软件项目开发全程实录

Python Web 项目开发全程实录

明日科技 编著

清华大学出版社
北京

内 容 简 介

本书精选 10 个热门项目，涉及 Flask 和 Django 两大框架及多领域应用，实用性非常强。具体项目包含：用户登录校验、员工信息审核模块、在线学习笔记、甜橙音乐网、乐购甄选在线商城、心灵驿站聊天室、站内全局搜索引擎、综艺之家、智慧校园考试系统、吃了么外卖网。从软件工程的角度出发，按照项目开发的顺序，系统、全面地讲解每一个项目的开发实现过程。体例上，每章一个项目，统一采用"开发背景→系统设计→技术准备→数据库设计/公共模块实现/各功能模块实现→项目运行→源码下载"的形式完整呈现项目，给读者明确的成就感，可以让读者快速积累实际项目经验与技巧，早日实现就业目标。

另外，本书配备丰富的 Python 在线开发资源库和电子课件，主要内容如下：

- ☑ 技术资源库：1456 个核心技术点
- ☑ 实例资源库：227 个应用实例
- ☑ 源码资源库：211 套项目与案例源码
- ☑ PPT 电子课件
- ☑ 技巧资源库：583 个开发技巧
- ☑ 项目资源库：44 个精选项目
- ☑ 视频资源库：598 集学习视频

本书可为 Python Web 入门自学者提供更广泛的项目实战场景，可为计算机专业学生进行项目实训、毕业设计提供项目参考，可作为计算机专业教师、IT 培训讲师的教学参考资料，还可作为软件工程师、IT 求职者、编程爱好者进行项目开发时的参考书。

本书封面贴有清华大学出版社防伪标签，无标签者不得销售。
版权所有，侵权必究。举报：010-62782989，beiqinquan@tup.tsinghua.edu.cn。

图书在版编目（CIP）数据

Python Web 项目开发全程实录 / 明日科技编著.
北京 : 清华大学出版社, 2024. 9. --（软件项目开发全程实录）. -- ISBN 978-7-302-67255-5

Ⅰ. TP312.8

中国国家版本馆 CIP 数据核字第 20240WE528 号

责任编辑：贾小红
封面设计：秦　丽
版式设计：文森时代
责任校对：马军令
责任印制：曹婉颖

出版发行：清华大学出版社
网　　址：https://www.tup.com.cn, https://www.wqxuetang.com
地　　址：北京清华大学学研大厦 A 座　　　邮　编：100084
社 总 机：010-83470000　　　　　　　　　邮　购：010-62786544
投稿与读者服务：010-62776969, c-service@tup.tsinghua.edu.cn
质量反馈：010-62772015, zhiliang@tup.tsinghua.edu.cn

印 装 者：三河市天利华印刷装订有限公司
经　　销：全国新华书店
开　　本：203mm×260mm　　　印　张：21.5　　　字　数：692 千字
版　　次：2024 年 10 月第 1 版　　　　　　　　印　次：2024 年 10 月第 1 次印刷
定　　价：89.80 元

产品编号：107421-01

如何使用本书开发资源库

本书赠送价值 999 元的"Python 在线开发资源库"一年的免费使用权限，结合图书和开发资源库，读者可快速提升编程水平和解决实际问题的能力。

1．VIP 会员注册

刮开并扫描图书封底的防盗码，按提示绑定手机微信，然后扫描右侧二维码，打开明日科技账号注册页面，填写注册信息后将自动获取一年（自注册之日起）的 Python 在线开发资源库的 VIP 使用权限。

Python 开发资源库

读者在注册、使用开发资源库时有任何问题，均可通过明日科技官网页面上提供的客服电话进行咨询。

2．开发资源库简介

Python 开发资源库中提供了技术资源库（1456 个核心技术点）、技巧资源库（583 个开发技巧）、实例资源库（227 个应用实例）、项目资源库（44 个精选项目）、源码资源库（211 套项目与案例源码）、视频资源库（598 集学习视频），共计六大类、3119 项学习资源。学会、练熟、用好这些资源，读者可在最短的时间内快速提升自己，从一名新手晋升为一名软件工程师。

3．开发资源库的使用方法

在学习本书的各项目时，可以通过 Python 开发资源库提供的大量技术点、技巧、热点实例等快速回顾或了解相关的知识和技巧，提升学习效率。

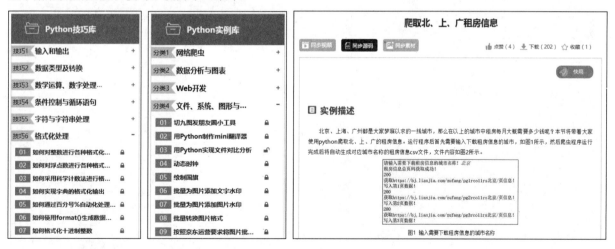

除此之外，开发资源库还配备了更多的大型实战项目，供读者进一步扩展学习，提升编程兴趣和信心，积累项目经验。

另外，利用页面上方的搜索栏，还可以对技术、技巧、实例、项目、源码、视频等资源进行快速查阅。

万事俱备后，读者该到软件开发的主战场上接受洗礼了。本书资源包中提供了 Python 的基础冲关 100 题以及企业面试真题，是求职面试的绝佳指南。读者可扫描图书封底的"文泉云盘"二维码获取。

前言

丛书说明："软件项目开发全程实录"丛书第 1 版于 2008 年 6 月出版，因其定位于项目开发案例、面向实际开发应用，并解决了社会需求和高校课程设置相对脱节的痛点，在软件项目开发类图书市场上产生了很大的反响，在全国软件项目开发零售图书排行榜中名列前茅。

"软件项目开发全程实录"丛书第 2 版于 2011 年 1 月出版，第 3 版于 2013 年 10 月出版，第 4 版于 2018 年 5 月出版。经过十六年的锤炼打造，不仅深受广大程序员的喜爱，还被百余所高校选为计算机科学、软件工程等相关专业的教材及教学参考用书，更被广大高校学子用作毕业设计和工作实习的必备参考用书。

"软件项目开发全程实录"丛书第 5 版在继承前 4 版所有优点的基础上，进行了大幅度的改版升级。首先，结合当前技术发展的最新趋势与市场需求，增加了程序员求职急需的新图书品种；其次，对图书内容进行了深度更新、优化，新增了当前热门的流行项目，优化了原有经典项目，将开发环境和工具更新为目前的新版本等，使之更与时代接轨，更适合读者学习；最后，录制了全新的项目精讲视频，并配备了更加丰富的学习资源与服务，可以给读者带来更好的项目学习及使用体验。

Python 凭借其简洁、易读的语法、强大的库支持和广泛的社区资源，在 Web 开发领域得到了广泛应用。常见的 Python Web 框架有 Flask 和 Django 等，这些框架为开发者提供了快速搭建 Web 应用的能力，同时也支持各种现代化的 Web 开发需求。本书以中小型项目为载体，带领读者切身感受软件开发的实际过程，可以让读者深刻体会 Python Web 技术在项目开发中的具体应用。全书内容不是枯燥的语法和陌生的术语，而是一步一步地引导读者实现一个个热门的项目，从而激发读者学习软件开发的兴趣，变被动学习为主动学习。另外，本书的项目开发过程完整，不但适合在学习软件开发时作为中小型项目开发的参考书，而且可以作为毕业设计的项目参考书。

本书内容

本书提供了采用 Flask 和 Diango 框架开发的项目，共 10 章，具体内容如下。

第 1 篇：**Flask 框架项目**。该篇主要通过"用户登录校验""员工信息审核模块""在线学习笔记""甜橙音乐网"和"乐购甄选在线商城"5 个功能完善的项目，帮助读者快速掌握使用 Flask 框架开发 Web 项目的关键技能，并让读者体验使用 Flask 框架开发 Python Web 项目的完整过程。

第 2 篇：**Django 框架项目**。该篇主要通过"心灵驿站聊天室""站内全局搜索引擎""综艺之家""智慧校园考试系统""吃了么外卖网"5 个功能完善的项目，帮助读者快速掌握使用 Django 框架开发 Web 项目的核心重点，并让读者全面体验使用 Django 框架开发 Python Web 项目的完整过程。

本书特点

- ☑ **项目典型**。本书精选 10 个热点项目，涉及 Flask 和 Django 两大框架及多领域应用。所有项目均从实际应用角度出发，可以让读者从项目学习中积累丰富的开发经验。
- ☑ **流程清晰**。本书项目从软件工程的角度出发，统一采用"开发背景→系统设计→技术准备→数据库设计/公共模块实现/各功能模块实现→项目运行→源码下载"的流程进行讲解，可以使项目的完整开发流程更加清晰。

- ☑ **技术新颖**。本书所有项目的实现技术均采用目前业内推荐使用的最新稳定版本，与时俱进，实用性极强。同时，项目全部配备"技术准备"环节，对项目中用到的基本技术点、高级应用、第三方模块等进行精要讲解，在 Python 基础和 Web 项目开发之间搭建了有效的桥梁，为仅有 Python 语言基础的初级编程人员参与 Web 项目开发扫清了障碍。
- ☑ **栏目精彩**。本书根据项目学习的需要，在每个项目讲解过程的关键位置添加"注意""说明"等特色栏目，点拨项目的开发要点和精华，以便读者能更快地掌握相关技术的应用技巧。
- ☑ **源码下载**。本书每个项目最后都安排了"源码下载"一节，读者在学习中能够通过扫描二维码下载对应项目的完整源码，方便学习。
- ☑ **项目视频**。本书为每个项目提供了开发及使用微视频，使读者能够更加轻松地搭建、运行、使用项目，并能够随时随地查看学习。

读者对象

- ☑ 初学 Web 编程的自学者
- ☑ 参与项目实训的学生
- ☑ 做毕业设计的学生
- ☑ 参加实习的初级程序员
- ☑ 高等院校的教师
- ☑ IT 培训机构的教师与学员
- ☑ 程序测试及维护人员
- ☑ 编程爱好者

资源与服务

本书提供了大量的辅助学习资源，同时还提供了专业的知识拓展与答疑服务，旨在帮助读者提高学习效率并解决学习过程中遇到的各种疑难问题。读者需要刮开图书封底的防盗码（刮刮卡），扫描并绑定微信，获取学习权限。

- ☑ **开发环境搭建视频**

搭建环境对于项目开发非常重要，它确保了项目开发在一致的环境下进行，减少了因环境差异导致的错误和冲突。通过搭建开发环境，可以方便地管理项目依赖，提高开发效率。本书提供了开发环境搭建讲解视频，可以引导读者快速准确地搭建本书项目的开发环境。扫描右侧二维码即可观看学习。

开发环境
搭建视频

- ☑ **项目精讲视频**

本书每个项目均配有对应的项目精讲微视频，主要针对项目的需求背景、应用价值、功能结构、业务流程、实现逻辑以及所用到的核心技术点进行精要讲解，可以帮助读者了解项目概要，把握项目要领，快速进入学习状态。扫描每章首页的对应二维码即可观看学习。

- ☑ **项目源码**

本书每章一个项目，系统全面地讲解了该项目的设计及实现过程。为了方便读者学习，本书提供了完整的项目源码（包含项目中用到的所有素材，如图片、数据表等）。扫描每章最后的二维码即可下载。

- ☑ **AI 辅助开发手册**

在人工智能浪潮的席卷之下，AI 大模型工具呈现百花齐放之态，辅助编程开发的代码助手类工具不断涌现，可为开发人员提供技术点问答、代码查错、辅助开发等非常实用的服务，极大地提高了编程学习和开发效率。为了帮助读者快速熟悉并使用这些工具，本书专门精心配备了电子版的《AI 辅助开发手册》，不仅为读者提供各个主流大语言模型的使用指南，而且详细讲解文心快码（Baidu Comate）、通义灵码、腾讯云 AI 代码助手、iFlyCode 等专业的智能代码助手的使用方法。扫描右侧二维码即可阅读学习。

AI 辅助
开发手册

☑ **代码查错器**

为了进一步帮助读者提升学习效率，培养良好的编码习惯，本书配备了由明日科技自主开发的代码查错器。读者可以将本书的项目源码保存为对应的 txt 文件，存放到代码查错器的对应文件夹中，然后自己编写相应的实现代码并与项目源码进行比对，快速找出自己编写的代码与源码不一致或者发生错误的地方。代码查错器配有详细的使用说明文档，扫描右侧二维码即可下载。

代码查错器

☑ **Python 开发资源库**

本书配备了强大的线上 Python 开发资源库，包括技术资源库、技巧资源库、实例资源库、项目资源库、源码资源库、视频资源库。扫描右侧二维码，可登录明日科技网站，获取 Python 开发资源库一年的免费使用权限。

Python 开发资源库

☑ **Python 面试资源库**

本书配备了 Python 面试资源库，精心汇编了大量企业面试真题，是求职面试的绝佳指南。扫描本书封底的"文泉云盘"二维码即可获取。

☑ **教学 PPT**

本书配备了精美的教学 PPT，可供高校教师和培训机构讲师备课使用，也可供读者做知识梳理。扫描本书封底的"文泉云盘"二维码即可下载。另外，登录清华大学出版社网站（www.tup.com.cn），可在本书对应页面查阅教学 PPT 的获取方式。

☑ **学习答疑**

在学习过程中，读者难免会遇到各种疑难问题。本书配有完善的新媒体学习矩阵，包括 IT 今日热榜（实时提供最新技术热点）、微信公众号、学习交流群、400 电话等，可为读者提供专业的知识拓展与答疑服务。扫描右侧二维码，根据提示操作，即可享受答疑服务。

学习答疑

致读者

本书由明日科技 Python 开发团队组织编写，主要编写人员有王国辉、王小科、张鑫、刘书娟、赵宁、高春艳、赛奎春、田旭、葛忠月、杨丽、李颖、程瑞红、张颖鹤等。明日科技是一家专业从事软件开发、教育培训以及软件开发教育资源整合的高科技公司，其编写的图书非常注重选取软件开发中的必需、常用内容，同时也很注重内容的易学性、方便性以及相关知识的拓展性，深受读者喜爱。其编写的图书多次荣获"全行业优秀畅销品种""全国高校出版社优秀畅销书"等奖项，多个品种长期位居同类图书销售排行榜的前列。

在编写本书的过程中，我们始终本着科学、严谨的态度，力求精益求精，但难免有疏漏和不当之处，敬请广大读者批评指正。

感谢您购买本书，希望本书能成为您的良师益友，成为您步入编程高手之路的踏脚石。

宝剑锋从磨砺出，梅花香自苦寒来。祝读书快乐！

编　者
2024 年 9 月

目录

第 1 篇　Flask 框架项目

第 1 章　用户登录校验 ... 2
—— Flask + PyMySQL + Flask-SQLAlchemy + Flask-Login

- 1.1　开发背景 ... 2
- 1.2　系统设计 ... 3
  - 1.2.1　开发环境 ... 3
  - 1.2.2　业务流程 ... 3
  - 1.2.3　功能结构 ... 3
- 1.3　技术准备 ... 4
  - 1.3.1　技术概览 ... 4
  - 1.3.2　数据存储技术 ... 4
  - 1.3.3　使用 Flask-Login 模块 ... 6
  - 1.3.4　使用哈希加盐技术进行密码加密 ... 7
- 1.4　数据库设计 ... 8
  - 1.4.1　创建数据库 ... 8
  - 1.4.2　创建数据表 ... 8
- 1.5　项目主文件 ... 9
- 1.6　功能设计 ... 9
  - 1.6.1　明日学院首页 ... 9
  - 1.6.2　登录与信息校验 ... 12
  - 1.6.3　修改密码 ... 15
  - 1.6.4　退出登录 ... 18
- 1.7　项目运行 ... 19
- 1.8　源码下载 ... 20

第 2 章　员工信息审核模块 ... 21
—— Flask + Flask-SQLAlchemy + PyMySQL

- 2.1　开发背景 ... 21
- 2.2　系统设计 ... 22
  - 2.2.1　开发环境 ... 22
  - 2.2.2　业务流程 ... 22
  - 2.2.3　功能结构 ... 22
- 2.3　技术准备 ... 23
- 2.4　数据库设计 ... 23
  - 2.4.1　创建数据库 ... 23
  - 2.4.2　创建数据表 ... 24
- 2.5　初始化项目 ... 25
  - 2.5.1　创建程序入口 ... 26
  - 2.5.2　初始化信息 ... 26
- 2.6　员工信息管理设计 ... 26
  - 2.6.1　实现显示个人信息 ... 26
  - 2.6.2　实现修改个人信息 ... 30
- 2.7　审核管理设计 ... 33
  - 2.7.1　查看已审核列表和待审核列表 ... 34
  - 2.7.2　实现通过审核功能 ... 36
- 2.8　权限管理设计 ... 37
- 2.9　项目运行 ... 38
- 2.10　源码下载 ... 40

第 3 章　在线学习笔记 ... 41
—— Flask + WTForms + passlib + PyMySQL

- 3.1　开发背景 ... 41
- 3.2　系统设计 ... 42
  - 3.2.1　开发环境 ... 42
  - 3.2.2　业务流程 ... 42
  - 3.2.3　功能结构 ... 42
- 3.3　技术准备 ... 43
  - 3.3.1　技术概览 ... 43
  - 3.3.2　使用 WTForms 模块 ... 44
  - 3.3.3　使用 passlib 模块进行加密 ... 46
- 3.4　数据库设计 ... 47
  - 3.4.1　数据库概要说明 ... 47
  - 3.4.2　创建数据表 ... 47
  - 3.4.3　数据表结构 ... 47

3.5	数据库操作类设计	48
3.6	用户管理模块设计	50
	3.6.1 实现用户注册功能	50
	3.6.2 实现用户登录功能	52
	3.6.3 实现退出登录功能	55
	3.6.4 实现用户权限管理功能	55
3.7	笔记管理模块设计	56
	3.7.1 实现笔记列表功能	56
	3.7.2 实现添加笔记功能	57
	3.7.3 实现编辑笔记功能	59
	3.7.4 实现删除笔记功能	60
3.8	项目运行	61
3.9	源码下载	62

第4章 甜橙音乐网 ... 63
——Flask + Flask-SQLAlchemy + Flask-WTF + jPlayer

4.1	开发背景	63
4.2	系统设计	64
	4.2.1 开发环境	64
	4.2.2 业务流程	64
	4.2.3 功能结构	65
4.3	技术准备	65
	4.3.1 技术概览	65
	4.3.2 jPlayer 插件	67
	4.3.3 蓝图	68
4.4	数据库设计	69
	4.4.1 数据库概要说明	69
	3.4.2 数据表结构	69
	4.4.3 数据表模型	70
4.5	首页设计	71
	4.5.1 首页概述	71
	4.5.2 实现热门歌手	72
	4.5.3 实现热门歌曲	73
	4.5.4 实现音乐播放	74
4.6	排行榜模块设计	75
	4.6.1 排行榜模块概述	75
	4.6.2 实现歌曲排行榜	76
	4.6.3 实现播放歌曲	78
4.7	曲风模块设计	78
	4.7.1 曲风模块概述	78
	4.7.2 实现曲风模块数据的获取	79

	4.7.3 实现曲风模块页面的渲染	79
	4.7.4 实现曲风列表的分页功能	81
4.8	发现音乐模块设计	81
	4.8.1 发现音乐模块概述	81
	4.8.2 实现发现音乐的搜索功能	82
	4.8.3 实现发现音乐模块页面的渲染	82
4.9	歌手模块设计	84
	4.9.1 歌手模块概述	84
	4.9.2 实现歌手列表	85
	4.9.3 实现歌手详情	85
4.10	我的音乐模块设计	86
	4.10.1 我的音乐模块概述	86
	4.10.2 实现收藏歌曲	87
	4.10.3 实现我的音乐	89
4.11	项目运行	91
4.12	源码下载	92

第5章 乐购甄选在线商城 ... 93
——Flask + SQLALchemy + MySQL

5.1	开发背景	93
5.2	系统设计	94
	5.2.1 开发环境	94
	5.2.2 业务流程	94
	5.2.3 功能结构	94
5.3	技术准备	95
5.4	数据库设计	96
	5.4.1 数据库概要说明	96
	5.4.2 数据表结构	97
	5.4.3 数据表模型	98
	5.4.4 数据表关系	101
5.5	会员注册模块设计	101
	5.5.1 会员注册模块概述	101
	5.5.2 会员注册页面	102
	5.5.3 验证并保存注册信息	106
5.6	会员登录模块设计	106
	5.6.1 会员登录模块概述	106
	5.6.2 创建会员登录页面	107
	5.6.3 保存会员登录状态	109
	5.6.4 会员退出功能	110
5.7	首页模块设计	110
	5.7.1 首页模块概述	110

5.7.2	实现显示最新上架商品功能	112
5.7.3	实现显示打折商品功能	113
5.7.4	实现显示热门商品功能	115
5.8	购物车模块设计	116
5.8.1	购物车模块概述	116
5.8.2	实现显示商品详细信息功能	118
5.8.3	实现添加购物车功能	119
5.8.4	实现查看购物车功能	120
5.8.5	实现保存订单功能	123
5.8.6	实现查看订单功能	125
5.9	后台功能模块设计	126
5.9.1	后台登录模块设计	126
5.9.2	商品管理模块设计	128
5.9.3	销量排行榜模块设计	135
5.9.4	会员管理模块设计	136
5.9.5	订单管理模块设计	137
5.10	项目运行	139
5.11	源码下载	141

第 2 篇 Django 框架项目

第 6 章 心灵驿站聊天室 144
——WebSocket + Django + Channels + Channels-Redis

6.1	开发背景	144
6.2	系统设计	145
6.2.1	开发环境	145
6.2.2	业务流程	145
6.2.3	功能结构	145
6.3	技术准备	145
6.3.1	技术概览	145
6.3.2	Django 框架的基本使用	146
6.3.3	Channels 模块的基本使用	154
6.3.4	在 Channels 项目中集成 Channels-Redis	155
6.4	创建项目	157
6.5	功能设计	157
6.5.1	进入房间	157
6.5.2	实时聊天	159
6.5.3	退出房间	165
6.6	项目运行	166
6.7	源码下载	168

第 7 章 站内全局搜索引擎 169
——Django + Django-Haystack + Whoosh + Jieba

7.1	开发背景	169
7.2	系统设计	169
7.2.1	开发环境	169
7.2.2	业务流程	170
7.2.3	功能结构	170
7.3	技术准备	170
7.3.1	技术概览	170
7.3.2	Django 框架的模型与数据库	171
7.3.3	Django-Haystack 模块的基本使用方法	174
7.3.4	使用 Whoosh 模块	175
7.3.5	使用 jieba 模块进行分词	176
7.4	数据库设计	178
7.4.1	数据库设计概要	178
7.4.2	数据表模型	179
7.5	创建项目	179
7.6	功能设计	180
7.6.1	全局搜索数据	180
7.6.2	分页显示搜索结果	183
7.7	项目运行	185
7.8	源码下载	187

第 8 章 综艺之家 188
——Django-Spirit + ECharts

8.1	开发背景	188
8.2	系统设计	189
8.2.1	开发环境	189
8.2.2	业务流程	189
8.2.3	功能结构	190
8.3	技术准备	190
8.3.1	技术概览	190
8.3.2	Django-Spirit 模块的基本使用方法	190
8.3.3	使用 ECharts 模块显示图表	193
8.4	数据库设计	195
8.4.1	数据库设计概要	195

8.4.2 数据表模型 195
8.4.3 数据表关系 197
8.5 综艺管理模块设计 198
   8.5.1 实现后台录入综艺信息和视频的功能 198
   8.5.2 实现前台首页展示功能 200
   8.5.3 实现综艺详情页展示功能 203
8.6 搜索功能模块设计 205
8.7 分类功能模块设计 208
8.8 社交管理模块设计 211
   8.8.1 实现发帖和回帖功能 212
   8.8.2 实现论坛后台管理功能 214
8.9 可视化展示模块设计 215
8.10 项目运行 219
8.11 源码下载 221

第9章 智慧校园考试系统 222
—— Django + MySQL + Redis + 文件上传技术 + xlrd

9.1 开发背景 222
9.2 系统设计 223
   9.2.1 开发环境 223
   9.2.2 业务流程 223
   9.2.3 功能结构 224
9.3 技术准备 224
   9.3.1 技术概览 224
   9.3.2 数据存储技术 224
   9.3.3 Django中的文件上传技术 227
   9.3.4 使用 xlrd 读取 Excel 227
9.4 数据库设计 228
   9.4.1 数据库设计概要 228
   9.4.2 数据表模型 229
9.5 登录与注册模块设计 231
   9.5.1 普通用户登录与注册模块概述 231
   9.5.2 使用 Django 默认授权机制实现普通登录 232
   9.5.3 机构注册功能的实现 238
9.6 核心答题功能设计 242
   9.6.1 答题首页设计 242

9.6.2 考试详情页面 244
9.6.3 答题功能的实现 247
9.6.4 提交答案与显示成绩排行榜 250
9.7 批量录入题库功能设计 252
9.8 项目运行 257
9.9 源码下载 260

第10章 吃了么外卖网 261
—— Django + MySQL + Redis

10.1 开发背景 261
10.2 系统设计 262
   10.2.1 开发环境 262
   10.2.2 业务流程 262
   10.2.3 功能结构 263
10.3 技术准备 263
10.4 数据库设计 264
   10.4.1 数据库设计概要 264
   10.4.2 数据表结构 265
   10.4.3 数据表关系 267
10.5 商品管理模块设计 269
   10.5.1 添加商品 269
   10.5.2 分页展示商品 272
10.6 店铺模块设计 278
   10.6.1 店铺首页 278
   10.6.2 店铺列表 283
   10.6.3 店铺详情页 286
10.7 购物车模块设计 293
   10.7.1 添加至购物车 293
   10.7.2 确认费用 298
   10.7.3 修改收货地址 303
10.8 订单模块设计 305
   10.8.1 订单生成 305
   10.8.2 订单追踪 310
   10.8.3 订单管理 312
   10.8.4 订单状态 320
   10.8.5 订单查询 322
10.9 项目运行 326
10.10 源码下载 330

第1篇

Flask 框架项目

Flask 是一个使用 Python 编写的轻量级 Web 应用框架。它的设计哲学是"核心小但易于扩展"。它的核心只包含了处理 HTTP 请求和响应的基本功能，以及一个简单的路由系统。这使得 Flask 非常灵活，可以用于从小型个人项目到大型企业级应用的各种场景。由于其简洁性和灵活性，Flask 已经成为 Python Web 开发中最受欢迎的框架之一。

本篇将使用 Flask 框架开发"用户登录校验""员工信息审核模块""在线学习笔记""甜橙音乐网"和"乐购甄选在线商城"5个流行的 Web 项目，带领读者全面体验使用 Flask 框架开发 Web 项目的过程，并积累实战项目开发经验。

第 1 章 用户登录校验

——Flask + PyMySQL + Flask-SQLAlchemy + Flask-Login

随着信息技术的快速发展和广泛应用，保护信息安全已经成为一个至关重要的议题。为了确保数据的安全无虞，防止未经授权的访问或修改，众多领域都已经采取措施，通过实施用户身份验证程序，来严把大门，只有经过授权的用户才能查看或更改相关信息。因此，设计一个有效的用户登录校验模块成为保护信息安全的重要手段。本章将使用 Flask 框架及 Flask-Login 模块开发一个用户登录校验项目。

项目微视频

本项目的核心功能及实现技术如下：

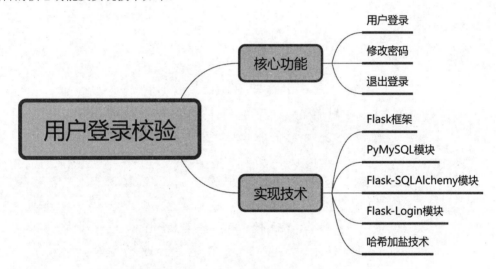

1.1 开发背景

在当今数字化高速发展的时代，信息安全问题变得越来越重要。以至于在多个领域都需要对用户进行身份验证，从而确保只有合法的用户可以访问或修改特定的数据。针对这一需求，可以设计一个有效的用户登录校验模块，为信息安全保驾护航。

Flask 框架是一种轻量级的 Python Web 框架，它具有灵活性强、扩展库丰富、易学易用以及出色的性能与安全性等优点。将 Flask 框架与 Flask-Login 模块搭配使用，就可以方便地开发常见的登录、注销和记住用户会话等任务。这种组合搭配是开发用户登录校验项目的理想选择。

本项目的实现目标如下：

- ☑ 安全性要求：在数据库中存储的密码应使用哈希加盐技术进行加密。
- ☑ 身份验证机制：系统应支持邮箱和密码登录。
- ☑ 用户体验：登录过程应简洁明了，对于输入错误应给予清晰的提示。
- ☑ 兼容性：系统应能兼容多种浏览器和设备。

☑ 性能要求：系统能够快速地响应用户的验证请求，即使在高流量下也能保持稳定的性能。

1.2 系统设计

1.2.1 开发环境

本项目的开发及运行环境如下：
☑ 操作系统：推荐 Windows 10、Windows 11 或更高版本。
☑ 开发工具：PyCharm 2024（向下兼容）。
☑ 开发语言：Python 3.12。
☑ 数据库：MySQL 8.0+PyMySQL 驱动。
☑ Python Web 框架：Flask 3.0。

1.2.2 业务流程

用户登录校验模块主要实现用户登录及密码修改等功能。用户可以输入邮箱和密码进行用户登录，登录后，可以对密码进行修改或者退出登录。本项目的业务流程如图 1.1 所示。

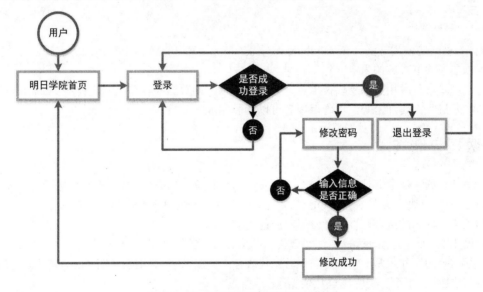

图 1.1　用户登录校验模块的业务流程

1.2.3 功能结构

本项目的功能结构已经在章首页中给出，其实现的具体功能如下：
☑ 用户登录：收集用户登录信息并校验登录信息的合法性。
☑ 修改密码：包括判断用户是否登录以及更改密码，在保存密码时，需要使用哈希加盐技术对密码进行加密。
☑ 退出登录：用于退出登录并进入用户登录页面。

1.3 技术准备

1.3.1 技术概览

本项目中应用的 Web 开发框架是 Flask 框架。Flask 是一个轻量级 Python Web 框架，它把 Werkzeug 和 Jinja 黏合在一起，能够很容易地生成安全的 HTML 页面。例如，本项目中，在项目主文件 run.py 中，首先创建 Flask 实例对象，然后创建并配置首页路由函数，在首页路由函数中，渲染首页文件，代码如下：

```python
from flask import Flask,render_template

app = Flask(__name__)              # 实例化 Flask 对象

@app.route('/')
def index():
    """
    首页
    """
    return render_template("index.html")
```

然后使用 run() 方法运行程序，代码如下：

```python
if __name__ == "__main__":
    app.run(debug=True)
```

有关 Flask 框架的使用方法，在《Python 从入门到精通（第 3 版）》中有详细的讲解，对该知识不太熟悉的读者可以参考该书对应的内容。

下面对实现本项目时用到的主要技术点进行必要介绍，如数据库存储技术、使用 Flask-Login 模块、使用哈希加盐技术进行密码加密等，以确保读者可以顺利完成本项目。

1.3.2 数据存储技术

本项目中的数据存储主要使用了 MySQL 数据库，其中操作 MySQL 数据库时使用了 PyMySQL 模块和 Flask-SQLAlchemy 模块，关于 PyMySQL 模块的知识在《Python 从入门到精通（第 3 版）》中有详细的讲解，对该知识不太熟悉的读者可以参考该书对应的内容。

下面对使用 Flask-SQLAlchemy 模块实现数据的增、查、改、删进行介绍。

数据库最常见的操作就是 CURD，它们分别代表插入（Create）、更新（Update）、读取（Retrieve）和删除（Delete）操作。下面分别介绍这几种操作。

1. 插入数据

在查询某些内容之前，必须插入一些数据。将数据插入数据库可以分为以下 3 个步骤。

（1）创建 Python 对象。
（2）将其添加到会话中。
（3）提交会话。

这里的会话不是 Flask 会话，而是 Flask-SQLAlchemy。它本质上是数据库事务的增强版本。下面通过一个示例介绍如何新增一个用户，示例代码如下：

```python
from models import User
```

```
me = User('admin', 'admin@qq.com')
db.session.add(me)
db.session.commit()
```

在将对象添加到会话之前，Flask-SQLAlchemy 不打算将其添加到事务中。此时，仍然可以放弃更改。add()方法可以将用户对象添加到会话中，但是并不会提交到数据库。commit()方法才能将会话提交到数据库。

2．读取数据

添加完数据以后，就可以从数据库中读取数据了。使用模型类提供的 query 属性，然后调用各种过滤方法及查询方法，即可从数据库中读取需要的数据。

通常，一个完整的查询语句格式如下：

`<模型类>.query.<过滤方法>.<查询方法>`

例如，查询 User 表中用户名为"mrsoft"的用户信息，示例代码如下：

`User.query.filter(username='mrsoft').get()`

上面的示例中，filter()是过滤方法，get()是查询方法。在 Flask-SQLAlchemy 中，常用的查询过滤器如表 1.1 所示，常用的查询方法如表 1.2 所示。

表 1.1　常用的 SQLAlchemy 查询过滤器

过滤器	说明
filter()	把过滤器添加到原查询上，返回一个新查询
filter_by()	把等值过滤器添加到原查询上，返回一个新查询
limit()	使用指定的值限制原查询返回的结果数量，返回一个新查询
offset()	偏移原查询返回的结果，返回一个新查询
order_by()	根据指定条件对原查询结果进行排序，返回一个新查询
group_by()	根据指定条件对原查询进行分组，返回一个新查询

表 1.2　常用的 SQLAlchemy 查询方法

方法	说明
all()	以列表形式返回查询的所有结果
first()	返回查询的第一个结果，如果没有结果，返回 None
first_or_404()	返回查询的第一个结果，如果没有结果，则终止请求，返回 404 错误响应
get()	返回指定主键对应的行，如果没有对应的行，返回 None
get_or_404()	返回指定主键对应的行，如果没有对应的行，则终止请求，返回 404 错误响应
count()	返回查询结果的数量
paginate()	返回一个 Paginate 对象，它包含指定范围内的结果。

在实际的开发过程中，使用的查询数据的方式比较多，下面介绍一些比较常用的查询方式。

（1）根据主键查询。在 get()方法中传递主键，示例代码如下：

`User.query.get(1)`

（2）精确查询。使用 filter_by()方法设置查询条件，示例代码如下：

`user = User.query.filter_by(username='mrsoft').first()`

使用 filter()方法设置查询条件，代码如下：

`user = User.query.filter(User.username='mrsoft').first()`

3. 更新数据

更新数据非常简单，直接赋值给模型类的字段属性就可以改变字段值，然后调用 commit()方法，提交会话即可。示例代码如下：

```
user = User.query.first()
user.username = 'guest'
db.session.commit()
```

4. 删除数据

删除数据也非常简单，只需要把插入数据的 add()替换成 delete()即可。示例代码如下：

```
user = User.query.first()
db.session.delete(user)
db.session.commit()
```

1.3.3 使用 Flask-Login 模块

Flask-Login 模块是 Flask 框架的一个插件，可以非常方便地管理用户对网站的访问。使用 Flask-Login 模块时，需要先安装该模块，命令如下：

```
pip install Flask-Login
```

Flask-Login 模块的常见操作如下。

（1）提供 user_loader()回调函数。使用 Flask-Login 时，需要为其提供一个 user_loader()回调函数。user_loader()函数主要是通过指定的用户 ID 获取 User 对象，并存储到 session 中。

（2）定义 User 类的属性和方法，如下所示。

- ☑ is_authenticated：用来判断是否已经授权，如果通过授权就会返回 true。
- ☑ is_active：判断是否已经激活。
- ☑ is_anonymous：判断是否是匿名用户。
- ☑ get_id()：返回用户的唯一标识。

这些属性和方法也可以直接继承于 UserMixin 的默认方法和属性，示例代码如下：

```
class User(db.Model,UserMixin):
    id = db.Column(db.Integer, autoincrement=True, primary_key=True)
    username = db.Column(db.String(125), nullable=False)
    email = db.Column(db.String(125), nullable=False)
    password = db.Column(db.String(255), nullable=False)
```

（3）自定义登录过程。当游客访问需要登录的页面时，应提示登录信息，并跳转到登录页面。Flask-Login 提供了配置属性，如下所示：

```
# 实例化 LoginManager 类
login_manager = LoginManager(app)
# 跳转的页面
login_manager.login_view = 'login'
# 提示信息
login_manager.login_message = "请先登录"
# 提示样式
login_manager.login_message_category = 'danger'
```

（4）"记住我"操作。默认情况下，当用户关闭浏览器时，Flask 会话被删除，用户注销。"记住我"可以防止用户在关闭浏览器时意外退出。这并不意味着在用户注销后记住或预先填写登录表单中的用户名或密码。只需将 remember = True 传递给 login_user()调用即可。示例代码如下：

```
login_user(user, remember=form.remember.data)
```

1.3.4 使用哈希加盐技术进行密码加密

Flask 框架的 werkzeug 库为我们提供了密码生成函数 generate_password_hash()和密码验证函数 check_password_hash()。这两个函数的使用说明如下所示。

（1）密码生成函数：generate_password_hash()。

密码生成函数的定义如下：

```
werkzeug.security.generate_password_hash(password, method='pbkdf2:sha1', salt_length=8)
```

generate_password_hash()是一个密码加盐哈希（hash）函数，用于对明文密码加盐，生成加密后的哈希字符串。生成的哈希字符串格式是这样的：

```
method$salt$hash
```

generate_password_hash()函数的参数说明如下：
- ☑ password：明文密码。
- ☑ method：哈希的方式（需要有 hashlib 库支持），格式如下：

```
pbpdf2:<method>[:iterations]
```

其中，method 为哈希的方式，一般为 SHA1；iterations 为可选参数，表示迭代次数，默认为 1000。
- ☑ salt_length：盐值的长度，默认为 8。

密码生成示例如下：

```
from werkzeug.security import generate_password_hash
print(generate_password_hash('123456'))
```

运行结果如下：

```
'pbkdf2:sha1:1000$X97hPa3g$252c0cca000c3674b8ef7a2b8ecd409695aac370'
```

因为盐值是随机的，所以就算是相同的密码，生成的哈希值也不会是一样的。

（2）密码验证函数：check_password_hash()。

check_password_hash()函数的定义如下：

```
werkzeug.security.check_password_hash(pwhash, password)
```

check_password_hash()函数用于验证经过 generate_password_hash()生成的哈希密码。如果密码匹配，则返回 True，否则返回 False。

check_password_hash()函数的参数说明如下：
- ☑ pwhash：generate_password_hash()生成的哈希字符串。
- ☑ password：需要验证的明文密码。

密码验证示例代码如下：

```
from werkzeug.security import check_password_hash
pwhash = 'pbkdf2:sha1:1000$X97hPa3g$252c0cca000c3674b8ef7a2b8ecd409695aac370'
print(check_password_hash(pwhash, '123456'))    # 输出为 True
```

运行结果如下：

```
True
```

1.4 数据库设计

1.4.1 创建数据库

本项目采用 MySQL 数据库，数据库名为 login。读者可以使用 MySQL 命令行方式或 MySQL 可视化管理工具（如 Navicat）创建数据库。使用命令行方式时输入如下命令：

```
create database login default character set utf8;
```

1.4.2 创建数据表

创建完数据库后，还需要创建数据表。在 login 数据库中，只有一张数据表，名称为 user，用于保存用户账号信息，其表结构如表 1.3 所示。

表 1.3 user 数据表的表结构

字段	类型	长度	是否允许为空	含义
id	INT	默认	否	数据编号，采用自动编号，是主键
username	VARCHAR	125	否	用户名
email	VARCHAR	125	否	邮箱（登录时使用）
password	VARCHAR	255	否	密码（需加密）

本项目使用 SQLAlchemy 模块进行数据库操作。SQLAlchemy 模块是一个常用的数据库抽象层和数据库关系映射包，并且需要一些设置才可以使用，因此，使用 Flask-SQLAlchemy 模块扩展来操作它。使用 SQLAlchemy 模块创建数据表的具体步骤如下：

（1）创建项目主文件 run.py，在该文件中创建数据表。首先导入所需的模块，然后实例化 Flask 对象，再定义数据库相关的配置信息，代码如下：

```python
from flask import Flask
from flask_login import UserMixin
from flask_sqlalchemy import SQLAlchemy
from werkzeug.security import generate_password_hash
app = Flask(__name__)                                      # 实例化 Flask 对象
# 基本配置
app.config["SECRET_KEY"] = "mrsoft"                        # 配置通用密钥
app.config['SQLALCHEMY_TRACK_MODIFICATIONS'] = True        # 是否跟踪数据库的变化
app.config['SQLALCHEMY_DATABASE_URI'] = (
    'mysql+pymysql://root:root@localhost/login'
)                                                          # 数据库基本配置信息
```

（2）实例化 SQLAlchemy 类，并创建用户模型类，然后使用用户表的模型类，在数据库中创建对应的数据表，代码如下：

```python
# 实例化 SQLAlchemy 类
db = SQLAlchemy(app)
# 用户表模型类
class User(db.Model,UserMixin):
    id = db.Column(db.Integer, autoincrement=True, primary_key=True)
    username = db.Column(db.String(125), nullable=False)
    email = db.Column(db.String(125), nullable=False)
```

```
password = db.Column(db.String(255), nullable=False)
if __name__ == "__main__":
    # 第一次运行程序时，先执行下面这段代码（只运行一次即可）
    with app.app_context():
        db.create_all()                             # 创建模型类中创建的数据表
```

（3）向数据表中插入一条测试数据，这里的密码需要加密，代码如下：

```
with app.app_context():                             # 向数据表中插入数据
    pwd = generate_password_hash('mrsoft')          # 将密码 mrsoft 转换为哈希密码
    user = User(username='mr', email='mingrisoft@qq.com', password=pwd)
    db.session.add(user)
    db.session.commit()                             # 提交
```

运行程序，将自动在数据库 login 中创建 user 数据表，并且在该数据表中添加如图 1.2 所示的数据。

图 1.2　插入用户数据

1.5　项目主文件

在开发用户登录校验模块时，需要先创建项目主文件，这里为 run.py，在该文件中，首先导入 Flask 类，然后实例化 Flask 对象，再使用 run()方法运行程序，代码如下：

```
from flask import Flask

app = Flask(__name__)              # 实例化 Flask 对象

if __name__ == "__main__":
    app.run(debug=True)            # 运行程序
```

上面的代码只是搭建了一个项目框架，还不能显示具体的网页。如果此时运行程序，在浏览器中将会显示如图 1.3 所示的 404 错误，还需要进行后续功能的开发。

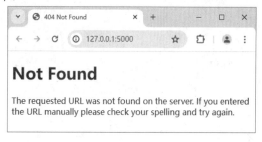

图 1.3　显示 404 错误

1.6　功 能 设 计

开发用户登录校验模块时，需要有一个主页，这里以明日学院首页为例进行开发。除了主页，还需要有用户登录、修改密码和退出登录等功能。针对这些需求，采用 Flask-Login 模块开发比较高效，所以本项目采用该插件开发。下面介绍具体的实现过程。

1.6.1　明日学院首页

在用户登录校验模块中，首页除了显示明日学院相关信息，还会根据用户登录状态显示不同的提示信息和功能按钮。实现首页时，主要是创建首页对应的模板文件 index.html。在项目的 templates 目录下创建名称为 index.html 的模板文件。在该文件中，首先包含 base.html 模板文件，然后编写涉及的 CSS 样式，以及根据登录状态显示的提示信息。index.html 模板文件的具体代码如下：

```
{% extends 'base.html' %}
{% block content%}
<style>
    .jumbotron {
        background-color: #563d7c;
        color:white
    }
</style>
<div class="jumbotron">
    <h1 class="display-4">
    {% if current_user.username %}
        Hello, {{ current_user.username }} !
    {% else %}
        Hello, world!
    {% endif %}
    </h1>
    <p class="lead">欢迎来到明日学院！</p>
    <hr class="my-4">
    <p>和我们一起开启编程学习之旅吧！</p>
    <a class="btn btn-success btn-lg" href="#" role="button">Learn more</a>
</div>
{% endblock %}
```

> **说明**
>
> base.html 文件中放置的是页面的公共内容，如导航栏和版权信息等。

在上面的代码中，包含了 base.html 模板文件，下面将创建这个模板文件。在该文件中，添加导航栏、功能按钮和底部版权信息等，关键代码如下：

```
<!doctype html>
<html lang="zh-CN">
<head>
    ……<!-- 此处省略了部分代码 -->
    <title>明日学院</title>
    <style>
        .nav-item {
            padding-left:20px;
        }
    </style>
</head>
<body>
<!-- 导航栏开始 -->
<nav class="navbar navbar-expand-lg navbar-dark bg-dark">
    <div class="container">
        <a class="navbar-brand" href="/">
            <span style="padding-left:5px">明日学院</span>
        </a>
        <div class="collapse navbar-collapse" id="navbarSupportedContent">
            <ul class="navbar-nav" style="margin:0 30px">
                <li class="nav-item active">
                    <a class="nav-link" href="/">首页 <span class="sr-only">(current)</span></a>
                </li>
                <li class="nav-item">
                    <a class="nav-link" href="/">全部课程</a>
                </li>
                <li class="nav-item">
                    <a class="nav-link" href="/">关于我们</a>
                </li>
            </ul>
        </div>
```

```html
        <ul class="navbar-nav">
            {% if current_user.is_authenticated %}
                <li class="nav-item dropdown">
                    <button class="btn btn-outline-success dropdown-toggle" href="#" id="navbarDropdown" role="button" data-toggle="dropdown" aria-haspopup="true" aria-expanded="false">
                        我的
                    </button>
                    <div class="dropdown-menu" aria-labelledby="navbarDropdown">
                        <a class="dropdown-item" href="/change_password">修改密码</a>
                        <div class="dropdown-divider"></div>
                        <a class="dropdown-item" href="/logout">退出登录</a>
                    </div>
                </li>
            {% else %}
                <li class="nav-item">
                    <a class="nav-link" href="/login" style="color:white;">
                        <button class="btn btn-outline-success" >
                            登录
                        </button>
                    </a>
                </li>
            {% endif %}
        </ul>
    </div>
</nav>
<!-- 导航栏结束 -->
{% block content %}

{% endblock %}
<!-- 底部信息开始 -->
<div class="footer">
    <div class="footer-left" >
        <p>明日学院 是明日科技公司旗下专注职业技能提升的在线学习平台。</p>
        <ul>
            <li>关于我们</li>
            <li>联系我们</li>
            <li>帮助中心</li>
        </ul>
    </div>
    <div class="footer-right">
        <img height="120px" width="120px" src="{{ url_for('static', filename='images/qrcode.jpg') }}">
        <p>关注微信公众平台</p>
    </div>
    <div>
        <p style="text-align:center;clear:both">©2001-2024 明日学院 版权所有</p>
    </div>
</div>
<!-- 底部信息结束 -->
</body>
</html>
```

模板文件创建完成后，还需要在项目主文件 run.py 中设置路由并渲染模板，关键代码如下：

```python
@app.route('/')
def index():
    """
    首页
    """
    return render_template("index.html")
```

运行程序，效果如图 1.4 所示。

图1.4 明日学院首页

1.6.2 登录与信息校验

在明日学院首页中，单击"登录"按钮，将打开用户登录界面。在该界面中实现用户登录及输入数据合法性的校验。具体的实现过程如下。

（1）在项目主文件 run.py 中，导入项目中用到的模块，代码如下：

```
from flask import Flask,render_template, request, redirect, url_for,flash
from forms import LoginForm,SettingForm
from werkzeug.security import check_password_hash
from flask_login import UserMixin,LoginManager,login_user,logout_user,current_user,login_required
from flask_sqlalchemy import SQLAlchemy
```

（2）实例化 LoginManager 类，并设置它的跳转页面、提示信息和提示样式，具体代码如下：

```
# 实例化 LoginManager 类
login_manager = LoginManager(app)
# 跳转的页面
login_manager.login_view = 'login'
# 提示信息
login_manager.login_message = "请先登录"
# 提示样式
login_manager.login_message_category = 'danger'
```

说明

LoginManager 类是 Flask-Login 模块中提供的。该类有多个方法和属性，用于保存用户登录的设置。

（3）由于涉及了用户信息，所以需要创建用户表，这里通过创建模型类 User 来实现。如果读者已经完成了 1.4.2 节介绍的创建数据表操作，那么，这里的创建用户表的代码则已经编写完成，就不需要再编写了。代码如下：

```python
class User(db.Model,UserMixin):
    id = db.Column(db.Integer, autoincrement=True, primary_key=True)
    username = db.Column(db.String(125), nullable=False)
    email = db.Column(db.String(125), nullable=False)
    password = db.Column(db.String(255), nullable=False)
```

（4）为 User 类设置回调函数，用于根据 ID 获取用户信息，并且为其添加装饰器，代码如下：

```python
@login_manager.user_loader
def load_user(user_id):
    return User.query.get(int(user_id))
```

（5）编写处理登录功能对应的业务逻辑的 login() 方法，并为其配置装饰器。在该方法中，判断用户是否已经登录，如果已经登录则跳转到首页，否则进行登录表单信息的验证。具体代码如下：

```python
@app.route('/login',methods=['GET','POST'])
def login():
    """
    登录
    """
    # 如果用户已经登录，访问登录页面会跳转到首页
    if current_user.is_authenticated:
        return redirect(url_for('index'))
    form = LoginForm() # 实例化 LoginForm()类
    # 验证表单
    if form.validate_on_submit():
        # 判断邮箱是否存在，如果不存在提示错误信息，
        # 如果存在，继续判断邮箱密码是否匹配
        # 如果匹配，跳转到上一页或首页，否则，提示错误信息
        user = User.query.filter_by(email=form.email.data).first()
        if not user:
            flash('邮箱不存在', 'danger')
        elif check_password_hash(user.password, form.password.data):
            login_user(user, remember=form.remember.data)
            next_page = request.args.get('next')
            # if not next_page or url_parse(next_page).netloc != '':
            if not next_page :
                next_page = url_for('index')
            return redirect(next_page)
        else:
            flash('用户名和密码不匹配', 'danger')
    return render_template('login.html',form=form)
```

（6）在上面的代码中，实例化了与登录相关的 form 表单类 LoginForm。所以需要先创建一个 forms.py 的文件，用于编写 LoginForm 类。在 forms.py 文件中，首先导入所需的模块，然后创建 LoginForm 类，在该类中，创建获取邮箱、密码、是否记住密码和提交按钮等信息的表单元素，并且为这些表单元素添加数据验证功能，当用户输入的数据不符合要求时，给予相应的提示，具体代码如下：

```python
from flask_wtf import FlaskForm
from wtforms import StringField, PasswordField, SubmitField, BooleanField
from wtforms.validators import DataRequired, Length, Email, EqualTo
class LoginForm(FlaskForm):
    email = StringField('邮箱',
                        validators=[
                            DataRequired(message="邮箱不能为空"),
                            Email()])
    password = PasswordField('密码',
                        validators=[
                            DataRequired(message="密码不能为空"),
                            Length(min=6, max=25, message='密码长度为 6-25 个字符'),
                        ])
```

```
    remember = BooleanField('Remember Me')
    submit = SubmitField('Login')
```

（7）创建用户登录页面对应的模板文件 login.html，并将其保存在 templates 目录下。在该文件中，首先包含 base.html 文件并编写需要的 CSS 样式代码，然后包含 layout.html 模板文件（该文件用于显示提示框），并放置登录表单。具体代码如下：

```
{% extends 'base.html' %}
{% block content%}
<style>
    .login-container {
        width: 500px;
        margin: 100px auto;
        padding: 20px 10px;
        background-color: #eef1f4;
        border-radius: .5rem;
        padding: 20px;
    }
    .login-title {
        text-align: center;
    }
    .login-form {
        padding: 20px;
    }
</style>

{% include "layout.html" %}

<div class="login-container">
    <h2 class="login-title">账号密码登录</h2>
    <form class="login-form" method="post" action="{{ url_for('login',next=request.args.next) }}">
        <div class="form-group">
            {{ form.email.label() }}
            {% if form.email.errors %}
                {{ form.email(class="form-control is-invalid") }}
                <div class="invalid-feedback">
                    {% for error in form.email.errors %}
                        <span>{{ error }}</span>
                    {% endfor %}
                </div>
            {% else %}
                {{ form.email(class="form-control") }}
            {% endif %}
        </div>
        <div class="form-group">
            {{ form.password.label() }}
            {% if form.password.errors %}
                {{ form.password(class="form-control is-invalid") }}
                <div class="invalid-feedback">
                    {% for error in form.password.errors %}
                        <span>{{ error }}</span>
                    {% endfor %}
                </div>
            {% else %}
                {{ form.password(class="form-control") }}
            {% endif %}
        </div>
        <div class="form-check" style="padding-bottom:10px">
            <label class="form-check-label">
                <input class="form-check-input" type="checkbox" name="remember"> 记住我
            </label>
        </div>
```

```
            <button type="submit" class="btn btn-success btn-lg btn-block">登录</button>
            {{ form.hidden_tag() }}
        </form>
    </div>
{% endblock %}
```

（8）在 templates 目录下，创建 layout.html 文件，在该文件中将提示信息显示在提示框中。代码如下：

```
{% with messages = get_flashed_messages(with_categories=true) %}
    {% if messages %}
        {% for category , message in messages %}
            <div class="alert alert-{{ category }} alert-dismissible fade show" role="alert">
                {{ message }}
                <button type="button" class="close" data-dismiss="alert" aria-label="Close">
                    <span aria-hidden="true">&times;</span>
                </button>
            </div>
        {% endfor %}
    {% endif %}
{% endwith %}
```

（9）在 run.py 文件的底部添加本地运行程序的代码，这里设置允许调试，具体代码如下：

```
if __name__ == "__main__":
    app.run(debug=True)
```

运行程序，单击"登录"按钮，进入用户登录界面，输入正确的用户邮箱和密码（这里的邮箱为 mingrisoft@qq.com，密码为 mrsoft），如图 1.5 所示，单击"登录"按钮，将直接跳转到首页，同时，首页中的欢迎信息中将显示当前用户的用户名，如图 1.6 所示。如果输入错误的邮箱或密码，将给出相应的提示，如图 1.7 所示。

图 1.5　输入正确的登录信息

图 1.6　登录成功显示的欢迎信息

图 1.7　显示错误提示信息

1.6.3　修改密码

用户成功登录后，在明日学院首页中，单击右上角的"我的"菜单，将显示下拉列表，选择"修改密码"

列表项，将打开修改密码界面。在该界面中实现修改当前登录用户密码的功能。具体的实现过程如下。

（1）在主文件 run.py 中，编写处理修改密码功能对应的业务逻辑的 change_password()方法，并为其配置装饰器。在该方法中，判断是否为提交操作，如果是，则将新密码保存到数据库，否则将错误信息存入闪存。具体代码如下：

```python
@app.route('/change_password',methods=['GET','POST'])
@login_required
def change_password():
    """
    修改密码
    """
    form = SettingForm()
    if form.validate_on_submit():                                    # 如果是提交操作，则修改密码
        if check_password_hash(current_user.password, form.password.data):
            # 根据当前用户 id 获取用户信息
            user = User.query.filter_by(id=current_user.id).first()
            # 获取新密码
            new_password = form.new_password.data
            # 对新密码加密
            user.password = generate_password_hash(new_password)
            # 提交到数据库
            db.session.commit()
            # 将成功消息存入闪存
            flash('修改成功', 'success')
            # 保存成功后，跳转到登录页面
            return redirect(url_for('change_password'))
        else:
            # 将失败消息存入闪存
            flash('原始密码错误', 'danger')
    return render_template("change_password.html",form=form)         # 显示修改密码页面
```

（2）在上面的代码中，实例化了修改密码相关的 form 表单类 SettingForm()。所以需要编写 LoginForm 类。这里将保存在 forms.py 文件中，在该文件中，创建 LoginForm 类，在该类中，创建获取原始密码、新密码、确认新密码和提交按钮等信息的表单元素，并且为这些表单元素添加数据验证功能，当用户输入的数据不符合要求时，给予相应的提示，具体代码如下：

```python
class SettingForm(FlaskForm):
    password = PasswordField('原始密码',
                             validators=[
                                 DataRequired(message="原始密码不能为空"),
                                 Length(min=6, max=25, message='密码长度为 6-25 个字符'),
                             ])
    new_password = PasswordField('新密码',
                                 validators=[
                                     DataRequired(message="新密码不能为空"),
                                     Length(min=6, max=25, message='密码长度为 6-25 个字符')])
    confirm_password = PasswordField('确认密码',
                                     validators=[
                                         DataRequired(message="确认密码不能为空"),
                                         EqualTo('new_password', message="2 次输入密码不一致")])
    submit = SubmitField('Login')
```

（3）创建用户修改密码页面对应的模板文件 change_password.html，并将其保存在 templates 目录下。在该文件中，首先包含 base.html 文件并编写需要的 CSS 样式代码，然后包含 layout.html 模板文件（该文件用于显示提示框），并放置修改密码表单。具体代码如下：

```html
{% extends 'base.html' %}
{% block content%}
<style>
```

```css
    .login-container {
        width: 500px;
        margin: 100px auto;
        padding: 20px 10px;
        background-color: #eef1f4;
        border-radius: .5rem;
        padding: 20px;
    }
    .login-title {
        text-align: center;
    }
    .login-form {
        padding: 20px;
    }
</style>

{% include "layout.html" %}

<div class="login-container">
    <h2 class="login-title">修改密码</h2>
    <form class="login-form" method="post" action="/change_password">
        <div class="form-group">
            {{ form.password.label() }}
            {% if form.password.errors %}
                {{ form.password(class="form-control is-invalid") }}
                <div class="invalid-feedback">
                    {% for error in form.password.errors %}
                        <span>{{ error }}</span>
                    {% endfor %}
                </div>
            {% else %}
                {{ form.password(class="form-control") }}
            {% endif %}
        </div>
        <div class="form-group">
            {{ form.new_password.label() }}
            {% if form.new_password.errors %}
                {{ form.new_password(class="form-control is-invalid") }}
                <div class="invalid-feedback">
                    {% for error in form.new_password.errors %}
                        <span>{{ error }}</span>
                    {% endfor %}
                </div>
            {% else %}
                {{ form.new_password(class="form-control") }}
            {% endif %}
        </div>
        <div class="form-group">
            {{ form.confirm_password.label() }}
            {% if form.confirm_password.errors %}
                {{ form.confirm_password(class="form-control is-invalid") }}
                <div class="invalid-feedback">
                    {% for error in form.confirm_password.errors %}
                        <span>{{ error }}</span>
                    {% endfor %}
                </div>
            {% else %}
                {{ form.confirm_password(class="form-control") }}
            {% endif %}
        </div>
        <button type="submit" class="btn btn-success btn-lg btn-block">提交</button>
        {{ form.hidden_tag() }}
```

```
            </form>
        </div>
{% endblock %}
```

运行程序,在成功登录后,单击"我的"/"修改密码"菜单项,将进入修改密码页面,在该页面中输入原始密码、新密码和确认新密码,单击"提交"按钮,如果输入的原始密码错误,将显示错误提示,如图 1.8 所示。如果输入的原始密码正确,并且输入的新密码和确认新密码一致,单击"提交"按钮,将显示修改成功,运行效果如图 1.9 所示。

图 1.8 原始密码错误

图 1.9 修改密码成功

1.6.4 退出登录

退出登录功能十分简单,只要清除设置的安全 cookie,然后重定向页面即可。这里需要注意的是:重定向的页面必须是用户当前所在的页面。实现方法是，直接调用 Flask-Login 模块的 logout_user()方法清除会话中的 cookie，并且将页面重定向到登录页面即可。实现代码如下：

```
@app.route('/logout')
def logout():
    """
    退出登录
    """
    logout_user()
    return redirect(url_for('login'))
```

运行程序，单击"我的"/"退出登录"菜单项，将退出登录，并进入用户登录页面。

1.7 项目运行

通过前述步骤，设计并完成了"用户登录校验"项目的开发。下面运行该项目，检验一下我们的开发成果。运行"用户登录校验"项目的步骤如下。

（1）打开 login\run.py 文件，根据自己的数据库账号和密码修改如下代码：

```
# 基本配置
app.config["SECRET_KEY"] = "mrsoft"                      # 配置通用密钥
app.config['SQLALCHEMY_TRACK_MODIFICATIONS'] = True       # 是否跟踪数据库的变化
app.config['SQLALCHEMY_DATABASE_URI'] = (
    'mysql+pymysql://root:root@localhost/login'
)                                                          # 数据库基本配置信息
```

（2）打开命令提示符对话框，进入 login 项目文件夹所在目录，在命令提示符对话框中输入如下命令来创建 venv 虚拟环境：

```
virtualenv venv
```

（3）在命令提示符对话框中输入如下命令来启动 venv 虚拟环境：

```
venv\Scripts\activate
```

（4）在命令提示符对话框中使用如下命令来安装 Flask 等依赖包：

```
pip install -r requirements.txt
```

（5）创建数据库。可以使用 MySQL 命令行方式或 MySQL 可视化管理工具（如 Navicat）创建数据库。使用命令行方式时输入如下命令：

```
create database login default character set utf8;
```

（6）在命令提示符对话框中执行 createdb.py 文件，用于创建数据表及添加默认数据。具体命令如下：

```
python createdb.py
```

（7）在 PyCharm 中打开项目文件夹 login，在其中选中 run.py 文件，单击鼠标右键，在弹出的快捷菜单中选择 Run 'run'命令，如图 1.10 所示。

（8）如果在 PyCharm 底部出现如图 1.11 所示的提示，说明程序运行成功。

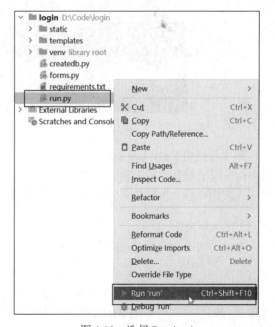

图 1.10　选择 Run 'run'

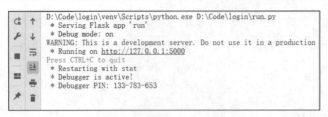

图 1.11　程序运行成功提示

（9）在浏览器中输入网址 http://127.0.0.1:5000/ 即可进入用户登录校验模块的首页，效果如图 1.12 所示。在该页面中，输入正确的邮箱和密码（例如，邮箱为 mingrisoft@qq.com；密码为 mrsoft）即可登录到明日学院首页，登录后，通过"我的"菜单可以实现修改密码和退出登录功能。

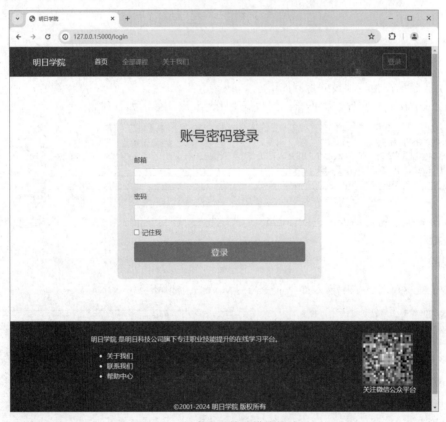

图 1.12　用户登录校验模块的首页

本项目使用 Flask-Login 模块实现了用户登录校验模块。通过项目的学习，能够使读者了解网站开发过程中用户登录校验功能涉及的相关技术及业务逻辑。希望读者通过本章的学习，可以学会借助 Flask-Login 模块开发用户登录校验相关的项目，提高开发效率。

1.8　源　码　下　载

本章虽然详细地讲解了如何编码实现"用户登录校验"项目的各个功能，但给出的代码都是代码片段，而非完整的源代码。为了方便读者学习，本书提供了该项目的完整源代码，读者可以通过扫描右侧的二维码进行下载。

第 2 章 员工信息审核模块

——Flask + Flask-SQLAlchemy + PyMySQL

在诸多的大、中、小企业中，都在使用员工信息管理系统，通过该系统可以更加方便地管理员工"入""离""调""转"全流程，快速寻找员工信息，第一时间了解各部门员工异动情况，以便于及时调整、维护组织架构。本章将使用 Flask 框架开发一个员工信息审核模块，从而实现员工自主审核个人信息，或者管理员确认员工信息正确后，通过审核。

项目微视频

本项目的核心功能及实现技术如下：

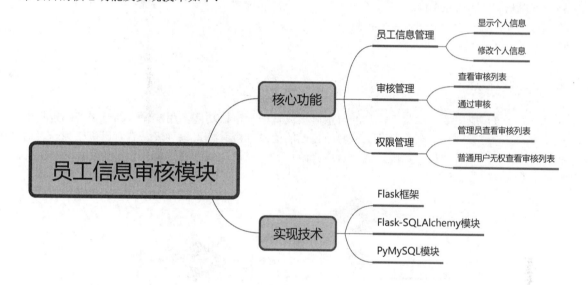

2.1 开发背景

在现代企业或组织中，员工信息管理是企业运营中不可或缺的一环。随着企业规模的不断扩大和业务的日益复杂化，传统的手工处理方式已经无法满足快速查询、更新和管理大量员工信息的需要。因此，开发一个高效、可靠的员工信息审核模块成为一项迫切的需求。

Flask 框架是一种轻量级的 Python Web 框架，它具有灵活性强、扩展库丰富、易学易用以及出色的性能与安全性等优点。将 Flask 框架与 Flask-SQLAlchemy 模块搭配使用，可以使开发过程更加高效、灵活且易于维护，同时 Flask-SQLAlchemy 模块采用 ORM 模式减少了 SQL 注入的风险，使数据更安全。

本项目的实现目标如下：

- ☑ 用户身份验证：系统应提供安全登录机制，确保只有授权人员可以访问。
- ☑ 员工自助服务：员工可以自行更新个人信息，减少 HR 的工作负担，同时提高数据的时效性和准确性。
- ☑ 权限分级管理：各司其职，确保只有授权人员才能进行相应的操作。

- ☑ 可用性：界面友好，易于使用，不需特殊培训即可使用系统。
- ☑ 兼容性：系统应能兼容多种浏览器和设备。
- ☑ 可维护性：系统设计要便于后期的维护和升级。

2.2 系统设计

2.2.1 开发环境

本项目的开发及运行环境如下：
- ☑ 操作系统：推荐 Windows 10、Windows 11 或更高版本。
- ☑ 开发工具：PyCharm 2024（向下兼容）。
- ☑ 开发语言：Python 3.12。
- ☑ 数据库：MySQL 8.0+PyMySQL 驱动。
- ☑ Python Web 框架：Flask 3.0。

2.2.2 业务流程

员工信息审核模块主要实现员工信息的自主审核或管理员审核功能。用户输入邮箱和密码进行用户登录，登录后，如果是管理员，则可以查看或修改自己的信息、查看待审核列表、查看已审核列表、通过审核或重新审核。如果是普通员工，则只能查看或修改自己的信息，或者通过审核自己的信息。员工信息审核模块的业务流程如图 2.1 所示。

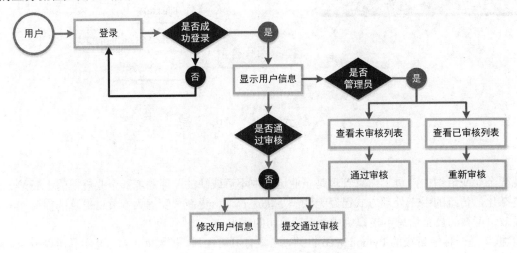

图 2.1 员工信息审核模块的业务流程

2.2.3 功能结构

本项目的功能结构已经在章首页中给出，其实现的具体功能如下：
- ☑ 审核管理：包括查看审核列表和通过审核功能。
- ☑ 员工信息管理：包括显示个人信息和修改个人信息功能。
- ☑ 权限管理：实现管理员可以查看审核列表，普通用户无权查看审核列表。

2.3 技术准备

实现员工信息审核模块时,主要使用了 Flask 框架技术、SQLAlchemy 操作 MySQL 数据库技术。基于此,这里将本项目所用的核心技术点及其具体作用简述如下。

☑ Flask 框架的使用:Flask 是一个轻量级 Python Web 框架,它把 Werkzeug 和 Jinja 黏合在一起,能够很容易地被扩展。例如,本项目中在__init__.py 初始化文件中创建 Flask 实例对象,并设置基本配置信息,代码如下:

```
from flask import Flask

app = Flask(__name__)
# 基本配置
app.config["SECRET_KEY"] = "mrsoft"
```

然后在项目启动文件 manage.py 中,导入路由文件,并使用 run()方法运行程序,代码如下:

```
from app import app

from app import routes    # 导入路由文件

if __name__ == "__main__":
    app.run(debug=True)
```

☑ 使用 Flask-SQLAlchemy 模块操作 MySQL 数据库:Flask-SQLAlchemy 是一个流行的 Python SQL 工具包和对象关系映射器(ORM),它采用简单的 Python 类来表示数据库表,并允许使用 Python 代码来创建、查询和更新数据库。使用 Flask-SQLAlchemy 模块操作 MySQL 数据库涉及以下关键步骤:安装必要的库(Flask-SQLAlchemy 和 MySQL 驱动库)、配置数据库连接、定义模型、创建表、创建会话、添加数据、执行查询以及更新和删除数据。例如,本项目在__init__.py 初始化文件中通过 SQLAlchemy 对象的 init_app()方法配置 MySQL 数据库连接,关键代码如下:

```
# 基本配置
app.config["SECRET_KEY"] = "mrsoft"
app.config['SQLALCHEMY_TRACK_MODIFICATIONS'] = True
app.config['SQLALCHEMY_DATABASE_URI'] = (
        'mysql+pymysql://root:root@localhost/online_check'
)
# 实例化 SQLAlchemy 类
db = SQLAlchemy(app)
```

有关 Flask 框架的使用方法,在《Python 从入门到精通(第 3 版)》中有详细的讲解,对该知识不太熟悉的读者可以参考该书对应的内容。关于使用 Flask-SQLAlchemy 模块操作 MySQL 数据库的相关知识,可以参考本书的 1.3.2 节。

2.4 数据库设计

2.4.1 创建数据库

本项目采用 MySQL 数据库,数据库名为 online_check。读者可以使用 MySQL 命令行方式或 MySQL 可视化管理工具（如 Navicat）创建数据库。使用命令行方式时输入如下命令：

```
create database online_check default character set utf8;
```

2.4.2 创建数据表

创建完数据库后，还需要创建数据表。在 online_check 数据库中，只有两张数据表，分别为 user 和 log。下面分别进行介绍。

用户表 user，用于保存用户登录信息。其表结构如表 2.1 所示。

表 2.1 user 数据表的表结构

字段	类型	长度	是否允许为空	含义
id	INT	默认	否	数据编号，采用自动编号，是主键
username	VARCHAR	125	否	用户名
email	VARCHAR	125	否	邮箱（登录时使用）
password	VARCHAR	255	否	密码（需加密）
is_admin	TINYINT	1	是	是否管理员（1 为管理员，0 为普通用户）
department	VARCHAR	255	否	所属部门
position	VARCHAR	255	否	职务
hiredate	DATE		否	入职日期
status	TINYINT	1	是	审核状态（1 为已审核，0 为待审核）

日志表 log，用于保存操作日志信息。其表结构如表 2.2 所示。

表 2.2 log 数据表的表结构

字段	类型	长度	是否允许为空	含义
id	INT	默认	否	数据编号，采用自动编号，是主键
user_id	INT	125	否	用户 ID
update_content	TEXT	125	否	日志内容
update_time	DATETIME	255	否	日期时间

本项目使用 Flask-SQLAlchemy 进行数据库操作，将所有的模型放置到一个单独的 models 模块中，使程序的结构更加明晰。使用 Flask-SQLAlchemy 模块创建数据表的具体步骤如下。

（1）创建模型文件 models.py，在该文件中创建数据表。首先导入所需的模块，然后实例化 Flask 对象，再定义数据库相关的配置信息，代码如下：

```python
from werkzeug.security import generate_password_hash
from app import db,login_manager
from flask_login import UserMixin
from datetime import datetime
from flask import Flask
from flask_sqlalchemy import SQLAlchemy
app = Flask(__name__)
# 基本配置
app.config["SECRET_KEY"] = "mrsoft"
app.config['SQLALCHEMY_TRACK_MODIFICATIONS'] = True
app.config['SQLALCHEMY_DATABASE_URI'] = (
    'mysql+pymysql://root:root@localhost/online_check'
)
```

（2）实例化 SQLAlchemy 类，并创建用户模型类，然后使用用户表的模型类，创建对应的数据表到数据库，代码如下：

```python
# 实例化 SQLAlchemy 类
db = SQLAlchemy(app)
@login_manager.user_loader
def load_user(user_id):
    return User.query.get(int(user_id))
class User(db.Model,UserMixin):
    id = db.Column(db.Integer, autoincrement=True, primary_key=True)
    username = db.Column(db.String(125), nullable=False)
    email = db.Column(db.String(125), nullable=False)
    password = db.Column(db.String(255), nullable=False)
    is_admin = db.Column(db.Boolean, default=0)
    department = db.Column(db.String(125), nullable=False)
    position = db.Column(db.String(125), nullable=False)
    hiredate = db.Column(db.Date,nullable=False)
    status = db.Column(db.Boolean, default=0)
    log = db.relationship("Log", backref="user")
class Log(db.Model):
    id = db.Column(db.Integer, autoincrement=True, primary_key=True)
    user_id = db.Column(db.Integer, db.ForeignKey('user.id'))
    update_content = db.Column(db.Text)
    update_time = db.Column(db.DateTime,default=datetime.now)
if __name__ == "__main__":
    with app.app_context():
        # db.create_all()                           # 创建模型类中创建的数据表
```

（3）向数据表 user 中插入两条测试数据，一条为管理员，另一条为普通员工。这里的密码需要加密，代码如下：

```python
# 向数据表中插入数据（管理员）
    pwd = generate_password_hash('mrsoft')          # 将密码 mrsoft 转换为哈希密码
    user = User(username='mr', email='mingrisoft@qq.com', password=pwd,is_admin=1,department='人资行政部',position='
        经理',hiredate='2001-01-01',status=1)
    db.session.add(user)
    db.session.commit()                             # 提交
# 向数据表中插入数据（普通员工）
    pwd = generate_password_hash('1234567')         # 将密码 mrsoft 转换为哈希密码
    user = User(username='wgh', email='wgh@qq.com', password=pwd, is_admin=0, department='开发部',position='程序员',
                hiredate='2010-01-01', status=0)
    db.session.add(user)
    db.session.commit()                             # 提交
```

运行程序，将自动在数据库 online_check 中创建 user 和 log 数据表，并且在数据表 user 中创建如图 2.2 所示的数据。

id	username	email	password	is_admin	department	position	hiredate	status
1	mr	mingrisoft@qq.com	scrypt:32768:8:1$kEw6Vb3cUWYi1wRP$a7969803134bc4be...	1	人资行政部	经理	2001-01-01	1
2	wgh	wgh@qq.com	scrypt:32768:8:1$52YT2nRKIP7ppPME$12d43e19972c595a9...	0	开发部	程序员	2007-10-12	0

图 2.2　插入用户数据

2.5　初始化项目

在开发项目前，需要先创建程序入口和初始化信息，以便规范代码。下面分别进行介绍。

2.5.1 创建程序入口

程序入口文件就是当前项目在运行时，从哪个文件开始运行。本项目的入口文件为 manage.py。在该文件中，首先导入 Flask 实例和路由文件，然后调用 run()方法运行程序。具体代码如下：

```python
from app import app
from app import routes

if __name__ == "__main__":
    app.run(debug=True)
```

2.5.2 初始化信息

在本项目中，单独创建一个 __init__.py 文件，用于设置初始化内容。在这里主要配置数据库连接信息、系统提示信息等内容，具体代码如下：

```python
from flask import Flask
from flask_sqlalchemy import SQLAlchemy
from flask_login import LoginManager
app = Flask(__name__)
# 基本配置
app.config["SECRET_KEY"] = "mrsoft"
app.config['SQLALCHEMY_TRACK_MODIFICATIONS'] = True
app.config['SQLALCHEMY_DATABASE_URI'] = (
         'mysql+pymysql://root:root@localhost/online_check'
         )
# 实例化 SQLAlchemy 类
db = SQLAlchemy(app)
# 实例化 LoginManager 类
login_manager = LoginManager(app)
# 跳转的页面
login_manager.login_view = 'login'
# 提示信息
login_manager.login_message = "请先登录"
# 提示样式
login_manager.login_message_category = 'danger'
```

2.6 员工信息管理设计

在员工信息审核模块中，员工信息管理主要分为两部分，一部分是显示个人信息，另一部分是修改个人信息，下面分别介绍具体的实现过程。

2.6.1 实现显示个人信息

当用户登录员工信息审核模块后，将显示该用户的个人员工信息。实现该功能时，首先是创建显示员工信息时对应的模板文件 info.html。在项目的 templates 目录下创建名称为 info.html 的模板文件，在该文件中，首先包含 base.html 模板文件，然后是根据登录状态显示的提示信息，以及动态显示的员工信息，这里需要对审核状态和是否为管理员进行相应判断，具体代码如下：

```
{% extends 'base.html' %}
```

```html
{% block content%}
{% include "layout.html" %}
<div class="container">
    <div class="card text-center" style="margin-top:20px">
        <div class="card-header" style="background-color: #28a345;">
            <span style="color:white">线上审核系统</span>
        </div>
        <div class="card-body">
            <table class="table col-sm-6 offset-sm-3">
                <tbody>
                    <tr>
                        <td style="text-align:right">姓名:</td>
                        <td>{{user.username}}</td>
                    </tr>
                    <tr>
                        <td style="text-align:right">部门:</td>
                        <td>{{user.department}}</td>
                    </tr>
                    <tr>
                        <td style="text-align:right">岗位名称:</td>
                        <td>{{user.position}}</td>
                    </tr>
                    <tr>
                        <td style="text-align:right">入职时间:</td>
                        <td>{{user.hiredate}}</td>
                    </tr>
                    <tr>
                        <td style="text-align:right">工龄:</td>
                        <td>{{working_age}}年</td>
                    </tr>
                    <tr>
                        <td style="text-align:right">工龄工资:</td>
                        <td>{{working_age * 10}}元</td>
                    </tr>
                </tbody>
            </table>
            {% if user.status == 0 %}
            <a href="{{url_for('update_status',id=user.id,type='checked')}}" class="btn btn-success" style="margin-right:20px">确 认</a>
            <a href="/edit" class="btn btn-danger" >修 改</a>
            {% endif %}
        </div>
        <div class="card-footer text-muted">
            <span style="color:#dc3545">
                {% if user.status == 1 %}
                当前状态：已审核
                {% else %}
                当前状态：待审核
                {% endif %}
            </span>
        </div>
    </div>
</div>
{% endblock %}
```

说明

base.html 文件中放置的是页面的公共内容，如导航栏、版权信息等。

在上面的代码中，包含了 base.html 模板文件，下面将创建这个模板文件。在该文件中，添加导航栏、功能按钮和底部版权信息等，关键代码如下：

```html
<!doctype html>
<html lang="zh-CN">
<head>
    … <!-- 此处省略部分代码 -->
</head>
<body>
<!-- 导航栏开始 -->
<nav class="navbar navbar-expand-lg navbar-dark bg-dark">
    <div class="container">
        <a class="navbar-brand" href="/">
            <span style="padding-left:5px">明日学院</span>
        </a>
        <div class="collapse navbar-collapse" id="navbarSupportedContent">
            <ul class="navbar-nav" style="margin:0 30px">
                <li class="nav-item active">
                    <a class="nav-link" href="/">首页 <span class="sr-only">(current)</span></a>
                </li>
                <li class="nav-item">
                    <a class="nav-link" href="/list/type/uncheck">待审核列表</a>
                </li>
                <li class="nav-item">
                    <a class="nav-link" href="/list/type/checked">已审核列表</a>
                </li>
            </ul>
        </div>
        <ul class="navbar-nav">
            {% if current_user.is_authenticated %}
            <li class="nav-item dropdown">
                <button class="btn btn-outline-success dropdown-toggle" href="#" id="navbarDropdown" role="button" data-toggle="dropdown" aria-haspopup="true" aria-expanded="false">
                    我的
                </button>
                <div class="dropdown-menu" aria-labelledby="navbarDropdown">
                    <a class="dropdown-item" href="/account">修改密码</a>
                    <div class="dropdown-divider"></div>
                    <a class="dropdown-item" href="/logout">退出登录</a>
                </div>
            </li>
            {% else %}
            <li class="nav-item">
                <a class="nav-link" href="/login" style="color:white;">
                    <button class="btn btn-outline-success" >
                        登录
                    </button>
                </a>
            </li>
            {% endif %}
        </ul>
    </div>
</nav>
<!-- 导航栏结束 -->
{% block content %}
{% endblock %}
<!-- 底部信息开始 -->
<div class="footer">
    <div class="footer-left" >
        <ul>
            <li style="display: inline; margin-right: 20px;">关于我们</li>
```

```
            <li style="display: inline; margin-right: 20px;">联系我们</li>
            <li style="display: inline; margin-right: 20px;">帮助中心</li>
        </ul>
    </div>
    <div>
        <p style="text-align:center;clear:both">©2001-2024 明日学院 版权所有</p>
    </div>
</div>
<!-- 底部信息结束 -->
<script>
    // 实现选中菜单功能
    var pathname = window.location.pathname;
    $(".nav-link").each(function(){
        // 删除原来选中菜单的选中状态
        $(this).parent().removeClass("active");
        // 为选中的菜单添加选中状态
        if(pathname == $(this).attr("href")){
            $(this).parent().addClass("active");
        }
    });
</script>
</body>
</html>
```

创建模板文件后，还需要在 route.py 文件中设置路由并渲染模板，关键代码如下：

```python
@app.route('/')
@app.route('/info')
@login_required
def index():
    """
    显示登录用户的信息
    :return:
    """
    # 获取用户信息
    user = User.query.filter_by(id=current_user.id).first()
    working_age = get_working_age(user.hiredate)
    # 渲染模板
    return render_template('info.html',user=user,working_age=working_age)
```

在显示员工信息时，需要自动计算工龄，所以还需要编写计算工龄的函数 get_working_age()，这里将其保存在 utils.py 文件中。在该函数中，主要是使用 datetime 模块中的 date 对象的 today() 函数获取当前的日期，再与入职日期相减得到天数差，并换算为年，则得到工龄。代码如下：

```python
import datetime
def get_working_age(hiredate):
    """
    计算工龄
    :param hiredate: datetime 类型数据
    :return:
    """
    # 计算时间间隔，结果为天数
    days = (datetime.date.today() - hiredate).days
    # 计算工龄，结果为年数
    working_age = days // 365
    return working_age
```

运行程序，将显示员工信息，如图 2.3 所示。

图 2.3 显示员工信息

2.6.2 实现修改个人信息

当员工信息处于待审核状态时，单击"修改"按钮，将打开修改员工信息界面。在该界面中，显示可修改员工信息的表单，修改相应信息后，单击"提交"按钮，可以保存所做的修改。具体的实现过程如下：

（1）在文件 routes.py 中，设置编辑用户信息所对应的路由并渲染模板，代码如下：

```python
@app.route('/edit')
@login_required
def edit():
    """
    编辑用户信息
    :return:
    """
    # 获取用户信息
    form = InfoForm()
    user = User.query.filter_by(id=current_user.id).first()
    return render_template('edit.html',user=user,form=form)
```

（2）在上面的代码中，实例化了用户相关的 form 表单类 InfoForm。所以需要先编写 InfoForm 类，这里创建一个 forms.py 的文件，在该文件中，首先导入所需的模块，然后创建 InfoForm 类，在该类中，创建获取姓名、部门、职务和入职时间等信息的表单元素对象，并且添加数据校验规则，在获取数据时，需要对数据进行验证，如果不符合要求，则给予相应的提示，具体代码如下：

```python
from flask_wtf import FlaskForm
from wtforms import StringField, PasswordField, SubmitField, BooleanField,DateField
from wtforms.validators import DataRequired, Length, Email, EqualTo
class InfoForm(FlaskForm):
    username = StringField('姓名',
                           validators=[
                               DataRequired(message="姓名不能为空"),
                               Length(min=2, max=60, message='密码长度为 2-25 个字符'),
```

```
                            ],
                            render_kw={'class': 'form-control','readonly': True}
                            )
    department = StringField('部门',
                            validators=[
                            DataRequired(message="部门不能为空"),
                            Length(min=2, max=60, message='密码长度为 2-25 个字符'),
                            ],
                            render_kw={'class': 'form-control'})
    position = StringField('职务',
                            validators=[
                            DataRequired(message="职务不能为空"),
                            Length(min=2, max=60, message='密码长度为 2-25 个字符'),
                            ],
                            render_kw={'class': 'form-control'})
    hiredate = DateField('入职时间',
                            validators=[
                            DataRequired(message="入职日期不能为空"),
                            ],
                            render_kw={'class': 'form-control'},
                            id="hiredate",
                            format='%Y-%m-%d'
                            )
```

（3）由于涉及了用户信息，所以需要创建用户表，这里通过创建模型类 User 来实现，具体代码如下：

```
class User(db.Model,UserMixin):
    id = db.Column(db.Integer, autoincrement=True, primary_key=True)
    username = db.Column(db.String(125), nullable=False)
    email = db.Column(db.String(125), nullable=False)
    password = db.Column(db.String(255), nullable=False)
    is_admin = db.Column(db.Boolean, default=0)
    department = db.Column(db.String(125), nullable=False)
    position = db.Column(db.String(125), nullable=False)
    hiredate = db.Column(db.Date,nullable=False)
    status = db.Column(db.Boolean, default=0)
    log = db.relationship("Log", backref="user")
```

说明

如果读者已经完成了 2.4.2 节介绍的创建数据表操作，那么，这里的创建用户表的代码则已经编写完成，就不需要再编写了。

（4）为 User 类设置回调函数，用于根据 ID 获取用户信息，并且为其添加装饰器，代码如下：

```
@login_manager.user_loader
def load_user(user_id):
    return User.query.get(int(user_id))
```

（5）编写修改员工信息功能对应的业务逻辑的 update()方法，并为其配置装饰器。在该方法中，首先根据用户 ID 获取该用户的员工信息，然后获取修改后的员工信息，并更新员工信息，再创建 Log 类的对象，并使用该对象保存日志信息，最后提交事务，将更改保存到数据库。具体代码如下：

```
@app.route('/update',methods=['POST'])
@login_required
def update():
    """
    更改用户信息
    :return:
    """
    # 获取用户信息
```

```python
user = User.query.filter_by(id=current_user.id).first()
form = InfoForm()
if form.validate_on_submit():
    content = {}
    content['username']   = f'{user.username  },{request.form["username"]  }'
    content['department'] = f'{user.department},{request.form["department"]}'
    content['position']   = f'{user.position  },{request.form["position"]  }'
    content['hiredate']   = f'{user.hiredate  },{request.form["hiredate"]  }'

    user.username   = request.form['username']
    user.department = request.form['department']
    user.position   = request.form['position']
    user.hiredate   = request.form['hiredate']
    # 保存日志信息
    log = Log()
    log.user_id = current_user.id                    # 将用户 ID 保存到 Log 对象中
    log.update_content = json.dumps(content)         # 将保存用户信息的字典转换为 JSON 字符串并保存到 Log 对象中
    try:
        db.session.add(log)                          # 向数据库中添加一条日志
        db.session.commit()                          # 提交
    except:
        db.session.rollback()                        # 事务回滚
    # 将成功消息存入闪存
    flash('修改成功', 'success')
    # 保存成功后，跳转到登录页面
    return redirect(url_for('index'))
return render_template('edit.html',form=form,user=user)
```

（6）创建修改员工信息页面对应的模板文件 edit.html，并将其保存在 templates 目录下。在该文件中，首先包含 base.html，然后创建一个用于显示员工信息的表单，表单元素中显示获取到的员工信息。具体代码如下：

```html
{% extends 'base.html' %}
{% block content%}
<script src="{{ url_for('static',filename='js/laydate/laydate.js') }}"></script>
<div class="container">
  <div class="card text-center" style="margin-top:20px">
    <div class="card-header" style="background-color: #28a345;">
      <span style="color:white">更改信息</span>
    </div>
    <div class="card-body">
        {% from "_formhelpers.html" import render_field %}
        <form class="offset-md-4" method="post" action="{{ url_for('update') }}" autocomplete="off">
            {{ form.csrf_token }}
            {{ render_field(form.username,user.username) }}
            {{ render_field(form.department,user.department) }}
            {{ render_field(form.position,user.position) }}

            <div class="form-group row">
                {{ form.hiredate.label(class="col-sm-2 col-form-label") }}
                <div class="col-sm-4">
                    {{ form.hiredate(value=user.hiredate.strftime("%Y-%m-%d")) }}
                </div>
            </div>
            <div class="col-sm-6">
                <button class="btn btn-success">提交</button>
                <a href="{{url_for('index')}}" class="btn btn-default">返回</a>
            </div>
        </form>
    </div>
  </div>
</div>
```

```
</div>
<script>
// 执行一个 laydate 实例
laydate.render({
    elem: '#hiredate' // 指定元素
});
</script>
{% endblock %}
```

运行程序，以普通用户身份登录系统（可以使用邮箱为 wgh@qq.com，密码为 123456 的用户进行登录），将显示员工信息，默认情况下，该用户为待审核状态，所以将显示如图 2.4 所示的界面，单击"修改"按钮，进入修改员工信息页面，如图 2.5 所示，修改某个信息后，单击"提交"按钮，即可将修改后的信息保存到数据库。

图 2.4　显示员工信息

图 2.5　修改员工信息

2.7　审核管理设计

在员工信息审核模块中，审核管理主要分为两部分，一部分是查看已审核列表和待审校列表，另一部分是通过审核（包括员工自主审核和管理员审核），下面分别介绍具体的实现过程。

2.7.1 查看已审核列表和待审核列表

以管理员身份登录系统后,单击导航栏上的待审核列表或已审校列表超链接,可以查看待审核信息或已审核信息。待审核信息和已审核信息通过数据表中的 status(状态)字段的值进行区分,值为 0 表示待审核,值为 1 表示已审核,在获取相应列表时,只需要根据 status 字段的值进行查询即可。下面介绍具体步骤。

(1)在 route.py 文件中,设置查看待审核列表或已审核列表的路由并渲染模板,代码如下:

```python
@app.route('/list/type/<type>')
@login_required
@is_admin
def list(type):
    """
    审核和待审核列表
    :param type:
    :return:
    """
    data = []
    if type == "uncheck":        # 待审核
        users = User.query.filter_by(status=0).all()
        data = get_uncheck_users(users)
    else:                         # 已审核
        users = User.query.filter_by(status=1).all()
        data = get_check_users(users)
    return render_template('list.html',data=data)
```

(2)在上面的代码中,获取待审核信息时调用了 get_uncheck_users() 函数,所以需要在 utils.py 文件中编写该函数。在 get_uncheck_users() 函数中,通过循环将获取到的待审核信息保存到一个列表中,值得注意的是:这里需要对工龄工资和入职时间进行单独处理。具体代码如下:

```python
def get_uncheck_users(users):
    data = []
    for user in users:
        # 入职时间转换为字符串
        user.hiredate_str = user.hiredate
        # 计算工龄工资
        user.working_age = get_working_age(user.hiredate)
        # 判断用户是否修改
        if user.log:
            content_json = user.log[-1].update_content
            content_dict = json.loads(content_json)
            for key, value in content_dict.items():
                old, new = value.split(',')
                if old != new :
                    # 入职时间需要单独处理
                    if key == "hiredate":
                        user.hiredate_str = f'{old}->{new}'
                    else:
                        exec(f'user.{key} = "{old}->{new}"')
        # 追加到列表
        data.append(user)
    return data
```

获取已审核信息调用了 get_check_users() 函数,所以需要在 utils.py 文件中编写该函数。在该函数中,通过循环将获取到的已审核信息保存到一个列表中,这里需要对工龄工资和入职时间进行相应转换。具体代码如下:

```python
def get_check_users(users):
```

```python
    data = []
    for user in users:
        # 入职时间转换为字符串
        user.hiredate_str = user.hiredate
        # 计算工龄工资
        user.working_age = get_working_age(user.hiredate)
        # 追加到列表
        data.append(user)
    return data
```

（3）创建查看待审核列表或已审核列表页面对应的模板文件 list.html，并将其保存在 templates 目录下。在该文件中，首先包含 base.html，然后以表格的形式将获取到的员工信息显示出来，同时根据审核状态添加相应的功能按钮。具体代码如下：

```html
{% extends 'base.html' %}
{% block content%}
<div class="container" style="min-height:500px">
  <div class="card text-center" style="margin-top:20px">
    <div class="card-header" style="background-color: #28a345;">
      <span style="color:white">审核列表</span>
    </div>
    <div class="card-body">
      <table class="table col-sm-12">
        <tbody>
          <thead>
            <tr>
              <th scope="col">姓名</th>
              <th scope="col">部门</th>
              <th scope="col">岗位名称</th>
              <th scope="col">入职时间</th>
              <th scope="col">工龄</th>
              <th scope="col">工龄工资</th>
              <th scope="col">状态</th>
              <th scope="col">操作</th>
            </tr>
          </thead>
          {% for user in data %}
            <tr>
              <td>{{user.username}}</td>
              <td>{{user.department}}</td>
              <td>{{user.position}}</td>
              <td>{{user.hiredate_str}}</td>
              <td>{{user.working_age}}年</td>
              <td>{{user.working_age * 10}}元</td>
              {% if user.status %}
                <td>已审核</td>
              {% else %}
                <td>待审核</td>
              {% endif %}
              <td>
                {% if user.status %}
                  <a href="/update_status/{{user.id}}/type/uncheck">
                    <button class="btn btn-info">重新审核</button>
                  </a>
                {% else %}
                  <a href="/update_status/{{user.id}}/type/checked">
                    <button class="btn btn-success">审核通过</button>
                  </a>
                {% endif %}
              </td>
            </tr>
          {% endfor %}
```

```
                </tbody>
            </table>
        </div>
    </div>
</div>
{% endblock %}
```

运行程序，以管理员身份登录，单击"待审核列表"超链接，将显示待审核列表页面，如图2.6所示。

图 2.6　待审核列表页面效果

单击"已审核列表"超链接，将显示已审核列表，效果如图2.7所示。

图 2.7　已审核列表页面效果

2.7.2　实现通过审核功能

在员工信息审核模块中，通过审核功能可以有两种情况：一种是普通用户在显示个人员工信息页面中，直接单击"确认"按钮完成；另一种情况是，管理员在待审核列表中，单击员工信息右侧的"审核通过"按钮完成。具体的实现过程如下。

（1）在审核列表模板文件中，找到"审核通过"按钮对应的超链接。从该超链接中，可以看出审核通过（即更改审核状态）对应的路由为"/update_status/{{user.id}}/type/checked"，代码如下：

```
<a href="/update_status/{{user.id}}/type/checked">
    <button class="btn btn-success">审核通过</button>
</a>
```

再打开显示个人员工信息页面的模板文件 info.html，找到"确认"按钮对应的超链接。从该超链接中，可以看出这里的路由和审核列表中更改审核状态对应的路由是一样的，所以我们配置一个即可。代码如下：

```
<a href="{{url_for('update_status',id=user.id,type='checked')}}" class="btn btn-success" style="margin-right:20px">确　认</a>
```

（2）在 route.py 文件中，设置更改审核状态的路由并渲染模板。这里主要根据用户 ID 获取相应员工的审核状态，并将其更改为1，表示已审核。代码如下：

```
@app.route("/update_status/<int:id>/type/<type>")
```

```python
@login_required
def update_status(id,type):
    """
    更改审核状态
    :param id:    用户ID
    :param type: 状态类型，uncked: 待审核，checked: 已审核
    :return:
    """
    # 员工只能更改自己的审核状态，管理员可以更改所有员工的审核状态
    if current_user.id != id and current_user.is_admin != 1:
        flash('没有修改权限',category='danger')
        redirect(url_for('index'))
    user = User.query.filter_by(id=id).first()
    if type == 'uncheck':
        user.status = 0
        db.session.commit()
        # 保存成功后，跳转到登录页面
        return redirect(url_for('list',type='uncheck'))
    elif type == 'checked':
        user.status = 1
        db.session.commit()
        # 如果是管理员账号，跳转到审核通过页面
        # 如果是员工账号，跳转到个人信息页面
        if current_user.is_admin:
            return redirect(url_for('list',type='checked'))
        else:
            return redirect(url_for('index'))
    return render_template('404.html'), 404
```

运行程序，以管理员身份登录后，单击"待审核列表"超链接，在显示的待审核列表页面中，单击某员工右侧的"审核通过"按钮，即可将该员工信息通过审核，然后，该员工信息将显示在已审核列表中。如果是普通员工登录，并且该员工的审核状态为待审核，则单击"确认"按钮，即将该员工信息通过审核，同时显示当前审核状态为已审核，如图2.8所示。

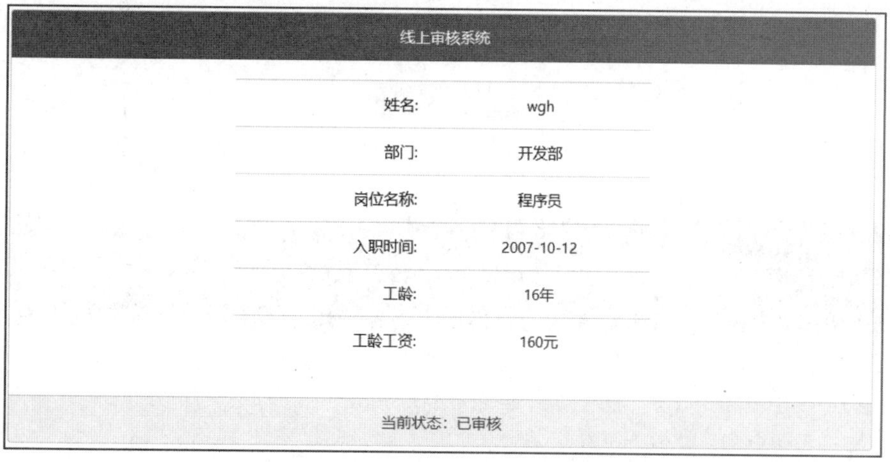

图2.8 通过审核的个人员工信息

2.8 权限管理设计

在员工信息审核模块中，只有管理员才可以查看已审核列表和待审核列表，并且审核通过员工信息，

实现代码如下：

```python
def is_admin(f):
    """
    管理员权限检测
    :param f:
    :return:
    """
    @wraps(f)
    def decorated_function(*args,**kwargs):
        if current_user.is_admin != 1:
            flash('非管理员账户无权访问', 'danger')
            return redirect(url_for('index'))
        return f(*args, **kwargs)
    return decorated_function
```

运行程序，以普通用户登录系统，单击"待审核列表"超链接，将显示提示信息，并且页面停留在显示个人员工信息页面，如图2.9所示。

图2.9　普通用户无权查看待审核列表

2.9　项目运行

通过前述步骤，设计并完成了"员工信息审核"项目的开发。下面运行该项目，检验一下我们的开发成果。运行"员工信息审核"项目的步骤如下。

（1）打开online_check__init__.py文件，根据自己的数据库账号和密码修改如下代码：

```
app.config["SECRET_KEY"] = "mrsoft"    # 配置通用密钥
```

```
app.config['SQLALCHEMY_TRACK_MODIFICATIONS'] = True        # 是否跟踪数据库的变化
app.config['SQLALCHEMY_DATABASE_URI'] = (
        'mysql+pymysql://root:root@localhost/online_check '
        )                                                  # 数据库基本配置信息
```

（2）打开命令提示符对话框，进入 online_check 项目文件夹所在目录，在命令提示符对话框中输入如下命令来创建 venv 虚拟环境：

```
virtualenv venv
```

（3）在命令提示符对话框中输入如下命令来启动 venv 虚拟环境：

```
venv\Scripts\activate
```

（4）在命令提示符对话框中使用如下命令来安装 Flask 等依赖包：

```
pip install -r requirements.txt
```

（5）创建数据库。可以使用 MySQL 命令行方式或 MySQL 可视化管理工具（如 Navicat）创建数据库。使用命令行方式时输入如下命令：

```
create database online_check default character set utf8;
```

（6）在命令提示符对话框中执行 createdb.py 文件，用于创建数据表及添加默认数据。具体命令如下：

```
python createdb.py
```

（7）在 PyCharm 中打开项目文件夹 online_check，在其中选中 manage.py 文件，单击鼠标右键，在弹出的快捷菜单中选择 Run 'manage' 命令，如图 2.10 所示。

（8）如果在 PyCharm 底部出现如图 2.11 所示的提示，说明程序运行成功。

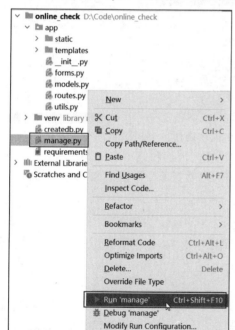

图 2.10　选择 Run 'manage'

图 2.11　程序运行成功提示

（9）在浏览器中输入网址 http://127.0.0.1:5000/ 即可进入员工信息审核模块的首页，效果如图 2.12 所示。输入正确的邮箱和密码即可登录到员工信息审核模块，管理员登录后，可以通过查看待审核列表和已审核列表来通过审核员工信息和重新审核已通过的员工信息；普通员工登录后，只能修改自己的信息和通过审核自己的信息。

图 2.12 员工信息审核模块的首页

> **说明**
>
> 以管理员身份登录时，输入邮箱为 mingrisoft@qq.com，密码为 mrsoft；以普通员工身份登录时，输入邮箱为 wgh@qq.com，密码为 1234567。

本项目使用 Flask 框架及 Flask-Sqlalchemy 模块实现了员工信息审核模块。通过项目的学习，能够使读者学会在网站开发过程中，如何对数据表进行操作，如查询数据、插入数据和修改数据等。希望读者通过本章的学习，可以学会借助 Flask 框架及 Flask-Sqlalchemy 模块开发与数据库操作相关的项目，提高开发效率。

2.10 源码下载

本章虽然详细地讲解了如何编码实现"员工信息审核模块"的各个功能，但给出的代码都是代码片段，而非完整的源代码。为了方便读者学习，本书提供了该项目的完整源代码，读者可以通过扫描右侧的二维码进行下载。

源码下载

第 3 章 在线学习笔记

——Flask + WTForms + passlib + PyMySQL

在当今数字化时代，阅读习惯正逐渐从传统的纸质书籍阅读方式转向电子化、网络化的阅读方式。我们也将面临一个新的挑战，那就是如何有效地筛选、整理和记忆每天接触到的大量信息。在此背景下，开发一个在线学习笔记项目就显得尤为迫切和重要。本章将使用 Python Web 框架 Flask 开发一个在线学习笔记项目。

项目微视频

本项目的核心功能及实现技术如下：

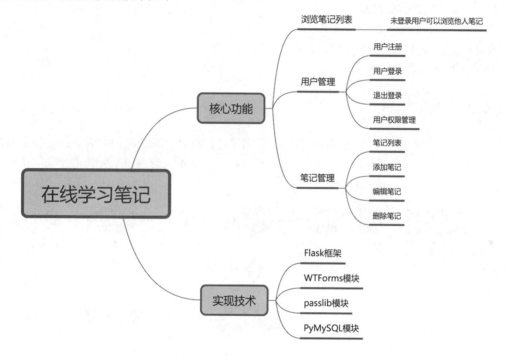

3.1 开发背景

随着互联网技术的飞速发展和智能移动设备的普及，人们越来越倾向于在碎片化的时间里通过手机、平板或计算机阅读书籍、文章与学习资料。然而，这一转变也带来了新的需求与挑战：如何高效地管理和整合个人的阅读笔记？这就需要有一个优秀的在线学习笔记项目。

Flask 框架是一种轻量级的 Python Web 框架，它具有灵活性强、扩展库丰富等特点。WTForms 模块提供了一套完整的机制来验证用户提交的数据，确保数据的合法性。将 Flask 框架与 WTForms 模块搭配使用，不仅可以提高开发效率，还能增强应用的安全性和用户体验。

本项目的实现目标如下：

☑ 笔记的创建与编辑：用户应能够轻松创建和编辑文本笔记。

- ☑ 用户体验：界面友好，操作直观简单，确保新用户能够快速上手。
- ☑ 响应式布局，用户在 PC 端和移动端都能达到较好的阅读体验。
- ☑ 性能：响应速度快，处理大量数据时仍能保持良好的性能。
- ☑ 兼容性：系统应能兼容多种浏览器和设备。

3.2 系统设计

3.2.1 开发环境

本项目的开发及运行环境如下：
- ☑ 操作系统：推荐 Windows 10、Windows 11 或更高版本。
- ☑ 开发工具：PyCharm 2024（向下兼容）。
- ☑ 开发语言：Python 3.12。
- ☑ 数据库：MySQL 8.0+PyMySQL 驱动。
- ☑ Python Web 框架：Flask 3.0。

3.2.2 业务流程

用户访问在线学习笔记时，可以使用游客的身份浏览笔记首页，以及笔记内容。但是如果需要管理笔记（如笔记列表、添加笔记、编辑笔记、删除笔记等），就必须先注册为网站会员，登录网站后才能执行相应的操作。系统业务流程如图 3.1 所示。

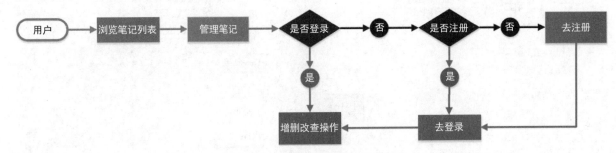

图 3.1　在线学习笔记的业务流程

> **说明**
> 在图 3.1 中，对笔记进行管理时进行的增、删、改、查操作是指添加笔记、删除笔记、编辑笔记和查看笔记列表。

3.2.3 功能结构

本项目的功能结构已经在章首页中给出，其实现的具体功能如下：
- ☑ 浏览笔记列表模块：未登录用户可以浏览他人写的笔记。
- ☑ 用户管理模块：包括用户注册、用户登录、退出登录和用户权限管理等功能。
- ☑ 笔记管理模块：包括笔记列表、添加笔记、编辑笔记、删除笔记等。

说明

由于浏览笔记列表模块与笔记管理模块中的笔记列表实现方法基本相同，这里不对其进行单独介绍。

3.3 技术准备

3.3.1 技术概览

在开发在线学习笔记时，应用的 Web 开发框架是 Flask 框架，操作数据库的模块是 PyMySQL。

1. Flask 框架

Flask 是一个轻量级 Python Web 框架，它把 Werkzeug 和 Jinja 黏合在一起，能够很容易地被扩展。例如，本项目中，在项目主文件 mange.py 中，首先创建 Flask 实例对象，然后创建并配置首页路由函数，在首页路由函数中，渲染首页文件，代码如下：

```
from flask import Flask,render_template

app = Flask(__name__)                          # 创建 Flask 实例对象
# 首页
@app.route('/')
def index():
    return render_template('home.html')        # 渲染模板
```

然后使用 run() 方法运行程序，代码如下：

```
if __name__ == '__main__':
    app.secret_key='secret123'
    app.run(debug=True)
```

有关 Flask 框架的使用方法，在《Python 从入门到精通（第 3 版）》中有详细的讲解，对该知识不太熟悉的读者可以参考该书对应的内容。

2. 使用 PyMySQL 模块操作数据库

由于 MySQL 服务器以独立的进程运行，并通过网络对外服务，所以，需要支持 Python 的 MySQL 驱动来连接到 MySQL 服务器。在 Python 中支持 MySQL 的数据库模块有很多，本项目选择使用简单方便的 PyMySQL 模块。使用 PyMySQL 模块时，大致可以分为 3 个步骤，下面分别进行介绍。

（1）安装 PyMySQL。

使用 pip 工具安装 PyMySQL 模块非常简单，只需要在 venv 虚拟环境下使用如下命令即可。

```
pip install PyMySQL
```

（2）连接 MySQL。

使用 PyMySQL 模块连接数据库。首先需要导入 PyMySQL 模块，然后使用 PyMSQL 的 connect() 方法来连接数据库。关键代码如下：

```
import pymysql
```

```
# 打开数据库连接，参数 1：主机名或 IP；参数 2：用户名；参数 3：密码；参数 4：数据库名称
db = pymysql.connect(host="localhost", user="root", password="root", db="flask")
……省略部分代码
# 关闭数据库连接
db.close()
```

在上述代码中，重点关注 connect()函数的参数：

```
db = pymysql.connect(
                    host='localhost',       # 主机名
                    user='root',            # 用户名
                    password='root',        # 密码
                    db='flask'              # 数据库名称
)
```

此外，connect()函数还有两个常用参数设置：

- ☑ charset:utf8：用于设置 MySQL 字符集为 UTF-8。
- ☑ cursorclass: pymysql.cursors.DictCursor：用于设置游标类型为字典类型，默认为元组类型。

（3）根据下面的操作流程使用 PyMySQL 模块操作数据库。操作 MySQL 的基本流程：连接 MySQL→创建游标→执行 SQL 语句→关闭连接。

有关 PyMySQL 模块的使用方法，在《Python 从入门到精通（第 3 版）》中有详细的讲解，对该知识不太熟悉的读者可以参考该书对应的内容。

下面对实现本项目时用到的其他主要技术点进行必要介绍，如使用 WTForms 模块、使用 passlib 模块进行加密等，以确保读者可以顺利完成本项目。

3.3.2 使用 WTForms 模块

由于 Flask 框架内部并没有提供全面的表单验证，所以通常搭配第三方模块来实现，例如，可以使用 WTForms 模块。下面将介绍在 Flask 项目中使用 WTForms 模块的基本方法。

1．安装 WTForms 模块

由于 WTForms 模块不是 Flask 框架自带的，在使用 WTForms 之前，需要确保已经安装了该模块。如果没有安装，可以通过 pip 工具来安装。例如，可以在 venv 虚拟环境下使用如下命令进行安装。

```
pip install wtforms
```

2．创建表单类

每个表单都应该被定义为一个 Python 类，这个类继承自 wtforms.Form，用于定义表单字段及其验证规则。在这个类里，可以定义多个字段，每个字段对应 HTML 表单中的一个输入域。例如，创建一个用户登录表单，包含用户名和密码两个字段，每个字段都有相应的验证器来确保数据的有效性。代码如下：

```
from wtforms import Form, StringField, PasswordField
from wtforms.validators import DataRequired

class LoginForm(Form):
    username = StringField(
        '用户名',
        validators=[
            DataRequired(message='请输入用户名')
        ]
    )
    password = PasswordField(
        '密码',
        validators = [
```

```
                DataRequired(message='密码不能为空')
        ]
)
```

3．在模板中渲染表单

在 HTML 模板中，可以直接使用表单实例的属性来渲染表单字段。例如，在模板文件中渲染上面创建的用户登录表单，代码如下：

```html
<!DOCTYPE html>
<html lang="zh-CN">
<head>
    <meta charset="UTF-8">
    <title>渲染表单</title>
</head>
<body>
    <form action="" method="POST">
        <div class="form-group">
            <label>用户名</label>
            <input type="text" name="username" class="form-control" value={{request.form.username}}>
        </div>
        <div class="form-group">
            <label>密码</label>
            <input type="password" name="password" class="form-control" value="">
        </div>
        <button type="submit" class="btn btn-primary">登录</button>
    </form>
</body>
</html>
```

4．处理表单提交

在 Flask 路由对应的视图函数中，首先实例化定义好的表单类，并将请求的数据传递给它。这通常是通过 request.form 来实现的。例如，处理用户登录表单，可以使用下面的代码。

```python
# 用户登录
@app.route('/login', methods=['GET', 'POST'])
def login():
    form = LoginForm(request.form)       # 实例化表单类
    if request.method == 'POST':          # 如果提交表单
        # 从表单中获取字段
        username = request.form['username']
        password_candidate = request.form['password']
        print('用户名：',username,' 密码：',password_candidate)
    return render_template('demo.html')
```

上面的代码实现了访问 http://127.0.0.1:5000/login 时显示用户登录页面，如图 3.2 所示。输入用户名和密码并单击"登录"按钮后，将提交表单，同时将获取到的用户名和密码输出到控制台，如图 3.3 所示。

图 3.2　用户登录页面　　　　　　　　图 3.3　获取提交的用户名和密码

除了上面代码中使用的 StringField 和 PasswordField HTML 标准字段，WTForms 还支持很多其他的 HTML 标准字段，如表 3.1 所示。

表 3.1　WTForms 支持的 HTML 标准字段

字段类型	说明	字段类型	说明
StringField	文本字段	BooleanField	复选框，值为 True 和 False
TextAreaField	多行文本字段	RadioField	一组单选框
PasswordField	密码文本字段	SelectField	下拉列表
HiddenField	隐藏文本字段	SelectMultipleField	下拉列表，可选择多个值
DateField	文本字段，值为 datetime.date 格式	FileField	文件上传字段
DateTimeField	文本字段，值为 datetime.datetime 格式	SubmitField	表单提交按钮
IntegerField	文本字段，值为整数	FormField	把表单作为字段嵌入另一个表单
DecimalField	文本字段，值为 decimal.Decimal	FieldList	一组指定类型的字段
FloatField	文本字段，值为浮点数		

WTForms 内置的验证函数如表 3.2 所示。

表 3.2　WTForms 支持的内置验证函数

字段类型	说明	字段类型	说明
Email	验证电子邮件地址	NumberRange	验证输入的值在数字范围内
EqualTo	比较两个字段的值；常用于要求输入两次密码进行确认的情况	Optional	无输入值时跳过其他验证函数
IPAddress	验证 IPv4 网络地址	Required	确保字段中有数据
Length	验证输入字符串的长度	Regexp	使用正则表达式验证输入值
NumberRange	验证输入的值在数字范围内	URL	验证 URL

3.3.3　使用 passlib 模块进行加密

passlib 模块是一个强大的 Python 密码哈希库，用于安全地处理用户密码。passlib 支持 30 多种密码散列算法，每种算法都有其特定用途和强度。例如，sha256_crypt 提供了一种基于 SHA-256 的密钥派生函数，该函数使用 SHA-256 算法，并且通常包含加盐（salt）和可配置的迭代次数（rounds）以增加安全性。加盐是为了防止彩虹表攻击，而迭代次数可以增加哈希计算的复杂度，从而抵御暴力破解。下面将介绍如何使用 passlib 模块的 sha256_crypt 进行密码处理。

1. 安装 passlib 模块

由于 passlib 模块是第三方模块，所以在使用该模块前，需要确保已经安装了该模块。如果没有安装，可以通过 pip 工具来安装。例如，可以在 venv 虚拟环境下使用如下命令进行安装：

```
pip install passlib
```

2. 对密码进行加密

通常情况下，程序中存储的用户密码不应该采用明文存储，而是存储其哈希值。这可以通过调用特定的散列算法（如 sha256_crypt）提供的方法来实现。例如，在 passlib 模块中，采用 sha256_crypt 算法时，对应的加密方法为 encrypt()，该方法会返回一个字符串，这个字符串是原始密码经过哈希处理后的结果。例如，通过下面的代码可以为给定的密码生成一个包含盐值和哈希值的字符串：

```
hashed_password = passlib.hash.sha256_crypt.hash('mrsoft')
```

加密码后的字符串类似下面的内容：

```
$5$rounds=535000$iWOTB4gpfZXV66xk$O0oYG7Q/knj.XfvVWFmSfZfNF6lnSaiXnApyCyKsP85
```

3. 验证密码

由于散列函数是一种单向函数，所以一旦数据被转换成哈希值，就几乎不可能从哈希值还原出原始数据。此时，想要验证密码，就需要将用户输入的密码与存储的哈希值进行比较，如果一样，就表示输入的密码正确。例如，通过下面的代码，可以验证用户提供的明文密码是否与之前存储的哈希值一致。

```
passlib.hash.sha256_crypt.verify('mrsoft', hashed_password)
```

3.4 数据库设计

3.4.1 数据库概要说明

本项目采用 MySQL 数据库，数据库名称为 notebook。读者可以使用 MySQL 命令行方式或 MySQL 可视化管理工具（如 Navicat）创建数据库。使用命令行方式时输入如下命令：

```
create database notebook default character set utf8;
```

3.4.2 创建数据表

因为本项目主要涉及用户和笔记两部分，所以在 notebook 数据库中创建两个数据表，数据表名称及作用如下。

- ☑ users：用户表，用于存储用户信息。
- ☑ articles：笔记表，用于存储笔记信息。

在 MySQL 命令行下或 MySQL 可视化管理工具（如 Navicat）下执行以下 SQL 语句创建数据表：

```sql
DROP TABLE IF EXISTS 'users';
CREATE TABLE 'users' (
  'id' int(8) NOT NULL AUTO_INCREMENT,
  'username' varchar(255) DEFAULT NULL,
  'email' varchar(255) DEFAULT NULL,
  'password' varchar(255) DEFAULT NULL,
  PRIMARY KEY ('id')
) ENGINE=InnoDB DEFAULT CHARSET=utf8;

DROP TABLE IF EXISTS 'articles';
CREATE TABLE 'articles' (
  'id' int(8) NOT NULL AUTO_INCREMENT,
  'title' varchar(255) DEFAULT NULL,
  'content' text,
  'author' varchar(255) DEFAULT NULL,
  'create_date' datetime DEFAULT NULL,
  PRIMARY KEY ('id')
) ENGINE=InnoDB DEFAULT CHARSET=utf8;
```

3.4.3 数据表结构

users 用户表的表结构如表 3.3 所示。

表 3.3　users 用户表的表结构

字　段	类　型	长　度	是否允许为空	含　义
id	int	默认	否	主键，编号
username	varchar	255	是	用户名
email	varchar	255	是	邮箱
password	varchar	255	是	密码

articles 笔记表的表结构如表 3.4 所示。

表 3.4　articles 笔记表的表结构

字　段	类　型	长　度	是否允许为空	含　义
id	int	默认	否	主键，编号
title	varchar	255	是	标题
content	text	默认	是	内容
author	varchar	255	是	作者
create_date	datetime	默认	是	记录时间

3.5　数据库操作类设计

本项目使用 PyMySQL 来驱动数据库，并实现对笔记的增、删、改、查功能。每次执行数据表操作都需要遵循如下流程：连接数据库→执行 SQL 语句→关闭数据库。

为了复用代码，单独创建一个 mysql_uitl.py 文件，该文件中包含一个 MysqlUtil 类，用于实现基本的增删改查方法。代码如下：

```python
import pymysql                                              # 引入 pymysql 模块
import traceback                                            # 引入 python 中的 traceback 模块，跟踪错误
import sys                                                  # 引入 Python 内置的 sys 模块

class MysqlUtil():
    def __init__(self):
        '''
            初始化方法，连接数据库
        '''
        host = '127.0.0.1'                                  # 主机名
        user = 'root'                                       # 数据库用户名
        password = 'root'                                   # 数据库密码
        database = 'notebook'                               # 数据库名称
        self.db = pymysql.connect(host=host,user=user,password=password,db=database)  # 建立连接
        self.cursor = self.db.cursor(cursor=pymysql.cursors.DictCursor)  # 设置游标，并将游标设置为字典类型

    def insert(self, sql):
        '''
        插入数据库
        sql:插入数据库的 SQL 语句
        '''
        try:
            # 执行 SQL 语句
            self.cursor.execute(sql)
            # 提交到数据库执行
            self.db.commit()
```

```python
        except Exception:                    # 方法一：捕获所有异常
            # 如果发生异常，则回滚
            print("发生异常", Exception)
            self.db.rollback()
        finally:
            # 最终关闭数据库连接
            self.db.close()

    def fetchone(self, sql):
        '''
        查询数据库：单个结果集
        fetchone(): 该方法获取下一个查询结果集。结果集是一个对象
        '''
        try:
            # 执行 SQL 语句
            self.cursor.execute(sql)
            result = self.cursor.fetchone()
        except:                              # 方法二：采用 traceback 模块查看异常
            # 输出异常信息
            traceback.print_exc()
            # 如果发生异常，则回滚
            self.db.rollback()
        finally:
            # 最终关闭数据库连接
            self.db.close()
        return result

    def fetchall(self, sql):
        '''
        查询数据库：多个结果集
        fetchall(): 接收全部的返回结果行
        '''
        try:
            # 执行 SQL 语句
            self.cursor.execute(sql)
            results = self.cursor.fetchall()
        except:                              # 方法三：采用 sys 模块回溯最后的异常
            # 输出异常信息
            info = sys.exc_info()
            print(info[0], ":", info[1])
            # 如果发生异常，则回滚
            self.db.rollback()
        finally:
            # 最终关闭数据库连接
            self.db.close()
        return results

    def delete(self, sql):
        '''
        删除结果集
        '''
        try:
            # 执行 SQL 语句
            self.cursor.execute(sql)
            self.db.commit()
        except:                              # 把这些异常保存到一个日志文件中，用来分析这些异常
            # 将错误日志输入目录文件中
            f = open("\log.txt", 'a')
            traceback.print_exc(file=f)
            f.flush()
```

```
            f.close()
            # 如果发生异常，则回滚
            self.db.rollback()
        finally:
            # 最终关闭数据库连接
            self.db.close()

    def update(self, sql):
        '''
        更新结果集
        '''
        try:
            # 执行 SQL 语句
            self.cursor.execute(sql)
            self.db.commit()
        except:
            # 如果发生异常，则回滚
            self.db.rollback()
        finally:
            # 最终关闭数据库连接
            self.db.close()
```

在使用 MysqlUtil 类时，只需要引入 MysqlUtil 类，实例化该类，并调用相应方法即可。

3.6 用户管理模块设计

用户管理模块主要包括 4 部分功能：用户注册、用户登录、退出登录和用户权限管理。这里的用户权限管理是指，只有登录后，用户才能访问某些页面（如控制台）。下面分别介绍每个功能的实现。

3.6.1 实现用户注册功能

用户注册功能是指在线学习笔记的注册新用户功能。在其页面中，需要填写用户名、邮箱、密码和确认密码。如果没有输入用户名、邮箱、密码或确认密码，系统将提示错误。此外，如果填写的格式错误也将提示错误。

1. 创建注册路由

实现用户注册时，需要先创建用户注册的路由。在 manage.py 入口文件中，创建一个名为 app 的 Flask 实例，然后调用 app.route()函数创建路由，关键代码如下：

```
app = Flask(__name__)                                     # 创建应用
# 用户注册
@app.route('/register', methods=['GET', 'POST'])
def register():
    form = RegisterForm(request.form)                     # 实例化表单类

    # 省略部分代码
    return render_template('register.html', form=form)    # 渲染模板
```

在上述代码中，@app.route()函数的第一个参数/register 是对应的 URL 的 path 部分；第二个参数 methods 是请求方式，这里使用列表指定接受 GET 和 POST 两种方式的请求。接下来，在 register()函数中实例化 RegisterForm 类，并使用 render_template()函数渲染模板。

2. 创建模板文件

因为 render_template()函数默认查找的模板文件路径为/templates/，所以需要在该路径下创建 register.html 模板文件。代码如下：

```html
{% extends 'layout.html' %}

{% block body %}
<div class="content">
  <h1 class="title-center">用户注册</h1>
  {% from "includes/_formhelpers.html" import render_field %}
  <form method="POST" action="">
    <div class="form-group">
      {{render_field(form.email, class_="form-control")}}
    </div>
    <div class="form-group">
      {{render_field(form.username, class_="form-control")}}
    </div>
    <div class="form-group">
      {{render_field(form.password, class_="form-control")}}
    </div>
    <div class="form-group">
      {{render_field(form.confirm, class_="form-control")}}
    </div>
    <p><input type="submit" class="btn btn-primary" value="注册"></p>
  </form>
</div>
{% endblock %}
```

在上述代码中，使用了 extends 标签来引入公共文件 layout.html，该文件包含了网站模板的基础框架，也称为父模板。由于网站页面包含很多通用的部分，如导航栏和底部信息等。将这些通用信息写入父模板，然后，使每个页面继承通用信息，并使用 block 标签来覆盖特有的信息。这样就简化了代码，达到了代码复用的目的。

此外，使用 WTForm 模块的 render_filed()函数来渲染表单中的字段。render_filed()函数的第一个参数是 Form 类的属性，该 Form 类是使用 render_tempalte()函数传递过来的，也就是 RegisterForm 类；第二个参数 _class 是模板中的 class 名称。

3. 实现注册功能

在 register.html 注册页面中，Form 表单的 action 属性值为空，即表示当用户单击"注册"按钮时，表单提交到当前页面。因此，需要在 manage.py 文件的 register()函数中继续编写提交表单的代码。register()函数的完整代码如下：

```python
# 用户注册
@app.route('/register', methods=['GET', 'POST'])
def register():
    form = RegisterForm(request.form)                              # 实例化表单类
    if request.method == 'POST' and form.validate():               # 如果提交表单，并且字段验证通过
        # 获取字段内容
        email = form.email.data
        username = form.username.data
        password = sha256_crypt.encrypt(str(form.password.data))   # 对密码进行加密

        db = MysqlUtil()                                           # 实例化数据库操作类
        sql = "INSERT INTO users(email,username,password) \
               VALUES ('%s', '%s', '%s')" % (email,username,password)  # user 表中插入记录
        db.insert(sql)
```

```
        flash('您已注册成功，请先登录', 'success')              # 闪存信息
        return redirect(url_for('login'))                      # 跳转到登录页面

    return render_template('register.html', form=form)         # 渲染模板
```

在上述代码的 if 语句中，先通过 request.method == 'POST' 来判断用户是否提交了表单。如果用户已经提交表单，就使用 form.validate() 判断是否通过 RegisterForm 类的全部验证规则。如果两个条件同时满足，则获取用户提交的注册信息，并对密码进行加密。接下来，实例化 MysqlUtil 类，将用户信息写入 users 表。最后，跳转到登录页面，并使用 flash() 闪存注册成功信息。如果用户没有提交表单或是字段验证失败，则执行 render_template() 函数显示注册页面。

用户注册失败的页面效果如图 3.4 所示，注册成功的页面效果如图 3.5 所示。

图 3.4　用户注册失败

图 3.5　用户注册成功

3.6.2　实现用户登录功能

在用户登录功能中，用户需要填写正确的用户名和密码，单击"登录"按钮，即可实现用户登录。如果没有输入用户名或者密码，将提示错误。另外，输入的账号和密码长度错误也将提示错误。

1．创建模板文件

在 /templates/ 路径下创建 login.html 模板文件。由于登录页面表单比较简单，只有两个字段，所以这里没有使用 WTForms 表单框架验证字段，而是直接通过 jQuery 来实现，具体代码如下：

```
{% extends 'layout.html' %}
```

```
{% block body %}
<div class="content">
  <h1 class="title-center">用户登录</h1>
  <form action="" method="POST" onsubmit="return checkLogin()">
    <div class="form-group">
      <label>用户名</label>
      <input type="text" name="username" class="form-control" value={{request.form.username}}>
    </div>
    <div class="form-group">
      <label>密码</label>
      <input type="password" name="password" class="form-control" value="">
    </div>
    <button type="submit" class="btn btn-primary">登录</button>
  </form>
</div>

<script>
  function checkLogin(){
    var username = $("input[name='username']").val()
    var password = $("input[name='password']").val()
    // 检测用户名长度
    if ( username.length < 2   || username.length > 25){
      alert('用户名长度在 2~25 个字符')
      return false;
    }
    // 检测密码长度
    if ( username.length < 2   || username.length > 25){
      alert('密码长度在 6~20 个字符之间')
      return false;
    }
  }
</script>

{% endblock %}
```

在上述代码中，由于需要验证账号长度、密码长度，所以在 Form 表单中，设置 onsubmit 属性验证表单。当单击"登录"按钮时，调用 checkLogin() 函数。如果 checkLogin() 函数返回 False，则表示验证没有通过，不提交表单。否则，正常提交表单。

2．实现登录功能

当用户填写登录信息后，还需要验证用户名是否存在，以及用户名和密码是否匹配等内容。如果验证全部通过，需要将登录标识和 username 写入 session 中，为后面判断用户是否登录做准备。此外，还需要在用户访问/login 路由时，判断用户是否已经登录，如果用户之前已经登录，那么不需要再次登录，而是直接跳转到控制台。具体代码如下：

```
# 用户登录
@app.route('/login', methods=['GET', 'POST'])
def login():
    if "logged_in" in session:                                    # 如果已经登录，则直接跳转到控制台
        return redirect(url_for("dashboard"))

    if request.method == 'POST':                                  # 如果提交表单
        # 从表单中获取字段
        username = request.form['username']
        password_candidate = request.form['password']
        sql = "SELECT * FROM users   WHERE username = '%s'" % (username)    # 根据用户名查找 user 表中记录
```

```
        db = MysqlUtil()                                    # 实例化数据库操作类
        result = db.fetchone(sql)                            # 获取一条记录
        if result :                                          # 如果查到记录
            password = result['password']                    # 用户填写的密码
            # 对比用户填写的密码和数据库中的记录密码是否一致
            if sha256_crypt.verify(password_candidate, password):  # 调用 verify 方法验证，如果为真，则验证通过
                # 写入 session
                session['logged_in'] = True
                session['username'] = username
                flash('登录成功！ ', 'success')
                return redirect(url_for('dashboard'))        # 跳转到控制台
            else:                                            # 如果密码错误
                error = '用户名和密码不匹配'
                return render_template('login.html', error=error)  # 跳转到登录页，并提示错误信息
        else:
            error = '用户名不存在'
            return render_template('login.html', error=error)
    return render_template('login.html')
```

在上述代码中，先判断 logged_in（登录标识）是否存在于 session 中。如果存在，则说明用户已经登录，直接跳转到控制台。如果不存在，则后续判断用户名和密码都正确时，通过 session['logged_in']=True 语句将 logged_in 标识存入 session，方便下次使用。

此外，还需要注意的是，在判断用户提交的密码和数据库中的密码是否匹配时，需要使用 sha256_crypt.verify()进行判断。verify()方法的第一个参数是用户输入的密码，第二个参数是数据库中保存的加密后的密码，如果返回 True，则表示密码相同，否则密码不同。

登录时，用户名不存在的页面效果如图 3.6 所示，用户名和密码不匹配的页面效果如图 3.7 所示，登录成功的页面效果如图 3.8 所示。

图 3.6　用户名不存在

图 3.7　用户名和密码不匹配

图 3.8　用户登录成功

3.6.3　实现退出登录功能

退出功能的实现比较简单，只是清空登录时保存在 session 中的值即可。使用 session.clear()函数来实现该功能。具体代码如下：

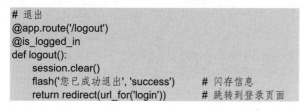

退出成功后，页面跳转到登录页。运行效果如图 3.9 所示。

3.6.4　实现用户权限管理功能

在线学习笔记项目中，需要用户登录后才能访问的路由及说明如下。

- ☑ /dashboard：控制台。
- ☑ /add_article：添加笔记。
- ☑ /edit_article：编辑笔记。
- ☑ /delete_article：删除笔记。
- ☑ /logout：退出登录。

对于这些路由，可以在每一个方法中都添加如下

图 3.9　退出登录页面效果

代码。

```
if 'logged_in' not in session:      # 如果用户没有登录
    return redirect(url_for('login'))   # 跳转到登录页面
```

如果需要用户登录才能访问的页面很多，显然这种方式不够优雅。在此，可以使用装饰器的方式来简化代码。在 manage.py 文件中实现一个 is_logged_in 装饰器。代码如下：

```
# 如果用户已经登录
def is_logged_in(f):
    @wraps(f)
    def wrap(*args, **kwargs):
        if 'logged_in' in session:      # 判断用户是否登录
            return f(*args, **kwargs)   # 如果登录，则继续执行被装饰的函数
        else:                           # 如果没有登录，则提示无权访问
            flash('无权访问，请先登录', 'danger')
            return redirect(url_for('login'))
    return wrap
```

定义完装饰器以后，就可以为需要用户登录的函数添加装饰器。例如，可以为 dashborad()函数添加装饰器，关键代码如下：

```
@app.route('/dashboard')
@is_logged_in
def dashboard():
    Pass
```

如果使用装饰器的方式，当执行 dashboard()函数时，优先执行 is_logged_in()函数判断用户是否登录。如果用户没有登录，则在浏览器中直接访问/dashboard，运行结果如图 3.10 所示。

图 3.10　未登录提示无权访问

3.7　笔记管理模块设计

笔记管理模块主要包括 4 部分功能：笔记列表、添加笔记、编辑笔记和删除笔记。因为用户必须登录后才能执行相应的操作，所以在每一个方法前添加@is_logged_in 装饰器来判断用户是否登录，如果没有登录，则跳转到登录页面。下面分别介绍每个功能的实现。

3.7.1　实现笔记列表功能

在控制台的笔记列表页面中，需要展示该用户的所有笔记信息。实现该功能的代码如下：

```
# 控制台
@app.route('/dashboard')
@is_logged_in
def dashboard():
    db = MysqlUtil()                    # 实例化数据库操作类
    sql = "SELECT * FROM articles WHERE author = '%s' ORDER BY create_date DESC" % (session['username'])
                                        # 根据用户名查找用户笔记信息，并根据时间降序排序
    result = db.fetchall(sql)           # 查找所有笔记
```

```
    if result:                                     # 如果笔记存在，则赋值给 articles 变量
        return render_template('dashboard.html', articles=result)
    else:                                          # 如果笔记不存在，则提示暂无笔记
        msg = '暂无笔记信息'
        return render_template('dashboard.html', msg=msg)
```

在上述代码中，需要注意使用 session()函数来获取用户名。如果用户登录成功，则使用 session['username'] = username 将 username 存入 session。所以，此时可以使用 session('username')来获取用户姓名。

接下来，使用 render_template()函数渲染模板文件。关键代码如下：

```
{% for article in articles %}
  <tr>
    <td>{{article.id}}</td>
    <td>{{article.title}}</td>
    <td>{{article.author}}</td>
    <td>{{article.create_date}}</td>
    <td><a href="edit_article/{{article.id}}" class="btn btn-default pull-right"> Edit</a></td>
    <td>
      <form action="{{url_for('delete_article', id=article.id)}}" method="post">
        <input type="hidden" name="_method" value="DELETE">
        <input type="submit" value="Delete" class="btn btn-danger">
      </form>
    </td>
  </tr>
{% endfor %}
```

在上述代码中，articles 变量表示所有笔记对象，使用 for 标签来遍历每一个笔记对象。

运行效果如图 3.11 所示。

图 3.11　笔记列表页面

3.7.2　实现添加笔记功能

在控制台列表页面中单击"添加笔记"按钮，即可进入添加笔记页面。在该页面中，用户需要填写笔

记标题和笔记内容。实现该功能的关键代码如下：

```python
# 添加笔记
@app.route('/add_article', methods=['GET', 'POST'])
@is_logged_in
def add_article():
    form = ArticleForm(request.form)                    # 实例化 ArticleForm 表单类
    if request.method == 'POST' and form.validate():    # 如果用户提交表单，并且表单验证通过
        # 获取表单字段内容
        title = form.title.data
        content = form.content.data
        author = session['username']
        create_date = time.strftime("%Y-%m-%d %H:%M:%S", time.localtime())
        db = MysqlUtil()                                # 实例化数据库操作类
        sql = "INSERT INTO articles(title,content,author,create_date) \
                VALUES ('%s', '%s', '%s','%s')" % (title,content,author,create_date)
        # 插入数据的 SQL 语句
        db.insert(sql)
        flash('创建成功', 'success')                    # 闪存信息
        return redirect(url_for('dashboard'))           # 跳转到控制台
    return render_template('add_article.html', form=form)   # 渲染模板
```

在上述代码中，接收表单的字段只包含标题和内容，此外，还需要使用 session()函数来获取用户名，使用 time 模块来获取当前时间。

在填写笔记内容时，使用了 CKEditor 编辑器替换普通的 textarea 文本框。CKEditor 编辑器和普通的 textarea 文本框的对比效果如图 3.12 所示。

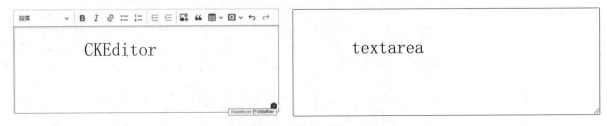

图 3.12　CKEditor 和 textarea 效果对比

在 add_article.html 模板中使用 CKEditor 的关键代码如下：

```html
{% extends 'layout.html' %}

{% block body %}
    <h1>添加笔记</h1>

    {% from "includes/_formhelpers.html" import render_field %}
    <form method="POST" action="">
        <div class="form-group">
            {{ render_field(form.title, class_="form-control") }}
        </div>
        <div class="form-group">
            <textarea id="editor" name="content"></textarea>
        </div>
        <p><input class="btn btn-primary" type="submit" value="提交"></p>
    </form>
<script src="{{ url_for('static', filename='js/ckeditor.js') }}"></script>
<script>
    // 初始化 CKEditor
    ClassicEditor.create(document.querySelector('#editor'))
        .then(editor => {
            console.log('CKEditor 5 初始化成功！');
```

```
        })
        .catch(error => {
            console.error('CKEditor 5 初始化失败:', error);
        });
</script>
{% endblock %}
```

在上述代码中，首先在 Form 表单的文本域中设置 id="editor"，然后引入 ckeditor.js，最后在 JavaScript 中通过 ClassicEditor.create()方法初始化 CKEditor。create()函数的参数就是表单中文本域字段，这里是通过 document.querySelector()方法根据文本域的 ID 值指定的。

添加笔记的运行效果如图 3.13 所示。

图 3.13　添加笔记

3.7.3　实现编辑笔记功能

在控制台列表中，单击笔记标题右侧的 Edit 按钮，即可根据笔记的 ID 进入该笔记的编辑页面。编辑页面和添加页面类似，只是编辑页面需要展示被编辑笔记的标题和内容。实现该功能的关键代码如下：

```
# 编辑笔记
@app.route('/edit_article/<string:id>', methods=['GET', 'POST'])
@is_logged_in
def edit_article(id):
    db = MysqlUtil()                                    # 实例化数据库操作类
    fetch_sql = "SELECT * FROM articles WHERE id = '%s' and author = '%s'" % (id,session
    ['username'])
    article = db.fetchone(fetch_sql)                    # 根据笔记 ID 查找笔记信息
    # 检测笔记不存在的情况                                # 查找一条记录
    if not article:
        flash('ID 错误', 'danger')
        return redirect(url_for('dashboard'))           # 闪存信息
    # 获取表单
    form = ArticleForm(request.form)
    if request.method == 'POST' and form.validate():    # 如果用户提交表单，并且表单验证通过
```

```
# 获取表单字段内容
title = request.form['title']
content = request.form['content']
update_sql = "UPDATE articles SET title='%s', content='%s' WHERE id='%s' and author = '%s'" % (title, content, id,session['username'])
db = MysqlUtil()                              # 实例化数据库操作类
db.update(update_sql)                         # 更新数据的 SQL 语句
flash('更改成功', 'success')                    # 闪存信息
return redirect(url_for('dashboard'))         # 跳转到控制台

# 从数据库中获取表单字段的值
form.title.data = article['title']
form.content.data = article['content']
# 渲染模板
return render_template('edit_article.html', form=form,mycontent=form.content.data)
```

在上述代码中,首先根据笔记的 ID 查找 articles 表中的笔记信息。如果 articles 表中没有此 ID,则提示错误信息。接下来,判断用户是否提交表单,并且表单验证通过。如果同时满足以上两个条件,则修改该 ID 的笔记信息,并跳转到控制台,否则,获取笔记信息后渲染模板。

编辑笔记的运行效果如图 3.14 所示。

图 3.14 编辑笔记

3.7.4 实现删除笔记功能

在控制台列表中,单击笔记标题右侧的 Delete 按钮,即可根据笔记的 ID 删除该笔记。删除成功后,页面跳转到控制台。实现该功能的关键代码如下:

```
# 删除笔记
@app.route('/delete_article/<string:id>', methods=['POST'])
@is_logged_in
def delete_article(id):
    db = MysqlUtil()                          # 实例化数据库操作类
    sql = "DELETE FROM articles WHERE id = '%s' and author = '%s'" % (id,session['username'])
```

```
# 执行删除笔记的 SQL 语句
db.delete(sql)
flash('删除成功', 'success')              # 闪存信息
return redirect(url_for('dashboard'))    # 跳转到控制台
```

在上述代码中，执行删除的 SQL 语句一定要添加 WHERE id 限定条件，否则，将删除所有笔记。

3.8 项目运行

通过前述步骤，设计并完成了"在线学习笔记"项目的开发。下面运行该项目，检验一下我们的开发成果。运行"在线学习笔记"项目的步骤如下。

（1）打开 Notebook\mysql_util.py 文件，根据自己的数据库账号和密码修改如下代码：

```
def __init__(self):
    """
        初始化方法，连接数据库
    """
    host = '127.0.0.1'                                          # 主机名
    user = 'root'                                               # 数据库用户名
    password = 'root'                                           # 数据库密码
    database = 'notebook'                                       # 数据库名称
    self.db = pymysql.connect(host=host,user=user,password=password,db=database)  # 建立连接
    self.cursor = self.db.cursor(cursor=pymysql.cursors.DictCursor)  # 设置游标，并将游标设置为字典类型
```

（2）打开命令提示符对话框，进入 Notebook 项目文件夹所在目录，在命令提示符对话框中输入如下命令来创建 venv 虚拟环境：

```
virtualenv venv
```

（3）在命令提示符对话框中输入如下命令来启动 venv 虚拟环境：

```
venv\Scripts\activate
```

（4）在命令提示符对话框中使用如下命令来安装 Flask 等依赖包：

```
pip install -r  requirements.txt
```

（5）创建数据库。可以使用 MySQL 命令行方式或 MySQL 可视化管理工具（如 Navicat）创建数据库。例如，在 MySQL 命令行窗口（MySQL 8.0 Command line Client）中输入登录密码后，使用下面的命令来实现。

```
create database notebook default character set utf8;
```

（6）执行 SQL 脚本文件 Code\Notebook\notebook.sql，创建数据表并插入数据。例如，在 MySQL 命令行窗口（MySQL 8.0 Command line Client）中输入登录密码后，使用下面的命令来实现。

```
use notebook
SOURCE D:\Code\Notebook\notebook.sql
```

（7）在 PyCharm 的左侧项目结构中展开"在线学习笔记"的项目文件夹 Notebook，在其中选中 manage.py 文件，单击鼠标右键，在弹出的快捷菜单中选择 Run 'manage'命令，如图 3.15 所示。

（8）如果在 PyCharm 底部出现如图 3.16 所示的提示，说明程序运行成功。

（9）在浏览器中输入网址 http://127.0.0.1:5000/即可进入在线学习笔记的首页，效果如图 3.17 所示。在该页面中，不登录时，可以浏览他人的笔记，注册并登录后，可以添加、编辑或删除自己的笔记。

本章主要使用 Flask 框架开发一个在线学习笔记的网站。在项目使用了很多开发中常用的模块和方法，例如，使用 WTFomrs 模块验证表单，使用 passlib 模块对密码加密，以及使用装饰器判断用户是否登录等。

通过本章的学习，希望读者能够了解 Flask 开发流程并掌握 Web 开发中常用的模块。

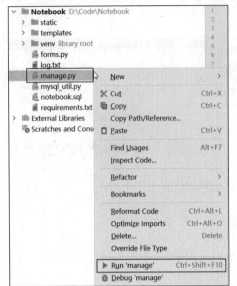

图 3.15　选择 Run 'manage'

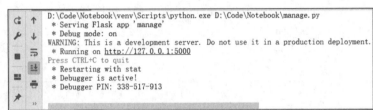

图 3.16　程序运行成功提示

图 3.17　在线学习笔记的首页

3.9　源码下载

源码下载

　　本章虽然详细地讲解了如何编码实现"在线学习笔记"的各个功能，但给出的代码都是代码片段，而非完整的源代码。为了方便读者学习，本书提供了该项目的完整源代码，读者可以通过扫描右侧的二维码进行下载。

第 4 章 甜橙音乐网

——Flask + Flask-SQLAlchemy + Flask-WTF + jPlayer

随着生活节奏的加快，人们的生活压力和工作压力也不断增加。为了缓解压力，现在的网络提供了许多娱乐项目，如网络游戏、网络电影和在线音乐等。听音乐可以放松心情，减轻生活或工作的压力。目前大多数的音乐网站都提供在线视听、音乐下载、在线交流、音乐收藏等功能。本章使用 Python Flask 框架开发一个在线音乐网站——甜橙音乐网。

项目微视频

本项目的核心功能及实现技术如下：

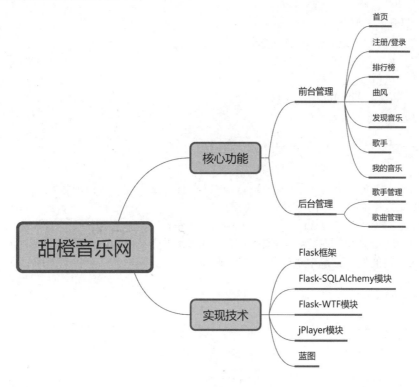

4.1 开发背景

在数字化浪潮席卷全球的今天，人们获取音乐的方式已从传统的 CD、广播转变为在线流媒体服务。人们对音乐的需求更加个性化、多样化，他们不仅希望随时随地能听到自己喜欢的歌曲，还希望能发现新的音乐、收藏音乐等。因此，甜橙音乐网——一个功能全面、操作便捷、内容丰富的音乐网站应运而生。

Flask 框架是一种轻量级的 Python Web 框架，它具有灵活性强、扩展库丰富的特点。Flask-WTF 模块能够自动渲染表单，并且支持服务器端数据验证，另外，提供了 CSRF 保护功能，能够保护所有表单免受跨站请求伪造（CSRF）的攻击。总之，将 Flask 与 Flask-WTF 搭配使用，可以极大地简化 Web 表单的处理流程，提高开发效率，同时还能增强应用的安全性和用户体验。

本项目的实现目标如下：
- ☑ 界面友好性：用户界面需简洁直观，易于导航，确保不同年龄段和技术水平的用户都能轻松操作。
- ☑ 在线播放：支持在线播放，用户可以即时听到音乐。
- ☑ 搜索与发现：支持关键词、歌手、曲风等多种搜索方式，让用户能迅速找到想听的音乐。
- ☑ 实时更新：保持音乐库的时效性，及时上线新歌，紧跟音乐潮流。
- ☑ 访问权限控制：用户需要注册和登录才能使用全部功能。
- ☑ 兼容性：系统应能兼容多种浏览器和设备。

4.2 系统设计

4.2.1 开发环境

本项目的开发及运行环境如下：
- ☑ 操作系统：推荐 Windows 10、Windows 11 或更高版本。
- ☑ 开发工具：PyCharm 2024（向下兼容）。
- ☑ 开发语言：Python 3.12。
- ☑ 数据库：MySQL 8.0+PyMySQL 驱动。
- ☑ Python Web 框架：Flask 3.0。

4.2.2 业务流程

普通用户首先使用浏览器进入音乐网的首页，可以查看歌曲排行榜、按曲风浏览歌曲、发现音乐、按歌手浏览歌曲、在我的音乐中查看收藏的歌曲（需登录）。

甜橙音乐网管理员（用户名为 mr，密码为 mrsoft），单击"登录"超链接，首先进入登录页面，进行系统登录操作，如果登录失败，则继续停留在登录页面，如果登录成功，则重新进入首页，单击右上角的"您好，mr"，在弹出的快捷菜单中，选择"后台管理"，则进入网站后台的管理页面，可以进行歌手管理和歌曲管理。

系统业务流程如图 4.1 所示。

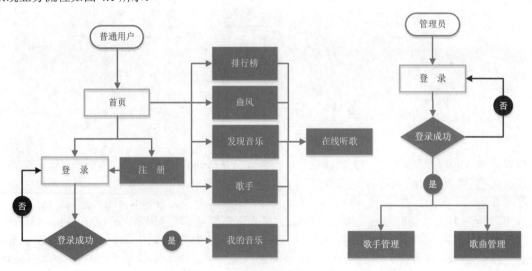

图 4.1 甜橙音乐网的业务流程

4.2.3 功能结构

本项目的功能结构已经在章首页中给出,其实现的具体功能如下:
- ☑ 前台管理:
  - ➤ 首页:包括热门歌手和热门歌曲。其中,单击热门歌曲名称可以播放该歌曲。
  - ➤ 登录/注册:包括用户注册、登录和退出等功能。
  - ➤ 排行榜:根据用户点击歌曲的次数进行排行。
  - ➤ 曲风:按曲风浏览歌曲。
  - ➤ 发现音乐:用户可根据歌曲名称搜索歌曲。
  - ➤ 歌手:按歌手浏览歌曲。
  - ➤ 我的音乐:用户登录后可以收藏歌曲。收藏完成后,点击"我的音乐",可以查看收藏的全部歌曲。
- ☑ 后台管理:
  - ➤ 歌手管理:包括添加歌手、修改歌手和删除歌手等功能。
  - ➤ 歌曲管理:包括添加歌曲、修改歌曲和删除歌曲等功能。

4.3 技术准备

4.3.1 技术概览

甜橙音乐网应用的 Web 开发框架是 Flask 框架,操作数据库的模块是 Flask-SQLAlchemy 模块。另外,还搭配了 Flask-WTF 模块来处理表单。下面分别介绍。

1. Flask 框架的使用

Flask 是一个轻量级 Python Web 框架,它把 Werkzeug 和 Jinja 黏合在一起,能够很容易地被扩展。例如,本项目中在 app__init__.py 初始化文件中创建 Flask 实例对象,并注册蓝图,代码如下:

```
# _*_ Coding:utf-8 _*_
from flask import Flask
from flask_sqlalchemy import SQLAlchemy
from flask_migrate import Migrate

db = SQLAlchemy()
def create_app():
    app = Flask(__name__)

    # 注册蓝图
    from app.home import home as home_blueprint
    app.register_blueprint(home_blueprint)
    return app
```

然后在 app.py 项目启动文件中使用 run()方法来运行程序,代码如下:

```
from app import create_app
```

```
if __name__ == "__main__":
    app = create_app()
    app.run(debug=True)
```

2．使用 Flask-SQLAlchemy 模块操作数据库

由于 MySQL 服务器以独立的进程运行，并通过网络对外服务，所以，需要支持 Python 的 MySQL 驱动来连接 MySQL 服务器。在采用 Flask 框架的 Python 程序中，通常使用 Flask-SQLAlchemy 模块操作数据库。使用 Flask-SQLAlchemy 模块时，通常涉及以下关键步骤：安装必要的库（Flask-SQLAlchemy 模块和 PyMySQL 模块）、配置数据库连接、定义模型、创建表、创建会话、添加数据、执行查询以及更新和删除数据。例如，本项目中在 app__init__.py 初始化文件中通过 SQLAlchemy 对象的 init_app()方法配置 MySQL 数据库连接，关键代码如下：

```
from flask_sqlalchemy import SQLAlchemy

db = SQLAlchemy()
def create_app():
    app = Flask(__name__)
    app.config['SQLALCHEMY_DATABASE_URI']='mysql+pymysql://root:root@127.0.0.1:3306/music'
    app.config['SQLALCHEMY_TRACK_MODIFICATIONS'] = True
    app.config['SECRET_KEY'] = 'mr'
    db.init_app(app)
```

3．使用 Flask-WTF 模块

由于 Flask 框架内部并没有提供全面的表单验证功能，所以通常搭配第三方模块来实现，例如，可以搭配 Flask-WTF 模块实现。在应用 Flask 框架的项目中，搭配 Flask-WTF 模块实现表单验证，大致步骤为：安装 Flask-WTF 模块（如果已经安装可以忽略）、创建表单类、在模板中渲染表单、处理表单提交。例如，在本项目中，将实现表单类的代码全部保存在名为 app\home\forms.py 的文件中。其中，用户登录表单类的关键代码如下：

```
from flask_wtf import FlaskForm
from wtforms import StringField, PasswordField, SubmitField

class LoginForm(FlaskForm):
    """
    登录功能
    """
    email = StringField(
        validators=[
            DataRequired("邮箱不能为空！")
        ],
        description="邮箱",
        render_kw={
            "type"         : "email",
            "placeholder": "请输入邮箱！",
        }
    )
    pwd = PasswordField(
        validators=[
            DataRequired("密码不能为空！")
        ],
        description="密码",
        render_kw={
            "type"         : "password",
            "placeholder": "请输入密码！",
        }
    )
```

```
        submit = SubmitField(
            '登录',
            render_kw={
                "class": "btn btn-primary",
            }
        )
```

有关 Flask 框架、Flask-SQLAlchemy 模块、Flask-WTF 模块的使用方法，在《Python 从入门到精通（第3版）》中有详细的讲解，对该知识不太熟悉的读者可以参考该书对应的内容。

下面对实现本项目时用到的其他主要技术点进行必要介绍，如使用 jPlayer 插件播放音乐、Flask 中的蓝图的使用等，以确保读者可以顺利完成本项目。

4.3.2 jPlayer 插件

jPlayer 是一个用 JavaScript 编写的完全免费和开源的 jQuery 多媒体库插件（现在也是一个 Zepto 插件）。jPlayer 可以让我们迅速编写跨平台的支持音频和视频播放的网页。jPlayer 丰富的 API 可以让我们创建个性化的多媒体应用，因此获得越来越多的社区成员的支持和鼓励。

1. 下载安装

jPlayer 插件的 github 网址为 https://github.com/jplayer/jPlayer。本项目使用 2.9.2 版本。

2. jPlayer 的基本使用

使用 jPlayer 时，需要先引入 jQuery 插件以及 jPlayer 的 CSS 文件和 JS 文件。接下来，在页面加载时，调用 $("#jquery_jplayer_1").jPlayer() 方法，并在 jPlayer() 方法内设置相应属性。参考代码如下：

```html
<!DOCTYPE html>
<html>
<head>
<meta charset="utf-8" />
<link href="../../dist/skin/blue.monday/css/jplayer.blue.monday.min.css" rel="stylesheet" type="text/css" />
<script type="text/javascript" src="../../lib/jquery.min.js"></script>
<script type="text/javascript" src="../../dist/jplayer/jquery.jplayer.min.js"></script>
<script type="text/javascript">
$(document).ready(function(){
    $("#jquery_jplayer_1").jPlayer({
        ready: function (event) {
            $(this).jPlayer("setMedia", {
                title: "Bubble",                                                  // 文件名称
                m4a: "http://jplayer.org/audio/m4a/Miaow-07-Bubble.m4a",         // 文件类型
                oga: "http://jplayer.org/audio/ogg/Miaow-07-Bubble.ogg"          // 文件类型
            });
        },
        swfPath: "../../dist/jplayer",              // 定义 jPlayer 的 jplayer.swf 文件的路径
        supplied: "m4a, oga",                        // 设置支持的文件类型
        wmode: "window",                             // 播放模式为 "window"
        useStateClassSkin: true,                     // 设置默认样式
        autoBlur: false,                             // GUI 交互时状态为 focus()
        smoothPlayBar: true,                         // 平滑过渡播放条
        keyEnabled: true,                            // 支持键盘
        remainingDuration: true,                     // 展示剩余时间
        toggleDuration: true                         // 单击 GUI 元素 duration 触发 jPlayer({remainingDuration}) 选项
    });
});
</script>
</head>
```

```
<body>
<div id="jquery_jplayer_1" class="jp-jplayer"></div>
<div id="jp_container_1" class="jp-audio" role="application" aria-label="media player">
    <!-- 省略部分代码 -->
</div>
</body>
</html>
```

在上述代码中，在 jPlayer()方法中只是设置了一部分参数，更多参数请参考 jPlayer 文档。运行结果如图 4.2 所示。

图 4.2　使用 jPlayer 插件播放音乐

4.3.3　蓝图

蓝图（Blueprint）是 Flask 中实现应用模块化的一种强大工具。它不仅有助于提高代码的可读性和可维护性，还能让应用的结构更加清晰，便于团队协作开发大型项目。

1．为什么使用蓝图

Flask 中的蓝图为以下这些情况设计。

☑　把一个应用分解为一个蓝图的集合。即一个项目可以实例化一个应用对象，初始化几个扩展，并注册一个集合的蓝图。这对大型应用是理想的。

☑　以 URL 前缀和/或子域名，在应用上注册一个蓝图。URL 前缀/子域名中的参数即成为这个蓝图下所有视图函数共同的视图参数（默认情况下）。

☑　在一个应用中，用不同的 URL 规则多次注册一个蓝图。

☑　通过蓝图提供模板过滤器、静态文件、模板和其他功能。一个蓝图不一定要实现应用或者视图函数。

☑　初始化一个 Flask 扩展时，在这些情况中注册一个蓝图。

Flask 中的蓝图不是即插应用，因为它实际上并不是一个应用——它是可以注册，甚至可以多次注册到应用上的操作集合。蓝图作为 Flask 层提供分割的替代，共享应用配置，并且在必要情况下可以更改所注册的应用对象。它的缺点是不能在应用创建后撤销注册一个蓝图而不销毁整个应用对象。

2．蓝图的设想

蓝图的基本设想是当它们注册到应用上时，它们记录将会被执行的操作。当分派请求和生成从一个端点到另一个端点的 URL 时，Flask 会关联蓝图中的视图函数。

3．创建蓝图

通常，将蓝图放在一个单独的包里。例如，创建一个 home 子目录，并创建一个空的"__init__.py"文件，表示它是一个 Python 的包。下面编写蓝图，将其保存在"home/__init__.py"文件中，代码如下：

```
from flask import Blueprint

home = Blueprint("home",__name__)

@home.route('/')
def index(name):
    return '<h1>Hello World!</h1>'
```

在上述代码中，创建了蓝图对象 home，它使用起来类似于 Flask 应用的 app 对象，它可以有自己的路由 home.route()。初始化 Blueprint 对象的第一个参数 home 指定了这个蓝图的名称，第二个参数指定了该蓝图所在的模块名,这里是当前文件。

4.注册蓝图

创建完蓝图后,需要注册蓝图。在 Flask 应用的主程序中,使用 app.register_blueprint()方法即可注册蓝图,代码如下:

```
from flask import Flask
from app.home import home as home_blueprint

app = Flask(__name__)
app.register_blueprint(home_blueprint, url_prefix='/home)

if __name__ == '__main__':
    app.run(debug=True)
```

在上述代码中,使用 app.register_blueprint()方法来注册蓝图,该方法的第一个参数是蓝图名称,第二个参数 url_prefix 是蓝图的 URL 前缀。也就是,当访问 http://localhost:5000/home/时就可以加载 home 蓝图的 index 视图了。

4.4 数据库设计

4.4.1 数据库概要说明

本项目采用 MySQL 数据库,数据库名称为 music,其中包含 4 张数据表,数据表名称及作用如表 4.1 所示。

表 4.1 数据库中的数据表及作用

表 名	含 义	作 用
user	用户表	用于存储用户的信息
song	歌曲表	用于存储歌曲信息
artist	歌手表	用于存储歌手信息
collect	收藏表	用于存储收藏歌曲信息

3.4.2 数据表结构

user 用户表的表结构如表 4.2 所示。

表 4.2 user 用户表的表结构

字 段	类 型	长 度	是否允许为空	含 义
id	INT	默认	否	主键,编号
username	VARCHAR	100	是	用户名
password	VARCHAR	255	是	密码
flag	TINYINT	1	是	用户标识,0:普通用户,1:管理员

song 歌曲表的表结构如表 4.3 所示。

表 4.3 song 歌曲表的表结构

字段	类型	长度	是否允许为空	含义
id	INT	默认	否	主键，编号
songName	VARCHAR	100	是	歌曲名称
singer	VARCHAR	100	是	歌手
fileURL	VARCHAR	100	是	歌曲图片路径
hits	INT	默认	是	点击数
style	INT	默认	是	歌曲类型，0：全部，1：流行，2：民歌，3：古典，4：摇滚，5：其他

artist 歌手表的表结构如表 4.4 所示。

表 4.4 artist 歌手表的表结构

字段	类型	长度	是否允许为空	含义
id	INT	默认	否	主键，编号
artistName	VARCHAR	100	是	歌手名
style	INT	默认	是	歌手类型
imgURL	VARCHAR	100	是	歌手图片路径
isHot	TINYINT	默认	是	是否热门

collect 收藏表的表结构如表 4.5 所示。

表 4.5 collect 收藏表的表结构

字段	类型	长度	是否允许为空	含义
id	INT	默认	否	主键，编号
song_id	INT	默认	是	所属歌曲
user_id	INT	默认	是	所属用户

4.4.3 数据表模型

本项目中使用 SQLAlchemy 进行数据库操作，将所有的模型放置到一个单独的 models 模块中，使程序的结构更加明晰，models 模块中的数据模型对应 MySQL 的每个数据表。

```python
from . import db

# 用户表
class User(db.Model):
    __tablename__ = "user"
    id = db.Column(db.Integer, primary_key=True)          # 编号
    username = db.Column(db.String(100))                  # 用户名
    pwd = db.Column(db.String(255))                       # 密码
    flag = db.Column(db.Boolean,default=0)                # 用户标识，0：普通用户，1：管理员

    def __repr__(self):
        return '<User %r>' % self.name

    def check_pwd(self, pwd):
        """
        检测密码是否正确
        :param pwd: 密码
        :return: 返回布尔值
```

```python
        """
        from werkzeug.security import check_password_hash
        return check_password_hash(self.pwd, pwd)

# 歌手表
class Artist(db.Model):
    __tablename__ = 'artist'
    id = db.Column(db.Integer, primary_key=True)          # 编号
    artistName = db.Column(db.String(100))                # 歌手名
    style = db.Column(db.Integer)                         # 歌手类型
    imgURL = db.Column(db.String(100))                    # 头像
    isHot = db.Column(db.Boolean,default=0)               # 是否热门

# 歌曲表
class Song(db.Model):
    __tablename__ = 'song'
    id = db.Column(db.Integer, primary_key=True)          # 编号
    songName = db.Column(db.String(100))                  # 歌曲名称
    singer = db.Column(db.String(100))                    # 歌手名称
    fileURL = db.Column(db.String(100))                   # 歌曲图片
    hits = db.Column(db.Integer,default=0)                # 点击量
    style = db.Column(db.Integer)         # 歌曲类型，0: 全部，1: 流行，2: 民歌，3: 古典，4: 摇滚，5: 其他
    collect = db.relationship('Collect', backref='song')  # 收藏外键关系关联

# 歌曲收藏
class Collect(db.Model):
    __tablename__ = "collect"
    id = db.Column(db.Integer, primary_key=True)          # 编号
    song_id = db.Column(db.Integer, db.ForeignKey('song.id'))  # 所属歌曲
    user_id = db.Column(db.Integer)                       # 所属用户
```

```sql
DROP TABLE IF EXISTS 'users';
CREATE TABLE 'users' (
  'id' int(8) NOT NULL AUTO_INCREMENT,
  'username' varchar(255) DEFAULT NULL,
  'email' varchar(255) DEFAULT NULL,
  'password' varchar(255) DEFAULT NULL,
  PRIMARY KEY ('id')
) ENGINE=InnoDB DEFAULT CHARSET=utf8;

DROP TABLE IF EXISTS 'articles';
CREATE TABLE 'articles' (
  'id' int(8) NOT NULL AUTO_INCREMENT,
  'title' varchar(255) DEFAULT NULL,
  'content' text,
  'author' varchar(255) DEFAULT NULL,
  'create_date' datetime DEFAULT NULL,
  PRIMARY KEY ('id')
) ENGINE=InnoDB DEFAULT CHARSET=utf8;
```

4.5 首页设计

4.5.1 首页概述

当用户访问甜橙音乐网时，首先进入的就是网站首页。在甜橙音乐网的首页中，用户可以浏览轮播图、

热门歌手和热门歌曲，同时通过菜单上的超链接也可以跳转到"排行榜""曲风""歌手"等页面。运行效果如图 4.3 所示。下面将重点介绍"热门歌手""热门歌曲"和"播放音乐"3 个主要功能模块。

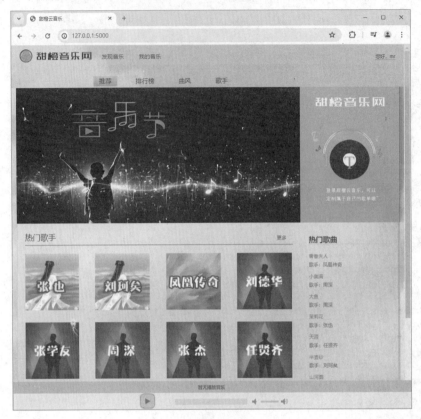

图 4.3　甜橙音乐网首页

4.5.2　实现热门歌手

1．获取热门歌手数据

热门歌手数据来源于 artist（歌手）表，该表中有 isHot（是否热门）字段。如果 isHot 字段的值为 1，则表示这条记录中的歌手是热门歌手；如果为 0，则表示非热门歌手。根据首页布局，从 user 表中筛选出 12 条 isHot 字段值为 1 的记录（不足 12 条时，则获取全部）。

使用 contentFrame()方法获取热门歌手数据，关键代码如下：

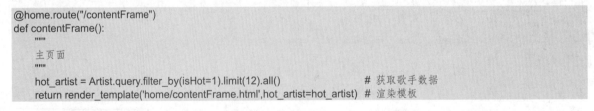

2．渲染热门歌手页面

在 contentFrame()方法中，使用 render_template()函数渲染模板，并将 hot_artist 变量赋值到模板，接下来，需要在 contentFrame.html 模板文件中展示数据。hot_artist 是所有热门歌手信息的集合，在模板中可以使用{%for%}标签来遍历数据，关键代码如下：

```
<div class="g-mn1">
    <div class="g-mn1c">
        <div class="g-wrap3">
            <div class="n-rcmd">
                <div class="v-hd2">
                    <a href="#" class="tit f-ff2 f-tdn">热门歌手</a>
                    <span class="more"><a href="{{url_for('home.artistList')}}"
                        class="s-fc3">更多</a><i class="cor s-bg s-bg-6"> </i> </span>
                </div>
                <ul class="m-cvrlst f-cb">
                    {% for artist in hot_artist %}
                    <li>
                        <div class="u-cover u-cover-1">
                            <a href="{{url_for('home.artist',id=artist.id)}}">
                                <img src="{{url_for('static',filename='images/artist/'+artist.imgURL)}}">
                            </a>
                        </div>
                    </li>
                    {% endfor %}
                </ul>
            </div>
        </div>
    </div>
</div>
```

首页中的热门歌手列表的运行结果如图 4.4 所示。

4.5.3 实现热门歌曲

1. 获取热门歌曲数据

热门歌曲数据来源于 song（歌曲）表，该表中有 hits（点击次数）字段。每当用户点击一次歌曲，该歌曲的 hits 字段值就加 1。根据首页布局，从 user 表中根据 hits 字段由高到低排序筛选出 10 条记录。使用 contentFrame()方法来获取热门歌曲数据，关键代码如下：

图 4.4 热门歌手列表的页面效果

```
@home.route("/contentFrame")
def contentFrame():
    """
    主页面
    """
    hot_song = Song.query.order_by(Song.hits.desc()).limit(10).all()    # 获取歌曲数据
    return render_template('home/contentFrame.html', hot_song=hot_song) # 渲染模板
```

2. 渲染热门歌曲页面

在 contentFrame()方法中，使用 render_template()函数渲染模板，并将 hot_song 变量赋值到模板，接下来，需要在 contentFrame.html 模板文件中展示数据。hot_song 是所有热门歌曲信息的集合，在模板中可以使用{%for%}标签来遍历数据，关键代码如下：

```
<div class="g-sd1">
    <div class="n-dj n-dj-1">
        <h1 class="v-hd3">
            热门歌曲
        </h1>
        <ul class="n-hotdj f-cb" id="hotdj-list">
            {% for song in hot_song %}
```

```html
<li>
    <div class="info">
        <p>
            <a onclick='playA("{{song.songName}}","{{song.id}}");'
                style="color: #1096A9">{{song.songName}} </a>
            <sup class="u-icn u-icn-1"></sup>
        </p>
        <p class="f-thide s-fc3">
            歌手：{{song.singer}}
        </p>
    </div>
</li>
{% endfor %}
    </ul>
</div>
```

首页中的热门歌曲栏目运行结果如图 4.5 所示。

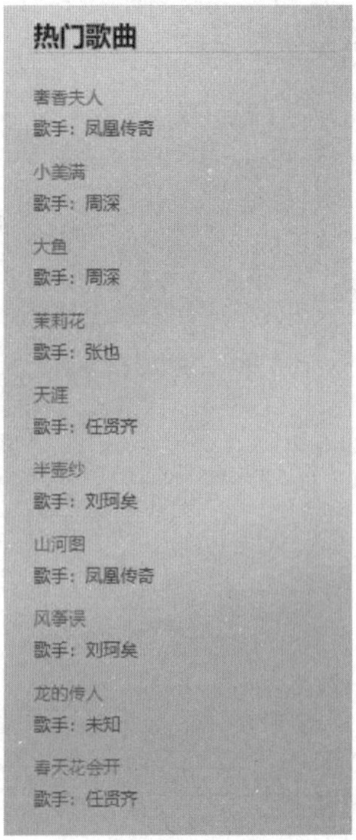

4.5.4 实现音乐播放

1．播放音乐

本项目使用 jPlayer 插件来实现播放音乐功能。使用 jPlayer 可以迅速编写一个跨平台的支持音频和视频播放的网页。

使用 jPlayer 前，需要先引入 jPlayer 对应的 JavaScript 文件和 CSS 文件，然后根据需求，编写相应的 JavaScript 代码。关键代码如下：

图 4.5 热门歌曲栏目的界面效果

```html
<link href="{{url_for('static',filename='css/jplayer.blue.monday.min.css')}}"
    rel="stylesheet" type="text/css" />
<script type="text/javascript"
        src="{{url_for('static',filename='js/jplayer/jquery.jplayer.min.js')}}"></script>
<script>
// 定义播放音乐的方法
function playMusic(name, id) {
    addMyList()                                              // 调用添加播放次数方法
    $("#jquery_jplayer").jPlayer( "destroy" );               // 销毁正在播放的音乐
    $("#jquery_jplayer").jPlayer({                           // 播放音乐
        ready: function(event) {                             // 准备音频
            $(this).jPlayer("setMedia", {
                title: name,                                 // 设置音乐标题
                mp3: "static/images/song/53.mp3"             // 设置播放音乐
            }).jPlayer( "play" );                            // 开始播放
        },
        swfPath: "dist/jplayer/jquery.jplayer.swf",          // 定义 jPlayer 的 jplayer.swf 文件的路径
        supplied: "mp3",                                     // 音乐格式为 mp3
        wmode: "window",                                     // 播放模式为 window
        useStateClassSkin: true,                             // 设置默认样式
        autoBlur: false,                                     // GUI 交互时状态为 focus()
        smoothPlayBar: true,                                 // 平滑过渡播放条
        keyEnabled: true,                                    // 支持键盘
        remainingDuration: true,                             // 显示剩余时间
        toggleDuration: true                                 // 点击 GUI 元素 duration 触发 jPlayer
    });
}
```

在上述代码中，首先需要对所有正在播放的音乐进行销毁处理，然后引入需要播放的音乐文件，设置音乐播放的题目，另外，还需要设置整个播放组件的相关参数信息，如是否支持图标、动画、进度条，最后播放音乐。

进入网站的首页后，单击热门歌曲中的任意一首，将会播放该歌曲。具体实现效果如图 4.6 所示。

图 4.6　音乐组件播放效果

2．统计播放次数

每次点击歌曲后，歌曲的点击次数都应该自动加 1。那么，在调用 playMusic()播放音乐时，可以调用一个自定义的 addMyList()方法，该方法使用 Ajax 异步提交的方式更改 song 表中 hits 字段的值。具体代码如下：

```
// 添加播放次数
function addMyList(id){
    $.ajax({
        url: "{{url_for('home.addHit')}}",         // 提交地址
        type: "get",                                // 提交类型
        data: {id: id},                             // 提交数据
        success: function(res) {                    // 回调函数
            console.log(res.message)
        }
    });
}
```

在上述代码中，使用 Ajax 将 id（歌曲 ID）使用 GET 方法提交到 home 蓝图下的 addHit()方法。因此，需要在 addHit()方法中，更改相应歌曲的点击次数。关键代码如下：

```
@home.route('/addHit')
def addHit():
    '''
    点击量加 1
    '''
    id = request.args.get('id')
    song = Song.query.get_or_404(int(id))
    if not song:
        res = {}
        res['status'] = -1
        res['message'] = '歌曲不存在'
    # 更改点击量
    else:
        song.hits += 1
        db.session.add(song)
        db.session.commit()
        res = {}
        res['status'] = 1
        res['message'] = '播放次数加 1'
    return jsonify(res)
```

在上述代码中，根据歌曲 ID 查找歌曲信息。如果歌曲存在，则令 hits 字段的值自增 1。最后使用 jsonify()函数返回 json 格式数据。

4.6　排行榜模块设计

4.6.1　排行榜模块概述

歌曲排行榜是音乐网站非常普遍的一个功能。从技术实现的原理上看，根据用户点击某歌曲的次数进

行排序，即形成了甜橙音乐网的歌曲排行榜。页面效果如图4.7所示。

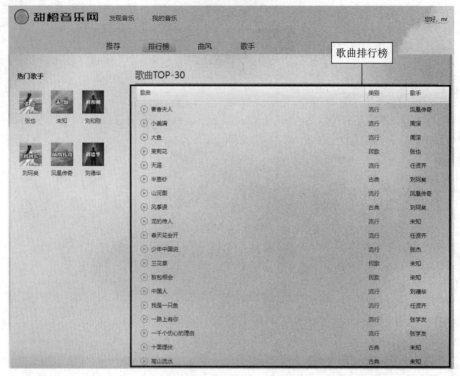

图4.7 排行榜的界面效果

4.6.2 实现歌曲排行榜

1. 获取排行榜数据

排行榜功能和首页的热门歌曲功能类似，区别在于首页热门歌曲只显示点击数量高的前10首歌曲的名称和歌手，而排行榜要显示排名前30首歌曲的详细信息。关键代码如下：

```
@home.route("/toplist")
def toplist():
    top_song = Song.query.order_by(Song.hits.desc()).limit(30).all()
    hot_artist = Artist.query.limit(6).all()
    return render_template('home/toplist.html', top_song=top_song, hot_artist=hot_artist)
```

2. 渲染歌曲排行榜页面

在 toplist()方法中，使用 render_template()函数渲染模板，并将 top_song 变量赋值给模板，接下来，需要在 toplist.html 模板文件中展示数据。top_song 是所有歌曲排行信息的集合，在模板中可以使用{%for%}标签来遍历数据，关键代码如下：

```
<div class="j-flag" id="auto-id-o5oRUwyIt22S4fpC">
    <table class="m-table m-table-rank">
        <thead>
            <tr>
                <th>
                    <div class="wp">
                        歌曲
                    </div>
                </th>
```

```html
                <th class="w2-1">
                    <div class="wp">
                        类别
                    </div>
                </th>
                <th class="w3">
                    <div class="wp">
                        歌手
                    </div>
                </th>
            </tr>
        </thead>
        <tbody>
        {% for song in top_song %}
            <tr class=" ">
                <td class="">
                    <div class="f-cb">
                        <div class="tt">
                            <span
                                onclick='playA("{{song.songName}}","{{song.id}}");'
                                class="ply "> </span>
                            <div class="ttc">
                                <span class="txt"><b>{{song.songName}} </b> </span>
                            </div>
                        </div>
                    </div>
                </td>
                <td class=" s-fc3">
                    <span class="u-dur ">
                        {% if song.style == 1 %}
                        流行
                        {% elif song.style == 2%}
                        民歌
                        {% elif song.style == 3%}
                        古典
                        {% elif song.style == 4%}
                        摇滚
                        {% elif song.style == 5%}
                        其他
                        {% endif %}
                    </span>
                    <div class="opt hshow">
                        <span onclick='addShow("{{song.id}}")' class="icn icn-fav" title="收藏">
                        </span>
                    </div>
                </td>
                <td class="">
                    <div class="text">
                        <span>{{song.singer}} </span>
                    </div>
                </td>
            </tr>
        {% endfor %}
        </tbody>
    </table>
</div>
```

在上述代码中，由于存储在数据库中的歌曲类型是数字，所以需要使用{%if%}标签判断 song.style 的值对应的歌曲类型数字。具体实现效果如图 4.8 所示。

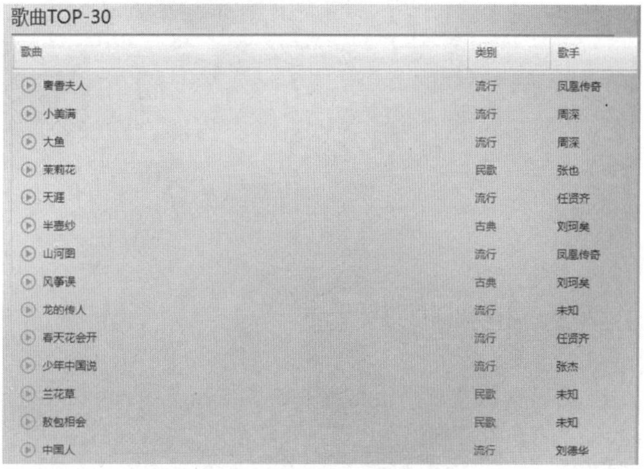

图 4.8 排行榜的页面实现效果

4.6.3 实现播放歌曲

在排行榜页面，单击歌曲名称左侧的播放图标，即可播放歌曲。由于在首页已经实现了歌曲的播放功能，所以在其他页面，可以共用首页的播放功能。在排行榜页面模板中，使用自定义函数 playA() 来调用父页面的播放功能。关键代码如下：

```
<script>
function playA(name,id){
    window.parent.playMusic(name,id);
}
</script>
```

在上述代码中，playA()函数接收两个参数：name 表示歌曲名称，用于在播放时显示播放歌曲名称；id 表示歌曲的 ID，用于播放歌曲时更改歌曲的点击量。playA()函数调用父页面的 playMusic()函数，继而实现播放歌曲的功能，运行结果如图 4.9 所示。

图 4.9　在排行榜中播放音乐

4.7　曲风模块设计

4.7.1　曲风模块概述

曲风模块主要是根据歌曲的风格浏览歌曲。在甜橙音乐网中，歌曲的曲风分成"全部""流行""民歌""古典""摇滚""其他"6 个子类。根据此分类标准，实现曲风模块的功能。曲风模块实现的效果如图 4.10 所示。

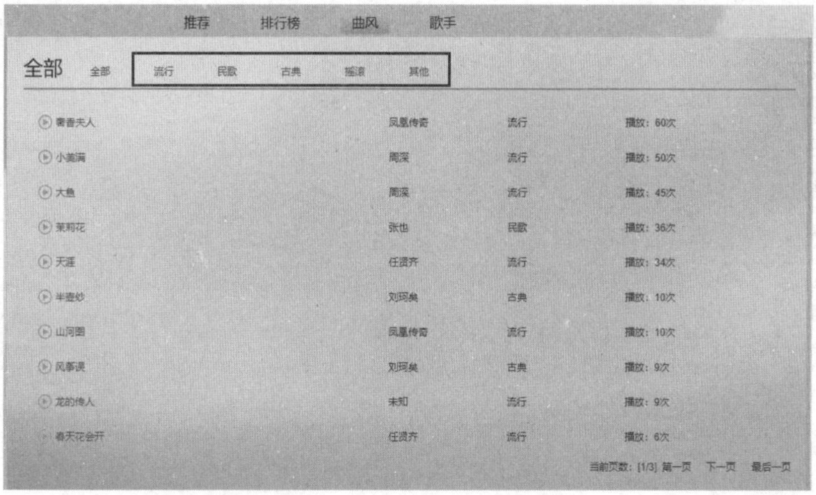

图 4.10　曲风模块的界面效果

4.7.2　实现曲风模块数据的获取

曲风模块功能和排行榜模块功能类似，区别在于排行榜模块只显示点击数量前 30 的歌曲信息，而曲风模块要显示所有的歌曲。为了更好地展示所有歌曲，还需要对歌曲进行分页。此外，还可以根据歌曲类型查找相应的歌曲。关键代码如下：

```python
@home.route('/style_list')
def styleList():
    """
    曲风
    """
    type = request.args.get('type',0,type=int)                    # 获取歌曲类型参数值
    page = request.args.get('page',type=int)                      # 获取 page 参数值
    if type:
        page_data = Song.query.filter_by(style=type).order_by(
                    Song.hits.desc()).paginate(page=page, per_page=10)
    else:
        page_data = Song.query.order_by(Song.hits.desc()).paginate(page=page, per_page=10)
    return render_template('home/styleList.html', page_data=page_data,type=type)    # 渲染模板
```

在上述代码中，首先判断 type 参数是否存在。如果 type 存在，则表示要筛选所有该类型的歌曲，否则，筛选所有类型的歌曲。

4.7.3　实现曲风模块页面的渲染

在 styleList()方法中，使用 render_template()函数渲染模板，并将 page_data 变量赋值到模板，接下来，需要在 styleList.html 模板文件中展示数据。page_data 是分页对象，page_data.items 则是所有歌曲信息的集合，在模板中可以使用{%for%}标签来遍历数据，关键代码如下：

```html
<div class="ztag j-flag" id="auto-id-oRFIQkCKNyCtcR5R">
    <div class="n-srchrst">
        <div class="srchsongst">
            {% for song in page_data.items %}
            <div class="item f-cb h-flag even ">
                <div class="td">
                    <div class="hd">
                        <a class="ply " title="播放"
```

```
                onclick='playA("{{song.songName}}","{{song.id}}");'></a>
            </div>
        </div>
        <div class="td w0">
            <div class="sn">
                <div class="text">
                    <b title="Lose Yourself "><span
                            class="s-fc7">{{song.songName}} </span></b>
                </div>
            </div>
        </div>
        <div class="td">
            <div class="opt hshow">
                <span onclick='addShow("{{song.id}}")' class="icn icn-fav" title=
                "收藏"></span>
            </div>
        </div>
        <div class="td w1">
            <div class="text">
                {{song.singer}}
            </div>
        </div>
        <div class="td w1">
            {% if song.style == 1 %}
                流行
            {% elif song.style == 2%}
                民歌
            {% elif song.style == 3%}
                古典
            {% elif song.style == 4%}
                摇滚
            {% elif song.style == 5%}
                其他
            {% endif %}
        </div>
        <div class="td">
            播放：{{song.hits}}次
        </div>
    </div>
    {% endfor %}
  </div>
 </div>
</div>
```

具体实现效果如图 4.11 所示。

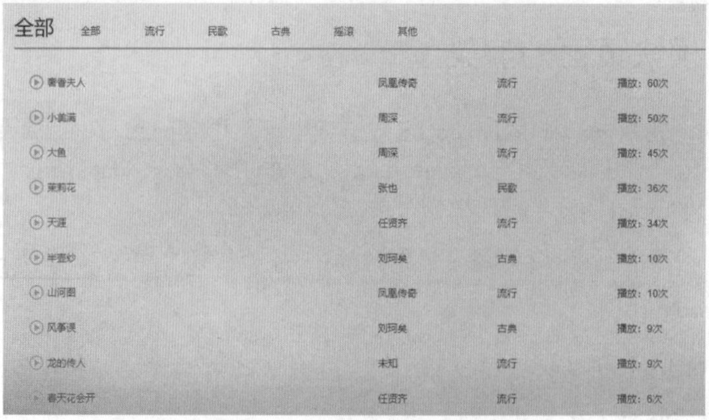

图 4.11　曲风列表的页面效果

4.7.4 实现曲风列表的分页功能

在曲风列表页中，由于歌曲数量较多，所以使用分页的方式展示歌曲数据。对于分页的处理，默认显示"第一页"和"最后一页"，然后再判断是否有上一页，如果有则显示超链接"上一页"；接下来再判断是否有下一页，如果有，则显示超链接"下一页"。实现分页功能的关键代码如下：

```
<table width="100%" border="0" cellspacing="0" cellpadding="0">
    <tr>
        <td height="24" align="right">
            当前页数：[{{page_data.page}}/{{page_data.pages}}] 
            <a href="{{ url_for('home.styleList',page=1,type=type) }}">第一页</a>
            {% if page_data.has_prev %}
                <a href="{{ url_for('home.styleList',page=page_data.prev_num,type=type) }}">
                上一页</a>
            {% endif %}
            {% if page_data.has_next %}
                <a href="{{ url_for('home.styleList',page=page_data.next_num,type=type) }}">
                下一页</a>
            {% endif %}
            <a href="{{ url_for('home.styleList',page=page_data.pages,type=type) }}">
            最后一页</a>
        </td>
    </tr>
</table>
```

在上述代码中，使用 page_data 分页类的相关属性。常用属性及说明如下。

- ☑ page_data.page：当前页数。
- ☑ page_data.pages：总页数。
- ☑ page_data.prev_num：上一页的页数。
- ☑ page_data.next_num：下一页的页数。

此外，在分页的链接中，传递了 type 参数，从而实现根据曲风类型进行分页。单击"曲风"菜单，实现效果如图 4.12 所示。

图 4.12　曲风列表分页组件的页面效果

4.8　发现音乐模块设计

4.8.1　发现音乐模块概述

发现音乐模块的功能实际上就是搜索音乐。在一般的音乐网站中，会提供根据歌手名、专辑名、歌曲名等检索条件进行搜索。本模块主要讲解根据歌曲名搜索歌曲的过程，其他搜索条件请读者自己尝试，实现原理都是相似的。发现音乐模块的页面效果如图 4.13 所示。

图 4.13 发现音乐的界面效果

4.8.2 实现发现音乐的搜索功能

当用户在搜索栏中输入歌曲名称并单击搜索按钮时,程序会根据用户输入的歌曲名进行模糊查询,这就需要使用 SQL 语句中的 like 语句。实现模糊查询功能的关键代码如下:

```python
@home.route('/search')
def search():
    keyword = request.args.get('keyword')           # 获取关键字
    page = request.args.get('page', type=int)       # 获取 page 参数值
    if keyword :
        keyword = keyword.strip()
        page_data = Song.query.filter(
                    Song.songName.like('%'+keyword+'%')).order_by(
                    Song.hits.desc()).paginate(page=page, per_page=10)
    else:
        page_data = Song.query.order_by(Song.hits.desc()).paginate(page=page, per_page=10)
    return render_template('home/search.html',keyword=keyword,page_data=page_data)
```

在上述代码中,首先接收用户输入的关键字 keyword,然后去除 keyword 左右空格,防止用户误输入空格。接着,使用 like 语句并结合分页功能实现分页查询。

4.8.3 实现发现音乐模块页面的渲染

发现音乐模块的主要功能是搜索音乐。对于搜索框,这里使用一个 form 表单,在该表单中,只包含一个 keyword 字段,当单击搜索按钮时,以 GET 方式提交表单。最后将获取到的结果显示在该页面。渲染页面的关键代码如下:

```html
<div class="g-bd" id="m-disc-pl-c">
    <div class="g-wrap n-srch">
        <div class="pgsrch f-pr j-suggest" id="auto-id-ErvdJrthwDvbXbzT">
            <form id="searchForm" action="" method="get">
                <input   type="text" name="keyword" class="srch j-flag" value=""
                    placeholder="请输入歌曲名称">
                <a hidefocus="true" href="javascript:document.getElementById('searchForm').submit();"
                    class="btn j-flag"
                    title="搜索" >搜索</a>
            </form>
        </div>
    </div>
    <div class="g-wrap p-pl f-pr">
        <div class="u-title f-cb">
            <h3>
                <span class="f-ff2 d-flag">搜索结果</span>
            </h3>
        </div>
        <div id="m-search">
            <div class="ztag j-flag" id="auto-id-oRFIQkCKNyCtcR5R">
                <div class="n-srchrst">
                    <div class="srchsongst">
                        {% for song in page_data.items%}
                        <div class="item f-cb h-flag even ">
                            <div class="td">
                                <div class="hd">
                                    <a class="ply " title="播放"
                                        onclick='playA("{{song.songName}}","{{song.id}}");'></a>
                                </div>
                            </div>
                            <div class="td w0">
                                <div class="sn">
                                    <div class="text">
                                        <b title="{{song.songName}}"><span
                                                class="s-fc7">{{song.songName}} </span></b>
                                    </div>
                                </div>
                            </div>
                            <div class="td">
                                <div class="opt hshow">
                                    <span onclick='addShow("{{song.id}}")' title="收藏"></span>
                                </div>
                            </div>
                            <div class="td w1">
                                <div class="text">
                                    {{song.singer}}
                                </div>
                            </div>
                            <div class="td w1">
                                {% if song.style == 1 %}
                                    流行
                                {% elif song.style == 2%}
                                    民歌
                                {% elif song.style == 3%}
                                    古典
                                {% elif song.style == 4%}
                                    摇滚
                                {% elif song.style == 5%}
                                    其他
                                {% else %}
                                    全部
                                {% endif %}
```

```
                </div>
                <div class="td">
                    播放：{{song.hits}}次
                </div>
            </div>
        {% endfor %}
        </div>
      </div>
    </div>
  </div>
</div>
```

在搜索框内，输入歌曲名的关键字，如"天"，然后单击搜索按钮，将会筛选出所有歌曲名称中包含"天"字的歌曲信息。运行效果如图4.14所示。

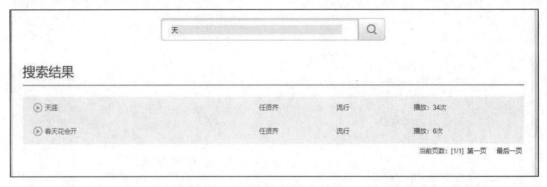

图4.14 搜索页面运行效果

4.9 歌手模块设计

4.9.1 歌手模块概述

歌手模块用于根据歌手浏览歌曲，即显示歌手所唱歌曲的列表。在甜橙音乐网中，歌手又根据风格属性划分成"全部""流行""民歌""古典""摇滚""其他"等类目，方便用户根据歌手风格查询相应的歌曲。歌手模块的运行效果如图4.15和图4.16所示。

图4.15 歌手列表页面

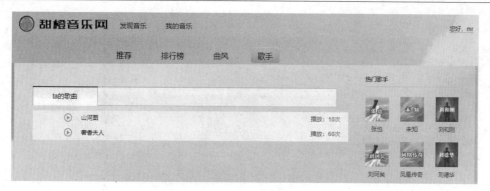

图 4.16 歌手详情页面

4.9.2 实现歌手列表

歌手模块的功能和曲风模块的功能类似，可以根据歌手的类型查看相关歌手。所以在获取歌手信息时，需要传递歌手类型参数。关键代码如下：

```python
@home.route('/artist_list')
def artistList():
    '''
    歌手列表
    '''
    type = request.args.get('type',0,type=int)
    page = request.args.get('page',type=int)    # 获取 page 参数值
    if type:
        page_data = Artist.query.filter_by(style=type).paginate(page=page, per_page=10)
    else:
        page_data = Artist.query.paginate(page=page, per_page=10)
    # 渲染模板
    return render_template('home/artistList.html', page_data=page_data,type=type)
```

歌手列表模板与曲风列表模板类似，这里不再赘述。具体实现效果如图 4.17 所示。

图 4.17 歌手列表的页面效果

4.9.3 实现歌手详情

在歌手列表页面单击歌手图片，可以根据歌手 ID 跳转到歌手详情页。然后，根据歌手 ID，联合查询

song（歌曲）表和 artist（歌手）表，获取该歌手的所有歌曲信息。关键代码如下：

```
@home.route("/artist/<int:id>")
def artist(id=None):
    """
    歌手页面
    """
    song = Song.query.join(Artist,Song.singer==Artist.artistName).filter(Artist.id== id).all()
    hot_artist = Artist.query.limit(6).all()
    # 渲染模板
    return render_template('home/artist.html',song=song,hot_artist=hot_artist)
```

在歌手列表页面，单击任意歌手项，将显示图 4.18 所示的运行效果。

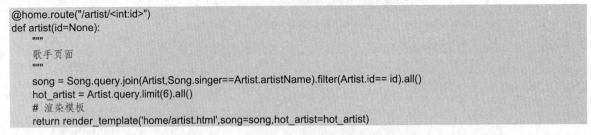

图 4.18　歌手详情页面的运行效果

4.10　我的音乐模块设计

4.10.1　我的音乐模块概述

用户在使用甜橙音乐网时，如果遇到喜欢的音乐可以单击收藏按钮进行收藏。程序会先判断该用户是否已经登录，如果已经登录，可以直接收藏，否则提示请先登录。

收藏的全部音乐可以在"我的音乐"列表中查看，如图 4.19 所示。

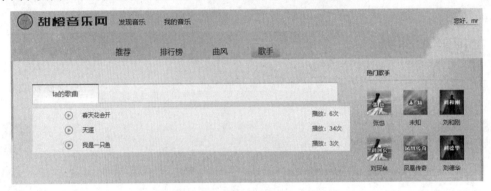

图 4.19　收藏列表

4.10.2 实现收藏歌曲

本项目中的多个页面都可以收藏歌曲,如排行榜页面、曲风页面、歌手详情页等。在这些页面中,当鼠标指针悬浮至歌曲的相应列时,即显示收藏按钮,如图 4.20 所示。

图 4.20　显示收藏按钮

以排行榜页面为例,当单击收藏按钮时,将调用 addShow()函数,关键代码如下:

```
<script>
function addShow(id){
    window.parent.addShow(id);
}
</script>
```

在上述代码中,addShow()函数接收一个 id 参数,即收藏歌曲的 ID。接下来,调用父页面的 addShow()函数,即在父页面实现收藏功能。关键代码如下:

```
// 添加收藏
function addShow(id){
    var username= '{{session['username']}}';
    if(username=="null" || username==""){
        layer.msg("收藏请先登录!",{icon:2,time:1000});
        return;
    }
    $.ajax({
        url: "{{url_for('home.collect')}}",
        type: "get",
        data: {
            id: id
        },
        success: function(res){
            if(res.status==1){
                layer.msg(res.message,{icon:1})
            }else{
                layer.msg(res.message,{icon:2})
            }
        }
    });
}
```

在上述代码中,先通过 session['username']判断用户是否登录,如果没有登录,则提示"收藏请先登录!",

运行效果如图 4.21 所示。

图 4.21 登录提示

如果用户已经登录，那么使用 Ajax 异步提交到 home 蓝图下的 collect() 方法，执行收藏的相关逻辑。关键代码如下：

```python
@home.route("/collect")
@user_login
def collect():
    """
    收藏歌曲
    """
    song_id = request.args.get("id", "")    # 接收传递的歌曲 ID 参数
    user_id = session['user_id']             # 获取当前用户的 ID
    collect = Collect.query.filter_by(       # 根据用户 ID 和歌曲 ID 判断是否应该收藏
        user_id =int(user_id),
        song_id=int(song_id)
    ).count()
    res = {}
    # 已收藏
    if collect == 1:
        res['status'] = 0
        res['message'] = '已经收藏'
    # 如未收藏，则收藏
    if collect == 0:
        collect = Collect(
            user_id =int(user_id),
            song_id=int(song_id)
        )
        db.session.add(collect)              # 添加数据
        db.session.commit()                  # 提交数据
        res['status'] = 1
        res['message'] = '收藏成功'
    return jsonify(res)                      # 返回 json 数据
```

在上述代码中，首先接收歌曲 ID 和登录用户 ID。接下来，根据歌曲 ID 和用户 ID 查找 collect 表，如果表中存在记录，那么，表示该用户已经收藏了这首歌曲，提示"已经收藏"，否则，将歌曲 ID 和登录用户 ID 写入 collect 表，最后使用 jsonify() 函数返回 json 数据。

登录账号后，在排行榜页面选中歌曲，单击收藏按钮，显示"收藏成功"，运行效果如图 4.22 所示。再次单击收藏按钮，收藏该歌曲，显示"已经收藏"，运行效果如图 4.23 所示。

图 4.22　提示收藏成功

图 4.23　提示已经收藏

4.10.3　实现我的音乐

用户收藏完歌曲后，可以单击"我的音乐"菜单查看所有收藏的音乐。收藏音乐信息来源于 collect 表，根据当前用户的 ID 查询该用户收藏的所有歌曲，关键代码如下：

```
@home.route("/collect_list")
@user_login
def collectList():
    page = request.args. get('page',type=int)        # 获取 page 参数值
    user_id = session['user_id']                      # 获取当前用户的 ID
    page_data = Collect.query.filter_by( user_id=int(user_id)).paginate(page=page, per_page=10)
    return render_template('home/collectList.html',page_data=page_data)
```

接下来，渲染我的音乐模板页面。关键代码如下：

```html
<div class="ztag j-flag" id="auto-id-oRFIQkCKNyCtcR5R">
    <div class="n-srchrst">
        <div class="srchsongst">
            {% for collect in page_data.items %}
            <div class="item f-cb h-flag even ">
                <div class="td">
                    <div class="hd">
                        <a class="ply " title="播放"  onclick='playA("{{collect.song.songName}}","{{collect.song.id}}");'></a>
                    </div>
                </div>
                <div class="td w0">
                    <div class="sn">
                        <div class="text">
                            <b title="Lose Yourself "><span class="s-fc7">{{collect.song.songName}}</span></b>
                        </div>
                    </div>
                </div>

                <div class="td w1">
                    <div class="text">
                        {{collect.song.singer}}
                    </div>
                </div>
                <div class="td w1">
                    {% if collect.song.style == 1 %}
                    流行
                    {% elif collect.song.style == 2%}
                    民歌
                    {% elif collect.song.style == 3%}
                    古典
                    {% elif collect.song.style == 4%}
                    摇滚
                    {% elif collect.song.style == 5%}
                    其他
                    {% endif %}
                </div>
                <div class="td">
                    播放：{{collect.song.hits}}次
                </div>
            </div>
            {% endfor %}
        </div>
    </div>
</div>
```

运行结果如图 4.24 所示。

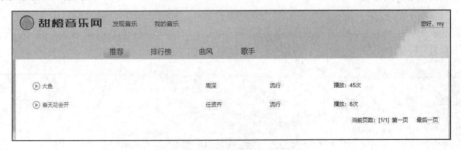

图 4.24　我的音乐页面效果

4.11 项目运行

通过前述步骤，设计并完成了"甜橙音乐网"项目的开发。下面运行该项目，检验一下我们的开发成果。运行"甜橙音乐网"项目的步骤如下。

（1）打开 OnlineMusic\app__init__.py 文件，根据自己的数据库账号和密码修改如下代码：

```
def create_app():
    app = Flask(__name__)
    app.config['SQLALCHEMY_DATABASE_URI']='mysql+pymysql://root:root@127.0.0.1:3306/music'
    app.config['SQLALCHEMY_TRACK_MODIFICATIONS'] = True
    app.config['SECRET_KEY'] = 'mr'
    db.init_app(app)
    migrate = Migrate(app, db)
```

（2）打开命令提示符对话框，进入 OnlineMusic 项目文件夹所在目录，在命令提示符对话框中输入如下命令来创建 venv 虚拟环境：

```
virtualenv venv
```

（3）在命令提示符对话框中输入如下命令来启动 venv 虚拟环境：

```
venv\Scripts\activate
```

（4）在命令提示符对话框中使用如下命令来安装 Flask 等依赖包：

```
pip install -r  requirements.txt
```

（5）创建数据库。可以使用 MySQL 命令行方式或 MySQL 可视化管理工具（如 Navicat）创建数据库。例如，在 MySQL 命令行窗口（MySQL 8.0 Command line Client）中输入登录密码后，使用下面的命令来实现。

```
create database music default character set utf8;
```

（6）在命令提示符对话框中使用如下命令创建数据表：

```
flask  db   init           # 创建迁移仓库,首次使用
flask  db   migrate        # 创建迁移脚本
flask  db   upgrade        # 把迁移应用到数据库中
```

（7）新增的数据表中数据为空，所以需要导入数据。将 OnlineMusic\music.sql 文件导入数据库中。

```
SOURCE D:\Code\OnlineMusic\music.sql
```

（8）在 PyCharm 的左侧项目结构中展开"甜橙音乐网"的项目文件夹 OnlineMusic，在其中选中 app.py 文件，单击鼠标右键，在弹出的快捷菜单中选择 Run 'app'命令，如图 4.25 所示。

（9）如果在 PyCharm 底部出现如图 4.26 所示的提示，说明程序运行成功。

（10）在浏览器中输入网址 http://127.0.0.1:5000/即可进入甜橙音乐网的首页，效果如图 4.27 所示。在甜橙音乐网的首页中，可以浏览热门歌手和热门歌曲，并且通过上面的导航还可以查看排行榜、发现音乐、收藏的音乐，以及根据曲风查找歌曲等。

本章运用软件工程的设计思想，通过开发一个完整的甜橙音乐网项目带领读者详细走完一个系统的开发流程。在程序开发过程中，采用了后端开发框架 Flask，并结合 iPlayer 组件等 Web 前端技术，使整个系统的视觉体验效果更加完美。通过本章的学习，读者不仅可以了解一般网站的开发流程，而且应该对前端技术有了比较深入的了解，掌握这些知识将对以后的开发工作大有裨益。

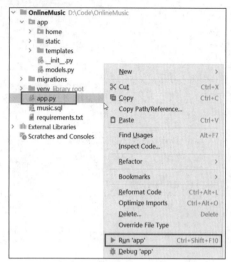

图 4.25　选择 Run 'app'

图 4.26　程序运行成功提示

图 4.27　甜橙音乐网的首页

4.12　源　码　下　载

源码下载

虽然本章详细地讲解了如何编码实现"甜橙音乐网"的各个功能，但给出的代码都是代码片段，而非完整的源代码。为了方便读者学习，本书提供了该项目的完整源代码，读者可以通过扫描右侧的二维码进行下载。

第 5 章 乐购甄选在线商城

——Flask + SQLALchemy + MySQL

网络购物已经不再是什么新鲜事物,当今无论是企业,还是个人,都可以很方便地在网上交易商品,批发零售。比如在淘宝上开网店、在微信上开微店、在抖音上开抖店等。本章使用 Python Web 框架 Flask 开发一个综合的在线商城项目——乐购甄选在线商城。

本项目的核心功能及实现技术如下:

项目微视频

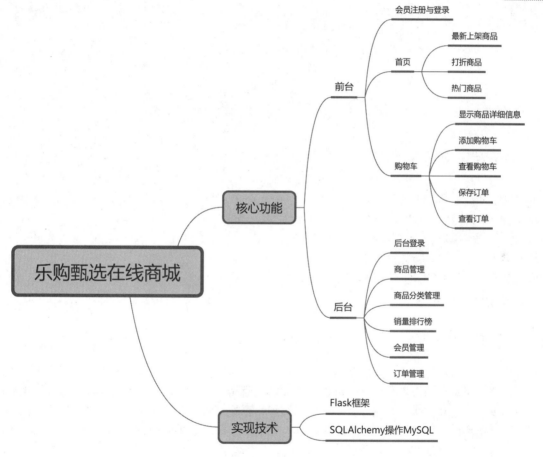

5.1 开发背景

随着电子商务的快速发展,越来越多的消费者选择在线购物,这为企业提供了巨大的市场机遇。乐购甄选在线商城旨在为消费者提供一个便捷、安全、丰富的在线购物平台,商城将集合各类优质商品,满足用户的多样化需求。

Flask 框架是一种轻量级的 Python Web 框架，它具有灵活性强、扩展库丰富、易学易用以及出色的性能与安全性等优点，是在线商城开发的理想选择。通过使用 Flask 框架，开发者可以更加高效、便捷地构建出功能丰富、性能稳定、安全可靠的在线购物平台。

本项目的实现目标如下：
- ☑ 用户能够方便地实现会员注册与登录。
- ☑ 用户能够方便地浏览和搜索商品，了解商品的详细信息。
- ☑ 用户能够方便地对自己的购物车进行管理。
- ☑ 用户能够安全地进行在线支付，并随时查看订单状态。
- ☑ 用户能够管理自己的个人信息和购物记录。
- ☑ 商城管理员能够方便地管理商品信息，包括添加、编辑和删除商品等。
- ☑ 商城管理员能够方便地对商品的分类进行管理，包括添加、删除、查看等。
- ☑ 商城管理员能够方便地查看商品销量排行榜。
- ☑ 商城管理员能够方便地查看用户会员注册信息并管理用户会员。
- ☑ 商城管理员能够方便地查看所有的用户订单。
- ☑ 商城需要保证数据安全，防止信息泄露和非法访问。

5.2 系统设计

5.2.1 开发环境

本项目的开发及运行环境如下：
- ☑ 操作系统：推荐 Windows 10、Windows 11 或更高版本。
- ☑ 开发工具：PyCharm 2024（向下兼容）。
- ☑ 开发语言：Python 3.12。
- ☑ 数据库：MySQL 8.0+PyMySQL 驱动。
- ☑ Python Web 框架：Flask 3.0。

5.2.2 业务流程

乐购甄选在线商城分为前台和后台两个部分。其中，前台首页默认显示商品列表，用户可以单击查看指定商品的详细信息，当用户登录后，可以选择和购买商品，并且可以对自己的购物车进行管理；而后台主要是对商城中的商品信息、商品分类信息进行添加、删除、修改等管理操作，并且可以查看商城的注册会员信息、订单信息以及相关商品的销量排行。本项目的业务流程如图 5.1 所示。

5.2.3 功能结构

本项目的功能结构已经在章首页中给出，其实现的具体功能如下：
- ☑ 前台：
  - ➢ 会员注册与登录功能。
  - ➢ 首页广告幻灯片展示功能。
  - ➢ 首页商品展示功能，包括展示最新上架商品、展示打折商品和展示热门商品等。

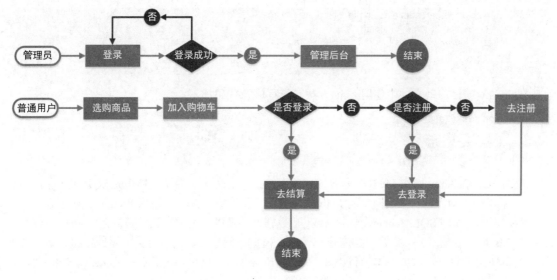

图 5.1　乐购甄选在线商城业务流程

- ➢ 查看商品详情功能，可以用于展示商品的详细信息。
- ➢ 购物车功能，用户可以将商品添加至购物车。
- ➢ 查看购物车，用户可以查看购物车中的所有商品，并且可以更改购买商品的数量、清空购物车等。
- ➢ 订单处理功能，用户可以填写地址信息，用于接收商品；提交订单后，显示支付宝收款码；完成订单后，可以查看订单详情。

☑ 后台：
- ➢ 管理员登录与注销功能。
- ➢ 商品管理功能，包括新增商品、编辑商品和删除商品等。
- ➢ 商品销量排行榜功能，根据商品销量显示商品排行榜。
- ➢ 商品分类管理功能，包括商品大分类、子分类的添加、删除、查看等。
- ➢ 会员管理功能，包括查看会员信息等。
- ➢ 订单管理功能，包括查看订单信息等。

5.3　技术准备

实现乐购甄选在线商城时，主要使用了 Flask 框架技术、SQLAlchemy 操作 MySQL 数据库技术。基于此，这里将本项目所用的核心技术点及其具体作用简述如下。

☑ Flask 框架的使用：Flask 是一个轻量级 Python Web 框架，它把 Werkzeug 和 Jinja 黏合在一起，能够很容易地被扩展。例如，本项目在 __init__.py 初始化文件中创建 Flask 实例对象，并注册蓝图，代码如下：

```
from flask import Flask

def create_app():
    app = Flask(__name__)
    # 注册蓝图
    from app.home import home as home_blueprint
```

```
from.app.admin   import admin as admin_blueprint
app.register_blueprint(home_blueprint)
app.register_blueprint(admin_blueprint,url_prefix="/admin")

return app
```

然后在 app.py 项目启动文件中使用 run()方法运行程序，代码如下：

```
from app import create_app

app = create_app()
app.run()
```

- ☑ 使用 SQLAlchemy 操作 MySQL 数据库：SQLAlchemy 是一个流行的 Python SQL 工具包和对象关系映射器（ORM），它采用简单的 Python 类表示数据库表，并允许使用 Python 代码创建、查询和更新数据库。使用 SQLAlchemy 操作 MySQL 数据库涉及以下关键步骤：安装必要的库（SQLAlchemy 和 MySQL 驱动库）、配置数据库连接、定义模型、创建表、创建会话、添加数据、执行查询以及更新和删除数据。例如，本项目中在__init__.py 初始化文件中通过 SQLAlchemy 对象的 init_app()方法配置 MySQL 数据库连接，关键代码如下：

```
db = SQLAlchemy()
def create_app():
    app = Flask(__name__)
    app.config['SQLALCHEMY_DATABASE_URI']='mysql+pymysql://root:root@127.0.0.1:3306/shop'
    app.config['SQLALCHEMY_TRACK_MODIFICATIONS'] = True
    app.config['SECRET_KEY'] = 'mr'
    db.init_app(app)
```

有关 Flask 框架的使用方法，在《Python 从入门到精通（第 3 版）》中有详细的讲解，对该知识不太熟悉的读者可以参考该书对应的内容。关于使用 SQLAlchemy 操作 MySQL 数据库的相关知识，可以参考本书 1.3.2 节。

5.4 数据库设计

5.4.1 数据库概要说明

本项目采用 MySQL 数据库，数据库名称为 shop，其中包含 8 张数据表，数据表名称及作用如表 5.1 所示。

表 5.1 数据库中的数据表及作用

表 名	含 义	作 用
admin	管理员表	用于存储管理员信息
user	用户表	用于存储用户的信息
goods	商品表	用于存储商品信息
cart	购物车表	用于存储购物车信息
orders	订单表	用于存储订单信息
orders_detail	订单明细表	用于存储订单明细信息
supercat	商品大分类表	用于存储商品大分类信息
subcat	商品子分类表	用于存储商品子分类信息

5.4.2 数据表结构

admin 管理员表的表结构如表 5.2 所示。

表 5.2 admin 管理员表的表结构

字 段	类 型	长 度	是否允许为空	含 义
id	int	默认	否	主键,编号
manager	varchar	100	是	管理员账号
password	varchar	100	是	管理员密码

user 用户表的表结构如表 5.3 所示。

表 5.3 user 用户表的表结构

字 段	类 型	长 度	是否允许为空	含 义
id	int	默认	否	主键,编号
username	varchar	100	是	用户名
email	varchar	100	是	邮箱
phone	varchar	11	是	手机号
addtime	datetime	默认	是	注册时间
password	text	默认	是	密码
consumption	decimal	10	是	消费额

goods 商品表的表结构如表 5.4 所示。

表 5.4 goods 商品表的表结构

字 段	类 型	长 度	是否允许为空	含 义
id	int	默认	否	主键,编号
name	varchar	255	是	商品名称
original_price	decimal	10	是	原价
current_price	decimal	10	是	现价
picture	varchar	255	是	商品图片
introduction	text	默认	是	商品简介
is_sale	tinyint	默认	是	是否特价
is_new	tinyint	默认	是	是否新品
addtime	datetime	默认	是	添加时间
views_count	int	默认	是	浏览次数
subcat_id	int	默认	是	所属子分类
supercat_id	int	默认	是	所属大分类

cart 购物车表的表结构如表 5.5 所示。

表 5.5 cart 购物车表的表结构

字 段	类 型	长 度	是否允许为空	含 义
id	int	默认	否	主键,编号
goods_id	int	默认	是	所属商品

续表

字段	类型	长度	是否允许为空	含义
user_id	int	默认	是	所属用户
number	int	默认	是	购买数量
addtime	datetime	默认	是	添加时间

orders 订单表的表结构如表 5.6 所示。

表 5.6 orders 订单表的表结构

字段	类型	长度	是否允许为空	含义
id	int	默认	否	主键，编号
user_id	int	默认	是	所属用户
recevie_name	varchar	255	是	收款人姓名
recevie_address	varchar	255	是	收款人地址
recevie_tel	varchar	255	是	收款人电话
remark	varchar	255	是	备注信息
addtime	datetime	默认	是	添加时间

orders_detail 订单明细表的表结构如表 5.7 所示。

表 5.7 orders_detail 订单明细表的表结构

字段	类型	长度	是否允许为空	含义
id	int	默认	否	主键，编号
goods_id	int	默认	是	所属商品
order_id	int	默认	是	所属订单
number	int	默认	是	购买数量

supercat 商品大分类表的表结构如表 5.8 所示。

表 5.8 supercat 商品大分类表的表结构

字段	类型	长度	是否允许为空	含义
id	int	默认	否	主键，编号
cat_name	varchar	100	是	大分类名称
addtime	datetime	默认	是	添加时间

subcat 商品子分类表的表结构如表 5.9 所示。

表 5.9 subcat 商品子分类表的表结构

字段	类型	长度	是否允许为空	含义
id	int	默认	否	主键，编号
cat_name	varchar	100	是	子分类名称
addtime	datetime	默认	是	添加时间
super_cat_id	int	默认	是	所属大分类

5.4.3 数据表模型

本项目中使用 SQLAlchemy 进行数据库操作，将所有的模型放置到一个单独的 models 模块中，使程序的结构更加明晰，models 模块中的数据模型对应 MySQL 的每个数据表。关键代码如下：

```python
from . import db
from datetime import datetime

# 用户数据模型
class User(db.Model):
    __tablename__ = "user"
    id = db.Column(db.Integer, primary_key=True)                          # 编号
    username = db.Column(db.String(100))                                  # 用户名
    password = db.Column(db.String(100))                                  # 密码
    email = db.Column(db.String(100), unique=True)                        # 邮箱
    phone = db.Column(db.String(11), unique=True)                         # 手机号
    consumption = db.Column(db.DECIMAL(10, 2), default=0)                 # 消费额
    addtime = db.Column(db.DateTime, index=True, default=datetime.now)    # 注册时间
    orders = db.relationship('Orders', backref='user')                    # 订单外键关系关联

    def __repr__(self):
        return '<User %r>' % self.name

    def check_password(self, password):
        """
        检测密码是否正确
        :param password: 密码
        :return: 返回布尔值
        """
        from werkzeug.security import check_password_hash
        return check_password_hash(self.password, password)

# 管理员
class Admin(db.Model):
    __tablename__ = "admin"
    id = db.Column(db.Integer, primary_key=True)                          # 编号
    manager = db.Column(db.String(100), unique=True)                      # 管理员账号
    password = db.Column(db.String(100))                                  # 管理员密码

    def __repr__(self):
        return "<Admin %r>" % self.manager

    def check_password(self, password):
        """
        检测密码是否正确
        :param password: 密码
        :return: 返回布尔值
        """
        from werkzeug.security import check_password_hash
        return check_password_hash(self.password, password)

# 大分类
class SuperCat(db.Model):
    __tablename__ = "supercat"
    id = db.Column(db.Integer, primary_key=True)                          # 编号
    cat_name = db.Column(db.String(100))                                  # 大分类名称
    addtime = db.Column(db.DateTime, index=True, default=datetime.now)    # 添加时间
    subcat = db.relationship("SubCat", backref='supercat')                # 外键关系关联
    goods = db.relationship("Goods", backref='supercat')                  # 外键关系关联

    def __repr__(self):
        return "<SuperCat %r>" % self.cat_name
```

```python
# 子分类
class SubCat(db.Model):
    __tablename__ = "subcat"
    id = db.Column(db.Integer, primary_key=True)                              # 编号
    cat_name = db.Column(db.String(100))                                      # 子分类名称
    addtime = db.Column(db.DateTime, index=True, default=datetime.now)        # 添加时间
    super_cat_id = db.Column(db.Integer, db.ForeignKey('supercat.id'))        # 所属大分类
    goods = db.relationship("Goods", backref='subcat')                        # 外键关系关联

    def __repr__(self):
        return "<SubCat %r>" % self.cat_name

# 商品
class Goods(db.Model):
    __tablename__ = "goods"
    id = db.Column(db.Integer, primary_key=True)                              # 编号
    name = db.Column(db.String(255))                                          # 名称
    original_price = db.Column(db.DECIMAL(10,2))                              # 原价
    current_price  = db.Column(db.DECIMAL(10,2))                              # 现价
    picture = db.Column(db.String(255))                                       # 图片
    introduction = db.Column(db.Text)                                         # 商品简介
    views_count = db.Column(db.Integer,default=0)                             # 浏览次数
    is_sale   = db.Column(db.Boolean(), default=0)                            # 是否特价
    is_new = db.Column(db.Boolean(), default=0)                               # 是否新品

    # 设置外键
    supercat_id = db.Column(db.Integer, db.ForeignKey('supercat.id'))         # 所属大分类
    subcat_id = db.Column(db.Integer, db.ForeignKey('subcat.id'))             # 所属子分类
    addtime = db.Column(db.DateTime, index=True, default=datetime.now)        # 添加时间
    cart = db.relationship("Cart", backref='goods')                           # 订单外键关系关联
    orders_detail = db.relationship("OrdersDetail", backref='goods')          # 订单外键关系关联

    def __repr__(self):
        return "<Goods %r>" % self.name

# 购物车
class Cart(db.Model):
    __tablename__ = 'cart'
    id = db.Column(db.Integer, primary_key=True)                              # 编号
    goods_id = db.Column(db.Integer, db.ForeignKey('goods.id'))               # 所属商品
    user_id = db.Column(db.Integer)                                           # 所属用户
    number = db.Column(db.Integer, default=0)                                 # 购买数量
    addtime = db.Column(db.DateTime, index=True, default=datetime.now)        # 添加时间
    def __repr__(self):
        return "<Cart %r>" % self.id

# 订单
class Orders(db.Model):
    __tablename__ = 'orders'
    id = db.Column(db.Integer, primary_key=True)                              # 编号
    user_id = db.Column(db.Integer, db.ForeignKey('user.id'))                 # 所属用户
    recevie_name = db.Column(db.String(255))                                  # 收款人姓名
    recevie_address = db.Column(db.String(255))                               # 收款人地址
    recevie_tel = db.Column(db.String(255))                                   # 收款人电话
    remark = db.Column(db.String(255))                                        # 备注信息
    addtime = db.Column(db.DateTime, index=True, default=datetime.now)        # 添加时间
    orders_detail = db.relationship("OrdersDetail", backref='orders')         # 外键关系关联
    def __repr__(self):
        return "<Orders %r>" % self.id
```

```
# 详细明细
class OrdersDetail(db.Model):
    __tablename__ = 'orders_detail'
    id = db.Column(db.Integer, primary_key=True)                    # 编号
    goods_id = db.Column(db.Integer, db.ForeignKey('goods.id'))     # 所属商品
    order_id = db.Column(db.Integer, db.ForeignKey('orders.id'))    # 所属订单
    number = db.Column(db.Integer, default=0)                       # 购买数量
```

5.4.4 数据表关系

本项目的数据表之间存在着多个数据关系，如一个大分类（supercat 表）对应着多个子分类（subcat 表），而每个大分类和子分类下又对应着多个商品（goods 表）。一个购物车（cart 表）对应着多个商品（goods 表），一个订单（orders 表）又对应着多个订单明细（orders_detail 表）。我们使用 ER 图来直观地展现数据表之间的关系，如图 5.2 所示。

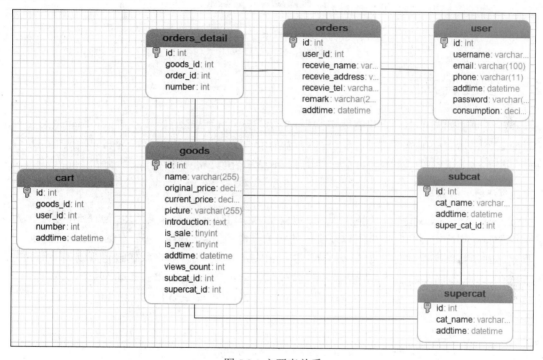

图 5.2　主要表关系

5.5　会员注册模块设计

5.5.1　会员注册模块概述

会员注册模块主要用于实现新用户注册成为网站会员的功能。在会员注册页面中，用户需要填写会员信息，然后单击"同意协议并注册"按钮，程序将自动验证输入的账户是否唯一。如果唯一，就把填写的会员信息保存到数据库；否则给出提示，账户需要修改为唯一的后，方可完成注册。另外，程序还将验证输入的信息是否合法，例如不能输入中文的账户名称等。会员注册页面运行结果如图 5.3 所示。

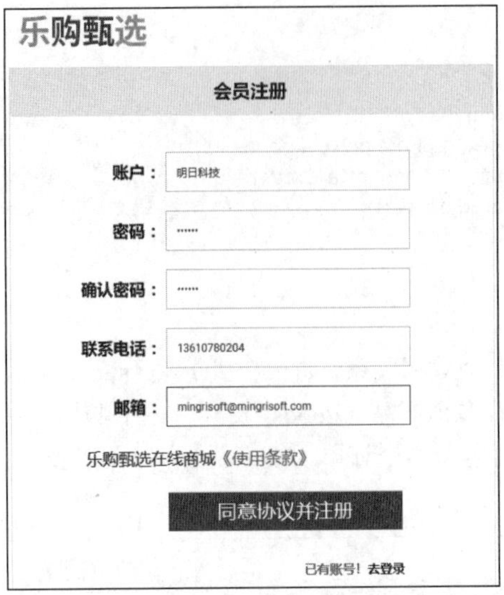

图 5.3 会员注册页面

5.5.2 会员注册页面

在会员注册页面的表单中，用户需要填写用户名、密码、确认密码、联系电话和邮箱信息。对于用户提交的信息，网站后台必须进行验证。验证内容包括用户名和密码是否为空，密码和确认密码是否一致，电话和邮箱格式是否正确等。在本项目中，使用 Flask-WTF 模块来创建表单。

1. 创建注册页面表单

在 app\home\forms.py 文件中，创建继承 FlaskForm 类的 RegiserForm 类。RegiserForm 类中，定义注册页面表单中的每个字段类型和验证规则以及字段的相关属性等信息。例如，定义 username 表示用户名，该字段类型是字符串型，所以需要从 wtforms 导入 StringField。对于用户名，我们设置规则为不能为空，长度在 3 至 50 之间。所以，将 validators 设置为一个列表，包含 DataRequired() 和 Length() 两个函数。而由于 Flask-并没有提供验证邮箱和验证手机号的功能，所以需要自定义 validate_email() 和 validate_phone() 这两个方法来实现。具体代码如下：

```python
from flask_wtf import FlaskForm
from wtforms import StringField, PasswordField, SubmitField, TextAreaField
from wtforms.validators import DataRequired, Email, Regexp, EqualTo, ValidationError,Length
class RegisterForm(FlaskForm):
    """
    用户注册表单
    """
    username = StringField(
        label="账户：",
        validators=[
            DataRequired("用户名不能为空！"),
            Length(min=3, max=50, message="用户名长度必须在3到50位之间")
        ],
        description="用户名",
        render_kw={
            "type"       : "text",
            "placeholder": "请输入用户名！",
            "class":"validate-username",
            "size" : 38,
```

```python
    }
)
phone = StringField(
    label="联系电话：",
    validators=[
        DataRequired("手机号不能为空！"),
        Regexp("1[34578][0-9]{9}", message="手机号码格式不正确")
    ],
    description="手机号",
    render_kw={
        "type": "text",
        "placeholder": "请输入联系电话！",
        "size": 38,
    }
)
email = StringField(
    label = "邮箱：",
    validators=[
        DataRequired("邮箱不能为空！"),
        Email("邮箱格式不正确！")
    ],
    description="邮箱",
    render_kw={
        "type": "email",
        "placeholder": "请输入邮箱！",
        "size": 38,
    }
)
password = PasswordField(
    label="密码：",
    validators=[
        DataRequired("密码不能为空！")
    ],
    description="密码",
    render_kw={
        "placeholder": "请输入密码！",
        "size": 38,
    }
)
repassword = PasswordField(
    label= "确认密码：",
    validators=[
        DataRequired("请输入确认密码！"),
        EqualTo('password', message="两次密码不一致！")
    ],
    description="确认密码",
    render_kw={
        "placeholder": "请输入确认密码！",
        "size": 38,
    }
)
submit = SubmitField(
    '同意协议并注册',
    render_kw={
        "class": "btn btn-primary login",
    }
)

def validate_email(self, field):
    """
    检测注册邮箱是否已经存在
    :param field: 字段名
```

```python
"""
email = field.data
user = User.query.filter_by(email=email).count()
if user == 1:
    raise ValidationError("邮箱已经存在！")
def validate_phone(self, field):
    """
    检测手机号是否已经存在
    :param field: 字段名
    """
    phone = field.data
    user = User.query.filter_by(phone=phone).count()
    if user == 1:
        raise ValidationError("手机号已经存在！")
```

> **说明**
> 自定义验证方法名称的格式为"validate_+字段名"，如自定义的验证手机号的函数名称为 validate_phone。

2. 显示注册页面

本项目中，所有模板文件均存储在 app/templates/ 路径下。如果是前台模板文件，则存放于 app/templates/home/ 路径下。在该路径下，创建 register.html 作为前台注册页面模板。接下来，需要使用 @home.route() 装饰器定义路由，并且使用 render_template() 函数来渲染模板。关键代码如下：

```python
@home.route("/register/", methods=["GET", "POST"])
def register():
    """
    注册功能
    """
    if "user_id" in session:
        return redirect(url_for("home.index"))
    form = RegisterForm()                                    # 导入注册表单
    # 省略部分代码
    return render_template("home/register.html", form=form)  # 渲染注册页面模板
```

上面代码中，实例化 LoginForm 类并赋值 form 变量，最后在 render_template() 函数中传递该参数。

我们已经使用了 Flask-Form 来设置表单字段，那么在模板文件中，直接可以使用 form 变量来设置表单中的字段，如用户名字段（username）就可以使用 form.username 来代替。关键代码如下：

```html
<form action="" method="post" class="form-horizontal">
    <fieldset>
        <div class="form-group">
            <div class="col-sm-4 control-label">
                {{form.username.label}}
            </div>
            <div class="col-sm-8">
                <!-- 账户文本框 -->
                {{form.username}}
                {% for err in form.username.errors %}
                <span class="error">{{ err }}</span>
                {% endfor %}
            </div>
        </div>
        <div class="form-group">
            <div class="col-sm-4 control-label">
                {{form.password.label}}
            </div>
```

```html
        <div class="col-sm-8">
            <!-- 密码文本框 -->
            {{form.password}}
            {% for err in form.password.errors %}
            <span class="error">{{ err }}</span>
            {% endfor %}
        </div>
    </div>
    <div class="form-group">
        <div class="col-sm-4 control-label">
            {{form.repassword.label}}
        </div>
        <div class="col-sm-8">
            <!-- 确认密码文本框 -->
            {{form.repassword}}
            {% for err in form.repassword.errors %}
            <span class="error">{{ err }}</span>
            {% endfor %}
        </div>
    </div>
    <div class="form-group">
        <div class="col-sm-4 control-label">
            {{form.phone.label}}
        </div>
        <div class="col-sm-8" style="clear: none;">
            <!-- 输入联系电话的文本框 -->
            {{form.phone}}
            {% for err in form.phone.errors %}
            <span class="error">{{ err }}</span>
            {% endfor %}
        </div>
    </div>
    <div class="form-group">
        <div class="col-sm-4 control-label">
            {{form.email.label}}
        </div>
        <div class="col-sm-8" style="clear: none;">
            <!-- 输入邮箱的文本框 -->
            {{form.email}}
            {% for err in form.email.errors %}
            <span class="error">{{ err }}</span>
            {% endfor %}
        </div>
    </div>
    <div class="form-group">
        <div style="float: right; padding-right: 216px;">
            乐购甄选在线商城<a href="#" style="color: #0885B1;">《使用条款》</a>
        </div>
    </div>
    <div class="form-group">
        <div class="col-sm-offset-4 col-sm-8">
            {{ form.csrf_token }}
            {{ form.submit }}
        </div>
    </div>
    <div class="form-group" style="margin: 20px;">
        <label>已有账号！<a
            href="{{url_for('home.login')}}">去登录</a>
        </label>
    </div>
</fieldset>
</form>
```

> **说明**
>
> 表单中使用{{form.csrf_token}}设置一个隐藏域字段csrf_token，该字段用于防止CSRF攻击。

渲染模板后，运行程序，当访问网址 http://127.0.0.1:5000/register 时，即可访问会员注册页面。

5.5.3 验证并保存注册信息

当用户填写完注册信息并单击"同意协议并注册"按钮时，程序将以POST方式提交表单。提交路径是form表单的action属性值。在register.html中，如果将action设置为空字符串，表示提交到当前URL。

在register()方法中，使用form.validate_on_submit()来验证表单信息，如果验证失败，则在页面返回相应的错误信息。验证全部通过后，将用户注册信息写入user表中。具体代码如下：

```python
@home.route("/register/", methods=["GET", "POST"])
def register():
    """
    注册功能
    """
    if "user_id" in session:
        return redirect(url_for("home.index"))
    form = RegisterForm()                                   # 导入注册表单
    if form.validate_on_submit():                           # 提交注册表单
        data = form.data                                    # 接收表单数据
        # 为 User 类属性赋值
        user = User(
            username = data["username"],                    # 用户名
            email = data["email"],                          # 邮箱
            password = generate_password_hash(data["password"]),  # 对密码加密
            phone = data['phone']
        )
        db.session.add(user)                                # 添加数据
        db.session.commit()                                 # 提交数据
        return redirect(url_for("home.login"))              # 注册成功，跳转到首页
    return render_template("home/register.html", form=form) # 渲染模板
```

在注册页面输入注册信息，当密码和确认密码不一致时，提示如图5.4所示的错误信息。当联系电话格式错误时，提示如图5.5所示错误信息。当验证通过后，则将注册用户信息保存到user表中，并且跳转到登录页面。

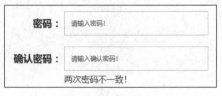

图5.4 密码不一致

图5.5 手机号码格式错误

5.6 会员登录模块设计

5.6.1 会员登录模块概述

会员登录模块主要用于实现网站的会员登录和退出功能，在该页面中，填写会员账户、密码和验证码

（如果验证码看不清楚，可以单击验证码图片刷新该验证码），单击"登录"按钮，即可实现会员登录。如果没有输入账户、密码或者验证码，将给予提示。另外，验证码输入错误也将给予提示。登录页面效果如图5.6所示。

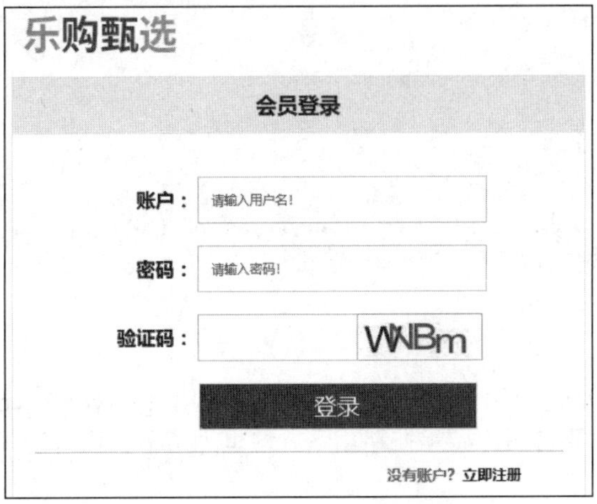

图5.6 会员登录页面

5.6.2 创建会员登录页面

在会员登录页面，需要用户填写用户名、密码和验证码。用户名和密码的表单字段与注册页面相同，这里不再赘述，我们重点介绍一下与验证码相关的内容。

1. 生成验证码

登录页面的验证码是一个图片验证码，也就是在一张图片上显示数字0~9、小写字母a~z、大写字母A~Z的随机组合。可以使用String模块的ascii_letters()和digits()方法，其中ascii_letters()方法用来生成所有字母，即a~z和A~Z。digits()方法用来生成所有数字0~9。最后使用PIL（图像处理标准库）来生成验证码图片。实现代码如下：

```
import random
import string
from PIL import Image, ImageFont, ImageDraw
from io import BytesIO

def rndColor():
    '''随机颜色'''
    return (random.randint(32, 127), random.randint(32, 127), random.randint(32, 127))

def gene_text():
    '''生成4位验证码'''
    return ''.join(random.sample(string.ascii_letters+string.digits, 4))

def draw_lines(draw, num, width, height):
    '''划线'''
    for num in range(num):
        x1 = random.randint(0, width / 2)
        y1 = random.randint(0, height / 2)
        x2 = random.randint(0, width)
        y2 = random.randint(height / 2, height)
        draw.line(((x1, y1), (x2, y2)), fill='black', width=1)

def get_verify_code():
```

```python
'''生成验证码图片'''
code = gene_text()
# 图片大小 120×50
width, height = 120, 50
# 新图片对象
im = Image.new('RGB',(width, height),'white')
# 字体
font = ImageFont.truetype('app/static/fonts/arial.ttf', 40)
# draw 对象
draw = ImageDraw.Draw(im)
# 绘制字符串
for item in range(4):
    draw.text((5+random.randint(-3,3)+23*item, 5+random.randint(-3,3)),
              text=code[item], fill=rndColor(),font=font )
return im, code
```

2．显示验证码

接下来，显示验证码。首先定义路由/code，在该路由下调用 get_verify_code()方法生成验证码，并生成一个 jpeg 格式的图片；然后需要将图片显示在路由下，这里为了节省内存空间，返回一张 gif 图片。具体代码如下：

```python
@home.route('/code')
def get_code():
    image, code = get_verify_code()
    # 图片以二进制形式写入
    buf = BytesIO()
    image.save(buf, 'jpeg')
    buf_str = buf.getvalue()
    # 把 buf_str 作为 response 返回前端，并设置首部字段
    response = make_response(buf_str)
    response.headers['Content-Type'] = 'image/gif'
    # 将验证码字符串储存在 session 中
    session['image'] = code
    return response
```

访问 http://127.0.0.1:5000/code，运行结果如图 5.7 所示。

图 5.7　生成验证码

最后，需要将验证码显示在登录页面上。这里，可以把模板文件中表示验证码图片的<img>标签的 src 属性设置为{{url_for('home.get_code')}}。此外，当单击验证码图片时，需要更新验证码图片，该功能可以通过 JavaScript 的 onclick 单击事件来实现。当点击图片时，设置使用 Math.random()来重新生成一个随机数。注意这里生成的随机数，其作用是为了重定向图片验证码链接，而不是验证码。关键代码如下：

```html
<div class="col-sm-8" style="clear: none;">
    <!-- 验证码文本框 -->
    {{form.verify_code}}
        <!-- 显示验证码 -->
        <img class="img_checkcode" src="{{url_for('home.get_code')}}" width="116"
            height="43" onclick="this.src='{{url_for('home.get_code')}}'+'?'+ Math.random()">
</div>
```

在登录页面中，单击验证码图片后，将会更新验证码，运行效果如图 5.8 所示。

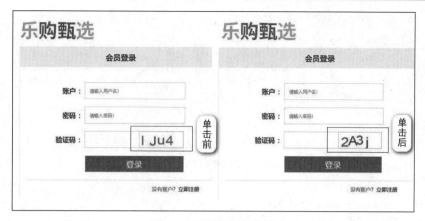

图 5.8　更新验证码效果

3. 检测验证码

在登录页面单击"登录"按钮后，程序会对用户输入的字段进行验证。那么对于验证码图片，该如何验证呢？其实，这里通过一种简单的方式将验证图片的功能进行了简化，在使用 get_code()方法生成验证码时，有如下代码：

```
session['image'] = code
```

上述代码将验证码的内容写入 session，因此只需要将用户输入的验证码和 session['image']进行对比即可。由于验证码内容包括英文大小写字母，所以在对比前，全部将其转换为英文小写字母，然后再对比。关键代码如下：

```
if session.get('image').lower() != form.verify_code.data.lower():
    flash('验证码错误',"err")
    return redirect(url_for("home.login"))                          # 跳转到登录页
```

在登录页面填写登录信息时，如果验证码错误，则提示错误信息，运行结果如图 5.9 所示。

图 5.9　验证码错误运行结果

5.6.3　保存会员登录状态

用户填写登录信息后，除了要判断验证码是否正确，还需要验证用户名是否存在，以及用户名和密码是否匹配等内容。如果验证全部通过，需要将 user_id 和 user_name 写入 session 中，为后面判断用户是否登录做准备。此外，还需要在用户访问/login 路由时，判断用户是否已经登录，如果用户之前已经登录，则不需要再次登录，而是直接跳转到商城首页。具体代码如下：

```python
@home.route("/login/", methods=["GET", "POST"])
def login():
    """
    登录
    """
    if "user_id" in session:                                    # 如果已经登录，则直接跳转到首页
        return redirect(url_for("home.index"))
    form = LoginForm()                                          # 实例化 LoginForm 类
    if form.validate_on_submit():                               # 如果提交
        data = form.data                                        # 接收表单数据
        # 判断用户名和密码是否匹配
        user = User.query.filter_by(username=data["username"]).first()# 获取用户信息
        if not user:
            flash("用户名不存在！", "err")                        # 输出错误信息
            return render_template("home/login.html", form=form) # 返回登录页
        # 调用 check_password()方法，检测用户名与密码是否匹配
        if not user.check_password(data["password"]):
            flash("密码错误！", "err")                            # 输出错误信息
            return render_template("home/login.html", form=form) # 返回登录页
        if session.get('image').lower() != form.verify_code.data.lower():
            flash('验证码错误','err')
            return render_template("home/login.html", form=form) # 返回登录页
        session["user_id"] = user.id                            # 将 user_id 写入 session，用于后面判断用户是否登录
        session["username"] = user.username                     # 将 username 写入 session，用于后面判断用户是否登录
        return redirect(url_for("home.index"))                  # 登录成功，跳转到首页
    return render_template("home/login.html",form=form)         # 渲染登录页面模板
```

5.6.4 会员退出功能

退出功能的实现比较简单，只是清空登录时保存在 session 中的 user_id 和 username 即可。使用 session.pop()函数实现该功能。具体代码如下：

```python
@home.route("/logout/")
def logout():
    """
    退出登录
    """
    # 重定向到 home 模块下的登录
    session.pop("user_id", None)
    session.pop("username", None)
    return redirect(url_for('home.login'))
```

当用户单击"退出"按钮时，执行 logout()方法，并且跳转到登录页。

5.7 首页模块设计

5.7.1 首页模块概述

当用户访问乐购甄选在线商城时，首先进入的便是前台首页。前台首页设计的美观程度将直接影响用户的购买欲望。在乐购甄选在线商城的前台首页中，用户不但可以查看最新上架、打折商品等信息，还可以及时了解大家喜爱的热门商品，以及商城推出的最新活动或者广告。乐购甄选在线商城前台首页的运行结果如图 5.10 所示。

图 5.10 乐购甄选在线商城前台首页

乐购甄选在线商城前台首页中，主要有 3 个部分需要添加动态代码，分别是热门商品、最新上架和打折商品，它们需要从数据库中读取 goods（商品表）中的数据，并通过循环显示在页面上。

5.7.2 实现显示最新上架商品功能

最新上架商品数据来源于 goods（商品表）中 is_new 字段为 1 的记录。由于数据可能会比较多，所以在商城首页中，根据商品的 addtime（添加时间）降序排序，筛选出 12 条记录。然后在模板中，遍历数据并显示。本项目中使用 Flask-SQLAlchemy 来操作数据库，查询最新上架商品的关键代码如下：

```python
@home.route("/")
def index():
    """
    首页
    """
    # 获取 12 个新品
    new_goods = Goods.query.filter_by(is_new=1).order_by(
                    Goods.addtime.desc()
                        ).limit(12).all()
    return render_template('home/index.html',new_goods=new_goods)          # 渲染模板
```

接下来渲染模板（index.html），关键代码如下：

```html
<div class="row">
    <!-- 循环显示最新上架商品：添加 12 条商品信息-->
    {% for item in new_goods %}
    <div class="product-grid col-lg-2 col-md-3 col-sm-6 col-xs-12">
        <div class="product-thumb transition">
            <div class="actions">
                <div class="image">
                    <a href="{{url_for('home.goods_detail',id=item.id)}}">
                        <img src="{{url_for('static',filename='images/goods/'+item.picture)}}" >
                    </a>
                </div>
                <div class="button-group">
                    <div class="cart">
                        <button class="btn btn-primary btn-primary" type="button"
                            data-toggle="tooltip"
                            onclick='javascript:window.location.href=
                                    "/cart_add/?goods_id={{item.id}}&number=1"; '
                            style="display: none; width: 33.3333%;"
                            data-original-title="加入到购物车">
                            <i class="fa fa-shopping-cart"></i>
                        </button>
                    </div>
                </div>
            </div>
            <div class="caption">
                <div class="name" style="height: 40px">
                    <a href="{{url_for('home.goods_detail',id=item.id)}}">
                        {{item.name}}
                    </a>
                </div>
                <p class="price">
                    价格：{{item.current_price}}元
                </p>
            </div>
        </div>
    </div>
    {% endfor %}
```

```
            <!-- // 循环显示最新上架商品：添加 12 条商品信息 -->
</div>
```

商城首页最新上架商品运行效果如图 5.11 所示。

图 5.11　最新上架商品

5.7.3　实现显示打折商品功能

打折商品数据来源于 goods（商品表）中 is_sale 字段为 1 的记录，由于数据可能比较多，所以在商城首页中，根据商品的 addtime（添加时间）降序排序，筛选出 12 条记录。然后在模板中，遍历数据并显示。查询打折商品的关键代码如下：

```
@home.route("/")
def index():
    """
    首页
    """
    # 获取 12 个打折商品
    sale_goods = Goods.query.filter_by(is_sale=1).order_by(
                    Goods.addtime.desc()
                    ).limit(12).all()
    return render_template('home/index.html',sale_goods=sale_goods)      # 渲染模板
```

接下来渲染模板（index.html），关键代码如下：

```
<div class="row">
    <!-- 循环显示打折商品：添加 12 条商品信息-->
    {% for item in sale_goods %}
    <div class="product-grid col-lg-2 col-md-3 col-sm-6 col-xs-12">
        <div class="product-thumb transition">
            <div class="actions">
                <div class="image">
                    <a href="{{url_for('home.goods_detail',id=item.id)}}">
                        <img src="{{url_for('static',filename='images/goods/'+item.picture)}}"
                             alt="{{item.name}}" class="img-responsive">
                    </a>
```

```html
            </div>
            <div class="button-group">
                <div class="cart">
                    <button class="btn btn-primary btn-primary" type="button"
                        data-toggle="tooltip"
                        onclick='javascript:window.location.href=
                                "/cart_add/?goods_id={{item.id}}&number=1"; '
                        style="display: none; width: 33.3333%;"
                        data-original-title="加入到购物车">
                        <i class="fa fa-shopping-cart"></i>
                    </button>
                </div>
            </div>
        </div>
        <div class="caption">
            <div class="name" style="height: 40px">
                <a href="{{url_for('home.goods_detail',id=item.id)}}" style="width: 95%">
                    {{item.name}}</a>
            </div>
            <div class="name" style="margin-top: 10px">
                <span style="color: #0885B1">分类：</span>{{item.subcat.cat_name}}
            </div>
            <span class="price"> 现价：{{item.current_price}} 元
            </span><br> <span class="oldprice">原价：{{item.original_price}}元
            </span>
        </div>
    </div>
{% endfor %}
<!-- 循环显示打折商品 ：添加 12 条商品信息-->
</div>
```

商城首页打折商品运行效果如图 5.12 所示。

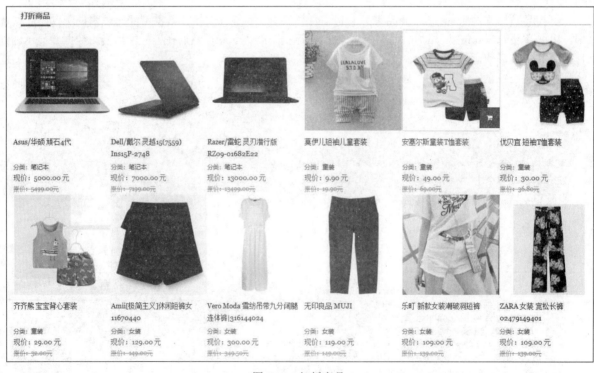

图 5.12　打折商品

5.7.4 实现显示热门商品功能

热门商品数据来源于 goods（商品表）中 view_count 字段值较高的记录。由于页面布局限制，我们只根据 view_count 字段降序筛选两条记录。然后在模板中，遍历数据并显示。查询热门商品的关键代码如下：

```python
@home.route("/")
def index():
    """
    首页
    """
    # 获取两个热门商品
    hot_goods = Goods.query.order_by(Goods.views_count.desc()).limit(2).all()
    return render_template('home/index.html', hot_goods=hot_goods)    # 渲染模板
```

接下来渲染模板（index.html），关键代码如下：

```html
<div class="box_oc">
    <!-- 循环显示热门商品：添加两条商品信息-->
    {% for item in hot_goods %}
    <div class="box-product product-grid">
        <div>
            <div class="image">
                <a href="{{url_for('home.goods_detail',id=item.id)}}">
                    <img src="{{url_for('static',filename='images/goods/'+item.picture)}}" >
                </a>
            </div>
            <div class="name">
                <a href="{{url_for('home.goods_detail',id=item.id)}}">{{item.name}}</a>
            </div>
            <!-- 商品价格 -->
            <div class="price">
                <span class="price-new">价格：{{item.current_price}} 元</span>
            </div>
            <!-- // 商品价格 -->
        </div>
    </div>
    {% endfor %}
    <!-- // 循环显示热门商品：添加两条商品信息-->
</div>
```

商城首页热门商品运行效果如图 5.13 所示。

图 5.13　热门商品

5.8 购物车模块设计

5.8.1 购物车模块概述

乐购甄选在线商城中，在首页单击某个商品可以进入显示商品详细信息的页面，如图 5.14 所示。在该页面中，单击"添加到购物车"按钮，即可将相应商品添加到购物车，然后填写物流信息，如图 5.15 所示。单击"结账"按钮，将弹出如图 5.16 所示的支付对话框。最后单击"支付"按钮，模拟提交支付并生成订单。

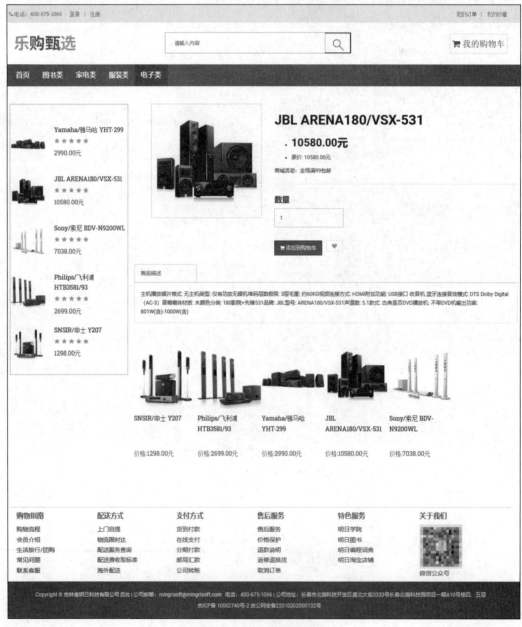

图 5.14 商品详细信息页面

图 5.15 查看购物车页面

图 5.16 支付对话框

5.8.2 实现显示商品详细信息功能

在首页单击任何商品名称或商品图片时，都将打开该商品的详细信息的页面。该页面中，除了显示商品的信息，还需要显示左侧的热门商品和底部的推荐商品。商品详细信息页面的实现关键点如下：

- ☑ 对于商品的详细信息，需要根据商品 ID，使用 get_or_404(id)方法来获取。
- ☑ 对于左侧热门商品，需要获取该商品的同一个子类别下的商品。例如，如果正在访问的商品子类别是音箱，那么左侧热门商品就是音箱相关的产品，并且根据浏览量从高到低排序，筛选出 5 条记录。
- ☑ 对于底部的推荐商品，与热门商品类似。只是根据商品添加时间从高到低排序，筛选出 5 条记录。
- ☑ 此外，由于要统计商品的浏览量，所以每当进入商品详情页时，需要更新 goods（商品）表中该商品的 view_count（浏览量）字段，将其值加 1。

实现商品详细信息页面主要功能的关键代码如下：

```
@home.route("/goods_detail/<int:id>/")
def goods_detail(id=None):                               # id 为商品 ID
    """
    详情页
    """
    user_id = session.get('user_id', 0)                  # 获取用户 ID，判断用户是否登录
    goods = Goods.query.get_or_404(id)                   # 根据商品 ID 获取数据，如果不存在返回 404
    # 浏览量加 1
    goods.views_count += 1
    db.session.add(goods)                                # 添加数据
    db.session.commit()                                  # 提交数据
    # 获取左侧热门商品
    hot_goods = Goods.query.filter_by(subcat_id=goods.subcat_id).order_by(
                    Goods.views_count.desc()).limit(5).all()
    # 获取底部相关商品
    similar_goods = Goods.query.filter_by(subcat_id=goods.subcat_id).order_by(
                    Goods.addtime.desc()).limit(5).all()
    return render_template('home/goods_detail.html',goods=goods,hot_goods=hot_goods,
                    similar_goods=similar_goods,user_id=user_id)    # 渲染模板
```

商品详情页运行结果如图 5.17 所示。

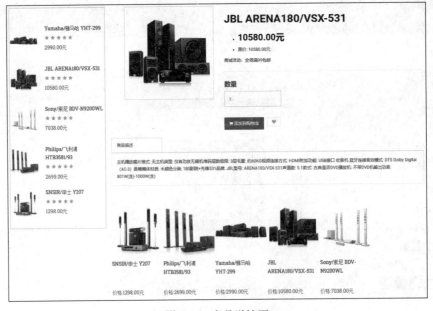

图 5.17 商品详情页

5.8.3 实现添加购物车功能

在乐购甄选在线商城中,有两种添加购物车的方法:商品详情页添加购物车和商品列表页添加购物车。它们之间的区别在于,在商品详情页中添加购物车时,可以选择购买商品的数量(大于或等于1);在商品列表页中添加购物车时,默认购买数量为1。

基于以上分析,可以通过设置<a>标签的方式来添加购物车。下面分别介绍这两种情况。

1. 商品详情页添加购物车功能

在商品详情页中,填写购买商品数量后,单击"添加到购物车"按钮时,首先判断用户是否登录。如果没有登录,页面跳转到登录页;如果已经登录,则执行加入购物车操作,这里需要注意的是,需要判断用户填写的购买数量,如果数量小于1,则提示错误信息。商品详情页模板(goods_detail.html)中实现添加购物车功能的关键代码如下:

```html
<button type="button" onclick="addCart()" class="btn btn-primary btn-primary">
    <i class="fa fa-shopping-cart"></i> 添加到购物车</button>

<script type="text/javascript">
function addCart() {
    var user_id = {{ user_id }};            // 获取当前用户的 ID
    var goods_id = {{ goods.id }}           // 获取商品的 ID
    if( !user_id){
        window.location.href = "/login/";   // 如果没有登录,跳转到登录页
        return ;
    }
    var number = $('#shuliang').val();      // 获取输入的商品数量
    // 验证输入的数量是否合法
    if (number < 1) {                       // 如果输入的数量不合法
        alert('数量不能小于1! ');
        return;
    }
    window.location.href = '/cart_add?goods_id='+goods_id+"&number="+number
    }
</script>
```

2. 商品列表页添加购物车功能

在商品列表页,当单击购物车图标时,执行添加购物车操作,并将添加的商品数量默认设置为1。在商品列表页(首页)模板(index.html)中实现添加购物车功能的关键代码如下:

```html
<button class="btn btn-primary btn-primary" type="button"
    data-toggle="tooltip"
    onclick='javascript:window.location.href="/cart_add/?goods_id={{item.id}}&number=1"; '
    style="display: none; width: 33.3333%;"
    data-original-title="加入到购物车">
    <i class="fa fa-shopping-cart"></i>
</button>
```

在以上两种情况下,添加购物车都执行链接/cart_add/并传递 goods_id 和 number 两个参数,而且分别将它们写入 cart(购物车表)中,关键代码如下:

```python
@home.route("/cart_add/")
@user_login
def cart_add():
    """
    添加购物车
```

```python
    """
    cart = Cart(
        goods_id = request.args.get('goods_id'),
        number = request.args.get('number'),
        user_id=session.get('user_id', 0)              # 获取用户 ID，判断用户是否登录
    )
    db.session.add(cart)                                # 添加数据
    db.session.commit()                                 # 提交数据
    return redirect(url_for('home.shopping_cart'))
```

5.8.4 实现查看购物车功能

在实现添加购物车时，将商品添加到购物车后，需要把页面跳转到查看购物车页面，用于显示已经添加到购物车中的商品。购物车中的商品数据来源于 cart（购物车表）和 goods（商品表）。由于 cart 表的 goods_id 字段与 goods 表的 id 字段关联，所以可以直接查找 cart 表中 user_id 为当前用户 ID 的记录。具体代码如下：

```python
@home.route("/shopping_cart/")
@user_login
def shopping_cart():
    user_id = session.get('user_id',0)
    cart = Cart.query.filter_by(user_id = int(user_id)).order_by(Cart.addtime.desc()).all()
    if cart:
        return render_template('home/shopping_cart.html',cart=cart)
    else:
        return render_template('home/empty_cart.html')
```

上面代码中，判断用户购物车中是否有商品，如果没有，则渲染 empty_cart.html 模板，否则渲染购物车列表页模板 shopping_cart.html，该模板中主要显示购物车商品列表，及用于输入收货信息的表单，关键代码如下：

```html
<div id="mr-content" class="mr-content col-xs-12">
    <div id="mrshop" class="mrshop common-home">
        <div class="container_oc">
            <div class="row">
                <div   class="col-sm-12">
                    <h1>我的购物车</h1>
                    <!-- 显示购物车中的商品 -->
                    <div class="table-responsive cart-info">
                        <table class="table table-bordered">
                            <thead>
                                <tr>
                                    <td class="text-center image">商品图片</td>
                                    <td class="text-left name">商品名称</td>
                                    <td class="text-left quantity">数量</td>
                                    <td class="text-right price">单价</td>
                                    <td class="text-right">总计</td>
                                </tr>
                            </thead>
                            <tbody>
                                <!-- 遍历购物车中的商品并显示 -->
                                {% for item in cart %}
                                    <tr>
                                        <td class="text-center image" width="20%">
                                            <a href="{{url_for('home.goods_detail',id=item.goods.id)}}">
                                                <img width="80px"
                                                 src="{{url_for('static',filename='images/goods/'+item.goods.picture)}}"> </a>
```

```html
                        </td>
                        <td class="text-left name">
                            <a href="{{url_for('home.goods_detail',id=item.goods.id)}}">
                                {{item.goods.name}}</a>
                        </td>
                        <td class="text-left quantity">{{item.number}}件</td>
                        <td class="text-right price">{{item.goods.current_price}}元</td>
                        <td class="text-right total" value="{{item.goods.current_price * item.number}}">
                            {{item.goods.current_price * item.number}}元
                        </td>
                    </tr>
                {% endfor %}
                <!-- // 遍历购物车中的商品并显示 -->
                </tbody>
            </table>
        </div>
        <!-- // 显示购物车中的商品 -->
        <!-- 显示总计金额 -->
        <div class="row cart-total">
            <div class="col-sm-4 col-sm-offset-8">
                <table class="table table-bordered">
                    <tbody>
                        <tr >
                            <span>
                                <strong>总计:</strong>
                                <p id="total_price"></p>
                            </span>
                        </tr>
                    </tbody>
                </table>
            </div>
        </div>
        <!-- // 显示总计金额  -->
    </div>
</div>
<!-- 填写物流信息 -->
<div class="row">
    <div id="content_oc" class="col-sm-12">
        <h1>物流信息</h1>
        <!-- 填写物流信息的表单 -->
        <form action="{{url_for('home.cart_order')}}" method="post" id="myform">
            <div class="table-responsive cart-info">
                <table class="table table-bordered">
                    <tbody>
                        <tr>
                            <td class="text-right" width="20%">收货人姓名：</td>
                            <td class="text-left quantity">
                                <div class="input-group btn-block" style="max-width: 400px;">
                                    <input type="text" id="recevieName" name="recevie_name"
                                        size="10" class="form-control">
                                </div>
                            </td>
                        </tr>
                        <tr>
                            <td class="text-right">收货人手机：</td>
                            <td class="text-left quantity">
                                <div class="input-group btn-block" style="max-width: 400px;">
                                    <input type="text" id="tel" name="recevie_tel"
                                        size="10" class="form-control">
                                </div>
                            </td>
                        </tr>
```

```html
                    <tr>
                        <td class="text-right">收货人地址：</td>
                        <td class="text-left quantity">
                            <div class="input-group btn-block" style="max-width: 400px;">
                                <input type="text" id="address" name="recevie_address"
                                    size="1" class="form-control">
                            </div>
                        </td>
                    </tr>
                    <tr>
                        <td class="text-right">备注：</td>
                        <td class="text-left quantity">
                            <div class="input-group btn-block" style="max-width: 400px;">
                                <input type="text" name="remark" size="1" class="form-control">
                            </div>
                        </td>
                    </tr>
                </tbody>
            </table>
        </div>
    </form>
    <!-- // 填写物流信息的表单 -->
</div>
</div>
<!-- // 填写物流信息 -->
<br />
<!-- 显示支付方式 -->
<div class="row">
    <div id="content_oc" class="col-sm-12">
        <h1>支付方式</h1>
        <div class="table-responsive cart-info">
            <table class="table table-bordered">
                <tbody>
                    <tr>
                        <td class="text-left">
                            <img src="{{url_for('static',filename='home/images/zhifubao.png')}}" /></td>
                    </tr>
                </tbody>
            </table>
        </div>
        <br /> <br />
        <div class="buttons">
            <div class="pull-left">
                <a href="{{url_for('home.index')}}" class="btn btn-primary btn-default">继续购物</a>
            </div>
            <div class="pull-left">
                <a href="{{url_for('home.cart_clear')}}" class="btn btn-primary btn-default">清空购物车</a>
            </div>
            <div class="pull-right">
                <a href="javascript:zhifu();" class="tigger btn btn-primary btn-primary">结账</a>
            </div>
        </div>
    </div>
</div>
<!-- // 显示支付方式 -->
</div>
</div>
</div>
```

购物车页面有商品时的效果如图 5.18 所示，购物车页面没有商品时的效果如图 5.19 所示。

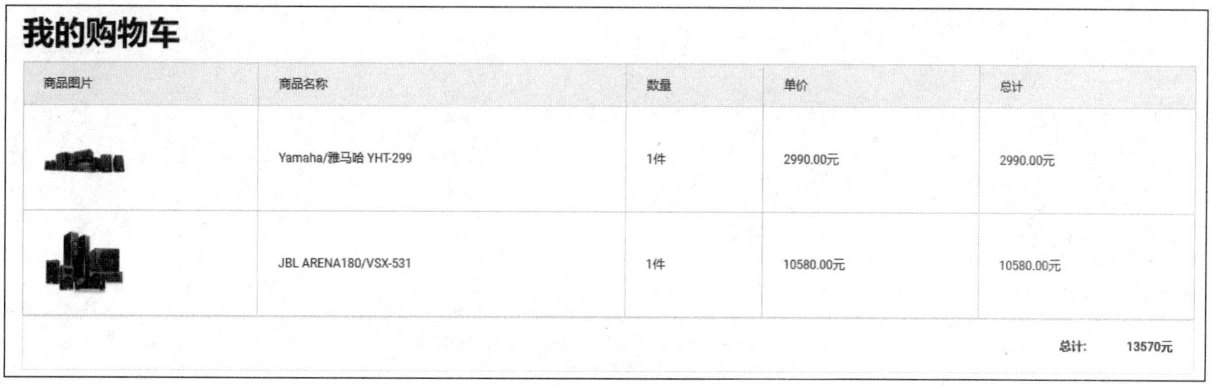

图 5.18 购物车页面

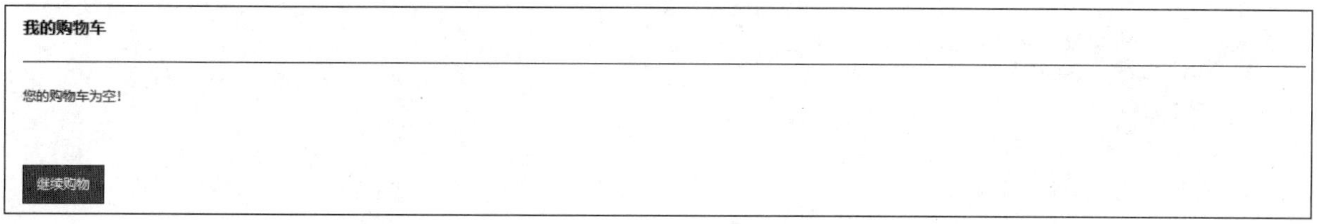

图 5.19 空购物车页面

5.8.5 实现保存订单功能

商品加入购物车后，需要填写物流信息，包括"收货人姓名""收货人手机"和"收货人地址"等。然后单击"结账"按钮，弹出支付二维码。因为调用支付宝接口，需要注册支付宝企业账户，并且完成实名认证，所以在本项目中只是模拟一下支付功能。模拟弹出支付二维码的功能是通过 jBox 插件实现的，jBox 插件是一款 JavaScript 多功能对话框插件，它适用于创建各种模态窗口、提示、通知等形式，还可以自定义提示框中的内容，这里主要使用 jBox 插件的 open() 函数打开一个对话框，并显示支付二维码，关键代码如下：

```
<!-- 使用 jBox 插件实现一个支付对话框 -->
<script type="text/javascript" src="{{url_for('static',filename='home/js/jBox/jquery-1.4.2.min.js')}}"></script>
<script type="text/javascript" src="{{url_for('static',filename='home/js/jBox/jquery.jBox-2.3.min.js')}}"></script>
<link type="text/css" rel="stylesheet" href="{{url_for('static',filename='home/js/jBox/Skins2/Pink/jbox.css')}}" />
<script type="text/javascript">
    // 获取总额
    $(document).ready(function(){
        var total_price = 0
        $('.total').each(function(){
            total_price += parseFloat($(this).attr('value'))
        })
        $('#total_price').text(total_price+"元")
    });
    function zhifu() {
        // 验证收货人姓名
        if ($('#recevieName').val() === "") {
            alert('收货人姓名不能为空！');
            return;
        }
        // 验证收货人手机
        if ($('#tel').val() === "") {
            alert('收货人手机不能为空！');
            return;
        }
        // 验证手机号是否合法
```

```javascript
        if (isNaN($('#tel').val())) {
            alert("手机号请输入数字");
            return;
        }
        // 验证收货人地址
        if ($('#address').val() === "") {
            alert('收货人地址不能为空！');
            return;
        }
        // 设置对话框中要显示的内容
        var html = '<div class="popup_cont">'
            + '<div style="width: 256px; height: 250px; text-align: center; margin:70px" >'
            + '<image src="/static/home/images/qr.png" width="256" height="256" />'
            + '<p style="color:red;padding-tope:30px">该页面仅为测试页面，并未实现支付功能</p></div>'
            + '</div>';
        var content = {
            state1 : {
                content : html,
                buttons : {
                    '取消' : 0,
                    '支付' : 1
                },
                buttonsFocus : 0,
                submit : function(v, h, f) {
                    if (v == 0) {                                           // 取消按钮的响应事件
                        return true;                                        // 关闭窗口
                    }
                    if (v == 1) {                                           // 支付按钮的响应事件
                        document.getElementById('myform').submit();         // 提交表单
                        return true;
                    }
                    return false;
                }
            }
        };
        $.jBox.open(content, '支付', 400, 450);                              // 打开支付窗口
    }
</script>
<!-- // 使用 jBox 插件实现一个支付对话框 -->
```

当单击弹窗右下角的"支付"按钮时，默认支付完成，此时需要保存订单。对于保存订单功能，需要结合 orders 表和 orders_detail 表来实现，它们之间是一对多的关系。例如，在一个订单中，可以有多个订单明细。orders 表用于记录收货人的姓名、电话和地址等信息，而 orders_detail 表用于记录该订单中的商品信息。所以，在添加订单时，需要同时添加到 orders 表和 orders_detail 表。实现代码如下：

```python
@home.route("/cart_order/",methods=['GET','POST'])
@user_login
def cart_order():
    if request.method == 'POST':
        user_id = session.get('user_id',0)                                  # 获取用户 ID
        # 添加订单
        orders = Orders(
            user_id = user_id,
            recevie_name = request.form.get('recevie_name'),
            recevie_tel = request.form.get('recevie_tel'),
            recevie_address = request.form.get('recevie_address'),
            remark = request.form.get('remark')
        )
        db.session.add(orders)                                              # 添加数据
        db.session.commit()                                                 # 提交数据
        # 添加订单详情
```

```python
        cart = Cart.query.filter_by(user_id=user_id).all()
        object = []
        for item in cart :
            object.append(
                OrdersDetail(
                    order_id=orders.id,
                    goods_id=item.goods_id,
                    number = item.number,)
            )
        db.session.add_all(object)
        # 更改购物车状态
        Cart.query.filter_by(user_id=user_id).update({'user_id': 0})
        db.session.commit()
    return redirect(url_for('home.index'))
```

上面代码中，在操作 orders_detail 表时，由于有多个数据，所以使用了 add_all()方法来批量添加。另外，需要注意的是，当添加完订单后，购物车就已经清空了，此时需要修改 cart（购物车）表的 order_id 字段，将其值更改为 0。这样，查看购物车时，购物车将没有数据。

5.8.6 实现查看订单功能

订单支付完成后，可以单击"我的订单"按钮来查看订单信息。订单数据信息来源于 orders 表和 orders_detail 表。查看订单功能的关键代码如下：

```python
@home.route("/order_list/",methods=['GET','POST'])
@user_login
def order_list():
    """
    我的订单
    """
    user_id = session.get('user_id',0)
    orders = OrdersDetail.query.join(Orders).filter(Orders.user_id==user_id).order_by(
             Orders.addtime.desc()).all()
    return render_template('home/order_list.html',orders=orders)
```

接下来渲染模板（order_list.html），关键代码如下：

```html
<div id="mr-content" class="mr-content col-xs-12">
    <div id="mrshop" class="mrshop common-home">
        <div class="container_oc">
            <div class="row">
                <div id="content_oc" class="col-sm-12">
                    <h1>我的订单</h1>
                    <div class="table-responsive cart-info">
                        <table class="table table-bordered">
                            <thead>
                                <tr>
                                    <td class="text-center image">订单号</td>
                                    <td class="text-center name">产品名称</td>
                                    <td class="text-center name">购买数量</td>
                                    <td class="text-center name">单价</td>
                                    <td class="text-center name">消费金额</td>
                                    <td class="text-center quantity">收货人姓名</td>
                                    <td class="text-center price">收货人手机</td>
                                    <td class="text-center total">下单日期</td>
                                </tr>
                            </thead>
                            <tbody>
                                {% for item in orders%}
                                    <tr>
```

```
                                        <td class="text-center image" width="10%">{{item.orders.id}}</td>
                                        <td class="text-center name">{{item.goods.name}}</td>
                                        <td class="text-center quantity">{{item.number}}件</td>
                                        <td class="text-center quantity">{{item.goods.current_price}}元</td>
                                        <td class="text-center quantity">
                                            {{item.number*item.goods.current_price}}元</td>
                                        <td class="text-center quantity">{{item.orders.recevie_name}}</td>
                                        <td class="text-center quantity">{{item.orders.recevie_tel}}</td>
                                        <td class="text-center quantity">{{item.orders.addtime}}</td>
                                    </tr>
                                    {% endfor %}
                                </tbody>
                            </table>
                        </div>
                </div>
            </div>
            <br /><br />
            <div class="row">
                <div    class="col-sm-12">
                    <br />
                    <br />
                    <div class="buttons">
                        <div class="pull-right">
                            <a href="{{url_for('home.index')}}" class="tigger btn btn-primary btn-primary">继续购物</a>
                        </div>
                    </div>
                </div>
            </div>
        </div>
    </div>
</div>
```

查看订单页面运行结果如图 5.20 所示。

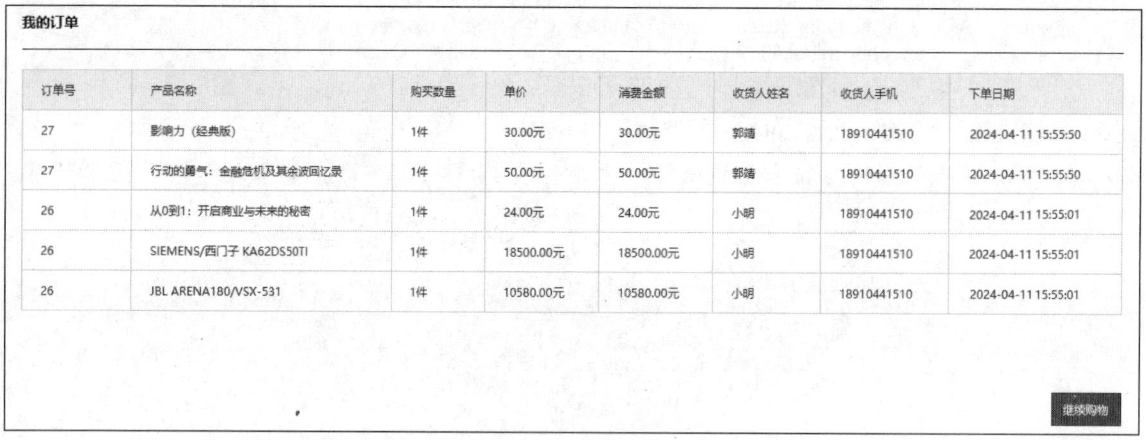
我的订单

订单号	产品名称	购买数量	单价	消费金额	收货人姓名	收货人手机	下单日期
27	影响力（经典版）	1件	30.00元	30.00元	郭靖	18910441510	2024-04-11 15:55:50
27	行动的勇气：金融危机及其余波回忆录	1件	50.00元	50.00元	郭靖	18910441510	2024-04-11 15:55:50
26	从0到1：开启商业与未来的秘密	1件	24.00元	24.00元	小明	18910441510	2024-04-11 15:55:01
26	SIEMENS/西门子 KA62DS50TI	1件	18500.00元	18500.00元	小明	18910441510	2024-04-11 15:55:01
26	JBL ARENA180/VSX-531	1件	10580.00元	10580.00元	小明	18910441510	2024-04-11 15:55:01

继续购物

图 5.20　查看订单页面

5.9　后台功能模块设计

5.9.1　后台登录模块设计

在网站前台首页的底部提供了后台管理员入口，通过该入口可以进入后台登录页面。在该页面，管理员通过输入正确的用户名和密码即可登录到网站后台。如果用户没有输入用户名或密码为空，系统都将进行判断并给予提示信息，否则进入管理员登录处理页验证用户信息。后台登录页面运行结果如图 5.21 所示。

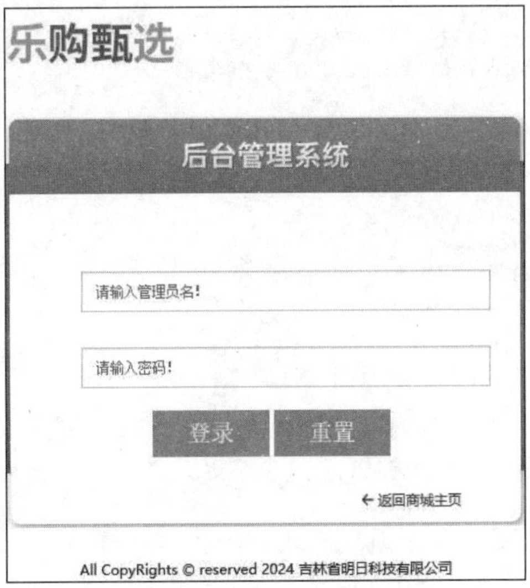

图 5.21　后台登录页面

后台登录功能主要是通过自定义的 login() 方法实现的。该方法中，对登录页面中的管理员账号和密码信息进行判断，如果输入正确，则进入后台主页。最后渲染登录模板页面 login.html。关键代码如下：

```python
@admin.route("/login/", methods=["GET","POST"])
def login():
    """
    登录功能
    """
    # 判断是否已经登录
    if "admin" in session:
        return redirect(url_for("admin.index"))
    form = LoginForm()                                          # 实例化登录表单
    if form.validate_on_submit():                               # 验证提交表单
        data = form.data                                        # 接收数据
        admin = Admin.query.filter_by(manager=data["manager"]).first()  # 查找 Admin 表数据
        # 密码错误时，则 not check_password(data["pwd"])为真
        if not admin.check_password(data["password"]):
            flash("密码错误!", "err")                           # 闪存错误信息
            return redirect(url_for("admin.login"))             # 跳转到后台登录页
        # 如果是正确的，就要定义 session 的会话进行保存
        session["admin"] = data["manager"]                      # 存入 session
        session["admin_id"] = admin.id  # 存入 session
        return redirect(url_for("admin.index"))                 # 返回后台主页
    return render_template("admin/login.html",form=form)
```

后台登录模板页面 login.html 的主要代码如下：

```html
<form   method="post" action="{{url_for('admin.login')}}" >
    <table width="448" height="345"  border="0" align="center"
        style="margin-top:170px;background:url('/static/admin/images/managerlogin_dialog.png') no-repeat"
        cellpadding="0" cellspacing="0">
        <tr>
            <td height="60" colspan="2" align="center"> </td>
        </tr>
        <tr>
            <td width="55" height="280" align="center" valign="top"> </td>
            <td width="436" align="left" valign="top">
```

```html
<table style="margin-top:30px" width="88%" height="240"  border="0" cellpadding="0" cellspacing="0">
    <tr>
        <td width="99%" height="74" align="center">
            {{ form.manager }}
            {% for err in form.manager.errors %}
                <p style="color: red;float:left">{{ err }}</p>
            {% endfor %}
        </td>
    </tr>
    <tr>
        <td height="30" align="center">
          <span class="word_white">
            {{ form.password }}
          </span></td>
    </tr>
    <tr>
        <td height="57" align="center">
          {{ form.csrf_token }}
          {{ form.submit }}
          <input   type="reset" class="login_reset" value="重置">
        </td>
    </tr>
    {% for msg in get_flashed_messages(category_filter=["err"]) %}
        <p class="login-box-msg" style="color: red">{{ msg }}</p>
    {% endfor %}
    {% for msg in get_flashed_messages(category_filter=["ok"]) %}
        <p class="login-box-msg" style="color: green">{{ msg }}</p>
    {% endfor %}
    <tr>
        <td height="35" align="right">
          <a href="/"><img src="{{url_for('static',filename='admin/images/back.png')}}"> 返回商城主页</a></td>
    </tr>
    </table>
    </td>
    </tr>
</table>
</form>
```

5.9.2 商品管理模块设计

乐购甄选在线商城的商品管理模块主要实现对商品信息的管理，包括分页显示商品信息、添加商品信息、修改商品信息、删除商品信息等功能。下面分别进行介绍。

1．分页显示商品信息

商品管理模块的首页是分页显示商品信息页，主要用于将商品信息表中的商品信息以列表的方式显示，并为之添加"修改"和"删除"功能，方便用户对商品信息进行修改和删除。商品管理模块首页的运行结果如图 5.22 所示。

分页显示商品信息是后台首页中的默认功能，该功能主要是通过自定义的 index()方法实现的。该方法中，根据 GET 请求中提交的页码和类型获取商品信息，并且分页显示；然后，渲染后台首页模板页面 index.html，在渲染页面时，将获取到的商品信息传递给模板页面，以便进行数据显示。关键代码如下：

```python
@admin.route("/")
@admin_login
def index():
    page = request.args.get('page', 1, type=int)  # 获取 page 参数值
    page_data = Goods.query.order_by(
        Goods.addtime.desc()
    ).paginate(page=page, per_page=10)
    return render_template("admin/index.html",page_data=page_data)
```

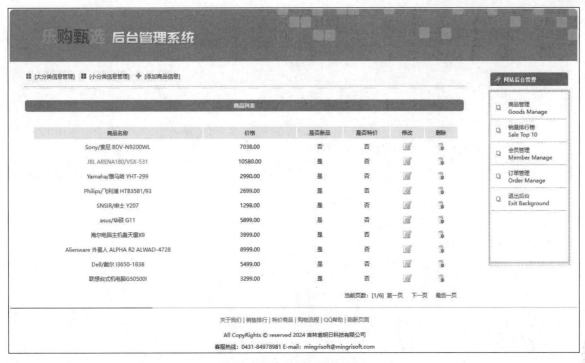

图 5.22　商品管理模块首页

后台首页模板页面 index.html 的主要代码如下：

```html
<table width="100%" height="60"  border="1" cellpadding="0" cellspacing="0"
    bordercolor="#FFFFFF" bordercolordark="#FFFFFF" bordercolorlight="#E6E6E6">
  <tr bgcolor="#eeeeee">
    <td width="40%" height="24" align="center">商品名称</td>
    <td width="22%" align="center">价格</td>
    <td width="11%" align="center">是否新品</td>
    <td width="11%" align="center">是否特价</td>
    <td width="8%" align="center">修改</td>
    <td width="8%" align="center">删除</td>
  </tr>
{% for v in page_data.items %}
    <tr style="padding:5px;">
      <td height="20" align="center">
        <a href="{{url_for('admin.goods_detail',goods_id=v.id)}}">{{ v.name }}</a>
      </td>
      <td align="center" >{{ v.current_price }}</td>
      <td align="center">{% if v.is_new %} 是 {% else %} 否 {% endif%}</td>
      <td align="center">{% if v.is_sale %} 是 {% else %} 否 {% endif%}</td>
      <td align="center">
        <a href="{{url_for('admin.goods_edit',id=v.id)}}">
          <img src="{{url_for('static',filename='admin/images/modify.gif')}}" width="19" height="19">
        </a>
      </td>
      <td align="center">
        <a href="{{url_for('admin.goods_del_confirm',goods_id=v.id)}}">
          <img src="{{url_for('static',filename='admin/images/del.gif')}}" width="20" height="20">
        </a>
      </td>
    </tr>
{% endfor %}
</table>
{% if page_data %}
```

```
<tbody>
  <tr>
    <td height="30" align="right">当前页数：[{{page_data.page}}/{{page_data.pages}}] 
      <a href="{{ url_for('admin.index',page=1) }}">第一页</a>
      {% if page_data.has_prev %}
        <a href="{{ url_for('admin.index',page=page_data.prev_num) }}">上一页</a>
      {% endif %}
      {% if page_data.has_next %}
        <a href="{{ url_for('admin.index',page=page_data.next_num) }}">下一页</a>
      {% endif %}
      <a href="{{ url_for('admin.index',page=page_data.pages) }}">最后一页 </a>
    </td>
  </tr>
</tbody>
{% endif %}
```

另外，在商品列表中单击删除图标时，会调整到确认删除商品页面，该页面中首先需要确认是否删除商品，如果确认删除，则调用 goods_del() 方法删除 goods 商品表中指定 ID 的商品，关键代码如下：

```python
@admin.route("/goods/del_confirm/")
@admin_login
def goods_del_confirm():
    '''确认删除商品'''
    goods_id = request.args.get('goods_id')
    goods = Goods.query.filter_by(id=goods_id).first_or_404()
    return render_template('admin/goods_del_confirm.html',goods=goods)

@admin.route("/goods/del/<int:id>/", methods=["GET"])
@admin_login
def goods_del(id=None):
    """
    删除商品
    """
    goods = Goods.query.get_or_404(id)        # 根据商品 ID 查找数据
    db.session.delete(goods)                   # 删除数据
    db.session.commit()                        # 提交数据
    return redirect(url_for('admin.index', page=1))   # 渲染模板
```

确认删除商品页面 goods_del_confirm.html 的主要代码如下：

```html
<form action="{{url_for('admin.goods_del',id=goods.id)}}" method="get" name="form1">
  <table width="94%"  border="0" align="right" cellpadding="-2" cellspacing="-2" bordercolordark="#FFFFFF">
    <tr>
      <td width="14%" height="27"> 商品名称：</td>
      <td height="27" colspan="3"> 
        <input name="ID" type="hidden" id="ID" value="24">
        {{goods.name}}  
      </td>
    </tr>
    <tr>
      <td height="27"> 所属大类：</td>
      <td width="31%" height="27"> {{goods.supercat.cat_name}}</td>
      <td width="13%" height="27">  所属小类：</td>
      <td width="42%" height="27"> {{goods.subcat.cat_name}}</td>
    </tr>
    <tr>
      <td height="16"> 图片文件：</td>
      <td height="27" colspan="3"> {{goods.picture}}</td>
    </tr>
    <tr>
      <td height="27" align="center">定      价：</td>
      <td height="27"> {{goods.original_price}}(元)</td>
```

```html
        <td height="27" align="center">现    价： </td>
        <td height="27"> {{goods.current_price}}(元)</td>
      </tr>
      <tr>
        <td height="45"> 是否新品： </td>
        <td> 
            {% if goods.is_new%}
                不是新品
            {% else %}
                是新品
            {% endif %}
        </td>
        <td> 是否特价： </td>
        <td>
            {% if goods.is_sale%}
                不是特价商品
            {% else %}
                是特价商品
            {% endif %}
        </td>
      </tr>
      <tr>
        <td height="103"> 商品简介： </td>
        <td colspan="3"><span class="style5">  </span>
           {{ goods.introduction }}
      </tr>
      <tr>
        <td height="38" colspan="4" align="center">
            <input name="Submit" type="submit" class="btn_bg_long1" value="确定删除">
             
            <input name="Submit3" type="button" class="btn_bg_short" value="返回" onClick="javascript:history.back()">
        </td>
      </tr>
    </table>
</form>
```

2．添加商品信息

在商品管理首页中单击"添加商品信息"即可进入添加商品信息页面。添加商品信息页面主要用于向数据库中添加新的商品信息。添加商品信息页面的运行结果如图 5.23 所示。

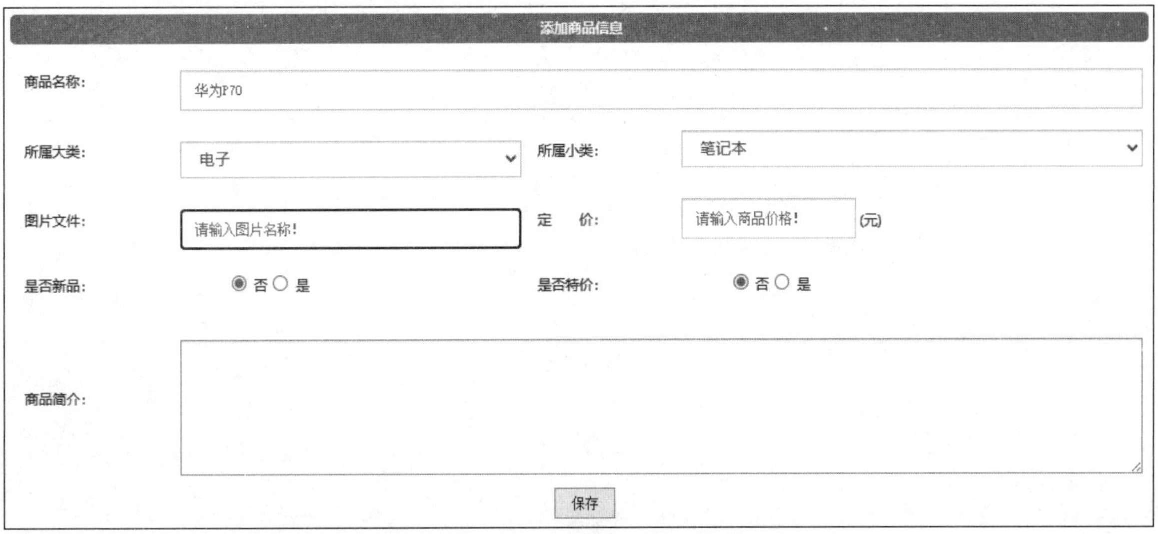

图 5.23 添加商品信息页面

添加商品信息功能主要是通过自定义的 goods_add()方法实现的。该方法中，首先从表单页面中获取提交的商品信息，然后对表单进行验证。如果验证成功，则向 goods 数据表中添加提交的商品信息，并返回商品列表页（即后台首页）。最后渲染添加商品模板页面 goods_add.html，并传递表单参数。关键代码如下：

```python
@admin.route("/goods/add/", methods=["GET", "POST"])
@admin_login
def goods_add():
    """
    添加商品
    """
    form = GoodsForm()                                                          # 实例化 form 表单
    supercat_list = [(v.id, v.cat_name) for v in SuperCat.query.all()]          # 获取所属大类列表
    form.supercat_id.choices = supercat_list                                    # 显示所属大类
    # 显示大类所包含的子类
    form.subcat_id.choices = [(v.id, v.cat_name) for v in SubCat.query.filter_by(super_cat_id=supercat_list[0][0]).all()]
    form.current_price.data = form.data['original_price']                       # 为 current_pirce 赋值
    if form.validate_on_submit():                                               # 添加商品情况
        data = form.data
        goods = Goods(
            name = data["name"],
            supercat_id = int(data['supercat_id']),
            subcat_id = int(data['subcat_id']),
            picture= data["picture"],
            original_price = Decimal(data["original_price"]).quantize(Decimal('0.00')),    # 转换为包含 2 位小数的形式
            current_price = Decimal(data["original_price"]).quantize(Decimal('0.00')),     # 转换为包含 2 位小数的形式
            is_new = int(data["is_new"]),
            is_sale = int(data["is_sale"]),
            introduction=data["introduction"],
        )
        db.session.add(goods)                                                   # 添加数据
        db.session.commit()                                                     # 提交数据
        return redirect(url_for('admin.index'))                                 # 页面跳转
    return render_template("admin/goods_add.html", form=form)                   # 渲染模板
```

添加商品模板页面 goods_add.html 的主要代码如下：

```html
<script>
$(document).ready(function(){
    $('#supercat_id').change(function(){
        super_id = $(this).children('option:selected').val()
        selSubCat(super_id);
    })
});
function selSubCat(val){
    $.get("{{ url_for('admin.select_sub_cat')}}",
            {super_id:val},
            function(result){
                html_doc = ''
                if(result.status == 1){
                    $.each(result.data,function(idx,obj){
                        html_doc += '<option value='+obj.id+'>'+obj.cat_name+'</option>'
                    });
                }else{
                    html_doc += '<option value=0>前选择子类</option>'
                }
                $("#subcat_id").html(html_doc);                                 // 显示获取到的小分类
    });
}
</script>
<table width="1280" height="288"    border="0" align="center" cellpadding="0" cellspacing="0" bgcolor="#FFFFFF">
    <!-- 省略部分代码 -->
```

```html
<form action="" method="post">
    <table width="94%"  border="0" align="center" cellpadding="0" cellspacing="0" bordercolordark="#FFFFFF">
      <tr>
        <td width="14%" height="27"> 商品名称：</td>
        <td height="27" colspan="3"> 
            {{ form.name }}
            {% for err in form.name.errors %}
                <span   style="float:left;padding-top:10px;color:red">{{ err }}</span>
            {% endfor %}
        </td>
      </tr>
      <tr>
        <td height="27"> 所属大类：</td>
        <td width="31%" height="27"> 
            {{ form.supercat_id }}
        </td>
        <td width="13%" height="27">  所属小类：</td>
        <td width="42%" height="27" id="subType">
            {{ form.subcat_id }}
        </td>
      </tr>
      <tr>
        <td height="41"> 图片文件：</td>
        <td height="41"> 
            {{ form.picture }}
        </td>
        <td height="41"> 定    价：</td>
        <td height="41">
            <span style="float:left;">
                {{form.original_price}}
            </span>
            <span   style="float:left;padding-top:10px;"> (元)</span>
            {% for err in form.original_price.errors %}
                <span   style="float:left;padding-top:10px;color:red">{{ err }}</span>
            {% endfor %}
        </td>
      </tr>
      <tr>
        <td height="45"> 是否新品：</td>
        <td> {{form.is_new}} </td>
        <td> 是否特价：</td>
        <td> {{form.is_sale}} </td>
      </tr>
      <tr>
        <td height="103"> 商品简介：</td>
        <td colspan="3">
            <span class="style5">  </span>
            {{ form.introduction }}
        </td>
      </tr>
      <tr>
        <td height="38" colspan="4" align="center">
            {{ form.csrf_token }}
            {{ form.submit }}
        </td>
      </tr>
    </table>
</form>
</td>
```

3. 修改商品信息

在商品管理首页中单击想要修改的商品信息后面的修改图标，即可进入修改商品信息页面。修改商品信息页面主要用于修改指定商品的基本信息。修改商品信息页面的运行结果如图 5.24 所示。

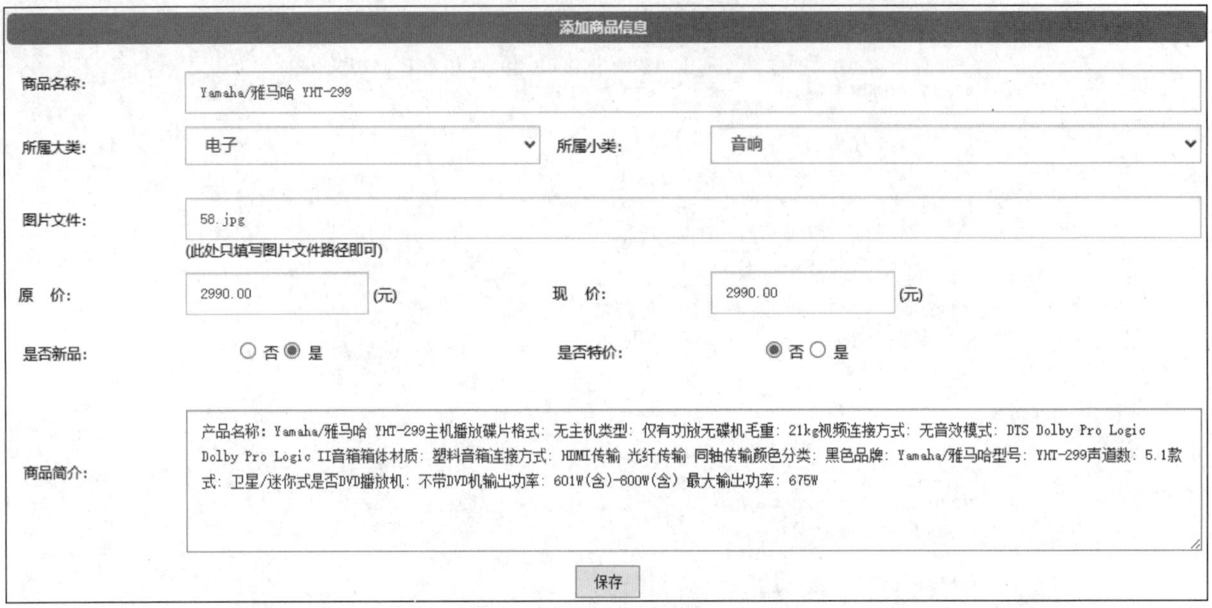

图 5.24　修改商品信息页面

修改商品信息功能主要是通过自定义的 goods_edit()方法实现的，该方法有一个参数，表示要修改的商品 ID。该方法中，首先根据 ID 参数的值获取要修改的商品信息，并显示在页面中。用户修改商品信息后，单击"保存"按钮，对表单信息验证。如果验证成功，则修改 goods 数据表中指定商品的信息。最后渲染修改商品模板页面 goods_edit.html，并传递表单参数。关键代码如下：

```python
@admin.route("/goods/edit/<int:id>", methods=["GET", "POST"])
@admin_login
def goods_edit(id=None):
    """
    修改商品
    """
    goods = Goods.query.get_or_404(id)
    form = GoodsForm()                    # 实例化 form 表单
    form.supercat_id.choices = [(v.id, v.cat_name) for v in SuperCat.query.all()]   # 获取所属大类
    # 获取所属子类
    form.subcat_id.choices = [(v.id, v.cat_name) for v in SubCat.query.filter_by(super_cat_id=goods.supercat_id).all()]

    if request.method == "GET":           # GET 请求，即在商品列表页中单击修改图标时跳转到该页面
        form.name.data = goods.name
        form.picture.data = goods.picture
        form.current_price.data = goods.current_price
        form.original_price.data = goods.original_price
        form.supercat_id.data = goods.supercat_id
        form.subcat_id.data = goods.subcat_id
        form.is_new.data = goods.is_new
        form.is_sale.data = goods.is_sale
        form.introduction.data = goods.introduction
    elif form.validate_on_submit():       # 提交操作，即修改指定的商品信息
        goods.name = form.data["name"]
        goods.supercat_id = int(form.data['supercat_id'])
```

```
            goods.subcat_id = int(form.data['subcat_id'])
            goods.picture= form.data["picture"]
            goods.original_price = Decimal(form.data["original_price"]).quantize(Decimal('0.00'))
            goods.current_price = Decimal(form.data["current_price"]).quantize(Decimal('0.00'))
            goods.is_new = int(form.data["is_new"])
            goods.is_sale = int(form.data["is_sale"])
            goods.introduction=form.data["introduction"]
            db.session.add(goods)                               # 添加数据
            db.session.commit()                                 # 提交数据
            return redirect(url_for('admin.index'))             # 页面跳转

    return render_template("admin/goods_edit.html", form=form)  # 渲染模板
```

修改商品模板页面 goods_edit.html 的代码与添加商品模板页面类似，这里不再赘述。

> **说明**
>
> 在乐购甄选在线商城的后台，除了对商品信息进行管理，还可以对商品大分类和子分类进行管理，它们的实现过程与商品管理模块类似，这里不再赘述，以下列出实现商品大分类和子分类管理模块主要用到的方法和渲染模板页面，方便读者查看源码。
> - ☑ 显示大分类列表：supercat_list()方法、supercat.html 页面。
> - ☑ 添加大分类：supercat_add()方法、supercat_add.html 页面。
> - ☑ 删除大分类：supercat_del()方法。
> - ☑ 显示子分类列表：subcat_list()方法、subcat.html 页面。
> - ☑ 添加子分类：subcat_add()方法、subcat_add.html 页面。
> - ☑ 删除子分类：subcat_del()方法。

5.9.3 销量排行榜模块设计

单击后台导航栏中的"销量排行榜"即可进入销量排行榜页面。在该页面中将以表格的形式对销量排在前 10 名的商品信息进行显示，以方便管理员及时了解各种商品的销量情况，从而根据该结果做出相应的促销活动。销量排行榜页面的运行效果如图 5.25 所示。

商品列表	
产品名称	销售数量（个）
商海悟道：商亦有道，大道至简	5
从0到1：开启商业与未来的秘密	1
商海悟道：商亦有道，大道至简	1
Bosch/博世 KAD63P70TI	1
Ronshen/容声 BCD-202M/TX6	1
Razer/雷蛇 灵刃潜行版 RZ09-01682E22	1
Asus/华硕 顽石4代	1
Apple/苹果 MacBook Pro MJLT2CH/A	1
asus/华硕 G11	1
从0到1：开启商业与未来的秘密	1

图 5.25 销量排行榜页面

商品销量排行榜功能的实现比较简单，主要是按照商品的销量（goods 表中的 number 字段）进行降序排序，关键代码如下：

```
@admin.route('/topgoods/', methods=['GET'])
```

```python
@admin_login
def topgoods():
    """
    销量排行榜(前 10 位)
    """
    orders = OrdersDetail.query.order_by(OrdersDetail.number.desc()).limit(10).all()
    return render_template("admin/topgoods.html", data=orders)
```

商品销量排行榜模板页面 topgoods.html 的主要代码如下：

```html
<table width="96%" height="48"   border="1" cellpadding="10" cellspacing="0"
    bordercolor="#FFFFFF" bordercolordark="#CCCCCC" bordercolorlight="#FFFFFF">
  <tr align="center">
     <td width="80%">产品名称</td>
     <td width="20%">销售数量（个）</td>
  </tr>
{% for v in data %}
<tr align="center">
    <td >{{v.goods.name}}</td>
    <td >{{v.number}}</td>
  </tr>
{% endfor %}
</table>
```

5.9.4　会员管理模块设计

单击后台导航栏中的"会员管理"，即可进入会员管理页面。对于会员信息的管理主要是查看会员基本信息。会员管理页面的运行效果如图 5.26 所示。

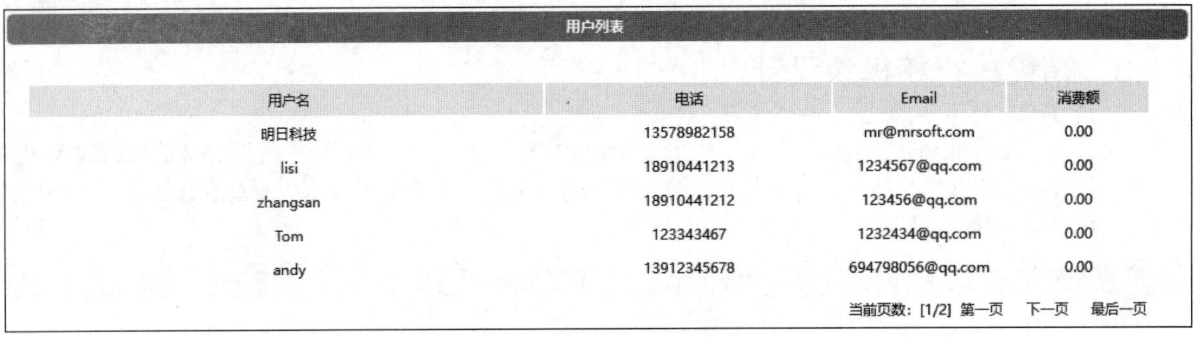

用户列表				
用户名		电话	Email	消费额
明日科技		13578982158	mr@mrsoft.com	0.00
lisi		18910441213	1234567@qq.com	0.00
zhangsan		18910441212	123456@qq.com	0.00
Tom		123343467	1232434@qq.com	0.00
andy		13912345678	694798056@qq.com	0.00
			当前页数：[1/2] 第一页 下一页 最后一页	

图 5.26　会员管理页面

查看会员基本信息功能主要是通过自定义的 user_list() 方法实现的。该方法中，主要分页显示所有注册该网站的会员的基本信息，每页显示 5 条记录。另外，可以根据姓名或者邮箱查询会员信息。关键代码如下：

```python
@admin.route("/user/list/", methods=["GET"])
@admin_login
def user_list():
    """
    会员列表
    """
    page = request.args.get('page', 1, type=int)         # 获取 page 参数值
    keyword = request.args.get('keyword', '', type=str)
    if keyword:
        # 根据姓名或邮箱查询
        filters = or_(User.username == keyword, User.email == keyword)
        page_data = User.query.filter(filters).order_by(
            User.addtime.desc()
        ).paginate(page=page, per_page=5)
```

```
    else:
        page_data = User.query.order_by(
            User.addtime.desc()
        ).paginate(page=page, per_page=5)

    return render_template("admin/user_list.html", page_data=page_data)
```

查看会员基本信息模板页面 user_list.html 的主要代码如下：

```html
<table width="100%" height="60"  border="1" cellpadding="0" cellspacing="0"
    bordercolor="#FFFFFF" bordercolordark="#FFFFFF" bordercolorlight="#E6E6E6">
  <tr bgcolor="#eeeeee">
    <td width="40%" height="24" align="center">用户名</td>
    <td width="22%" align="center">电话</td>
    <td width="11%" align="center">Email</td>
    <td width="11%" align="center">消费额</td>
  </tr>
{% if page_data.items %}
  {% for v in page_data.items %}
    <tr style="padding:5px;">
      <td height="20" align="center">{{ v.username }}</td>
      <td align="center" >{{ v.phone }}</td>
      <td align="center">{{ v.email }}</td>
      <td align="center">{{ v.consumption }}</td>
    </tr>
  {% endfor %}
{% else %}
    没有查找的信息
{% endif %}
</table>
{% if page_data %}
  <tbody>
    <tr>
      <td height="30" align="right">当前页数：[{{page_data.page}}/{{page_data.pages}}] 
        <a href="{{ url_for('admin.user_list',page=1) }}">第一页</a>
        {% if page_data.has_prev %}
          <a href="{{ url_for('admin.user_list',page=page_data.prev_num) }}">上一页</a>
        {% endif %}
        {% if page_data.has_next %}
          <a href="{{ url_for('admin.user_list',page=page_data.next_num) }}">下一页</a>
        {% endif %}
        <a href="{{ url_for('admin.user_list',page=page_data.pages) }}">最后一页 </a>
      </td>
    </tr>
  </tbody>
{% endif %}
```

5.9.5 订单管理模块设计

单击后台导航栏中的"订单管理"即可进入订单管理页面。对于订单的管理主要是显示订单列表，以及按照订单编号查询指定的订单。订单管理模块页面运行结果如图 5.27 所示。

订单管理功能主要是通过自定义的 orders_list()方法实现的。该方法中，主要分页显示所有网站订单信息，每页显示 10 条记录。另外，可以根据订单号查询订单信息。关键代码如下：

```python
@admin.route("/orders/list/", methods=["GET"])
@admin_login
def orders_list():
    """
    订单列表页面
    """
```

```python
keywords = request.args.get('keywords','',type=str)
page = request.args.get('page', 1, type=int) # 获取 page 参数值
if keywords :
    page_data = Orders.query.filter_by(id=keywords).order_by(
        Orders.addtime.desc()
    ).paginate(page=page, per_page=10)
else :
    page_data = Orders.query.order_by(
        Orders.addtime.desc()
    ).paginate(page=page, per_page=10)
return render_template("admin/orders_list.html", page_data=page_data)
```

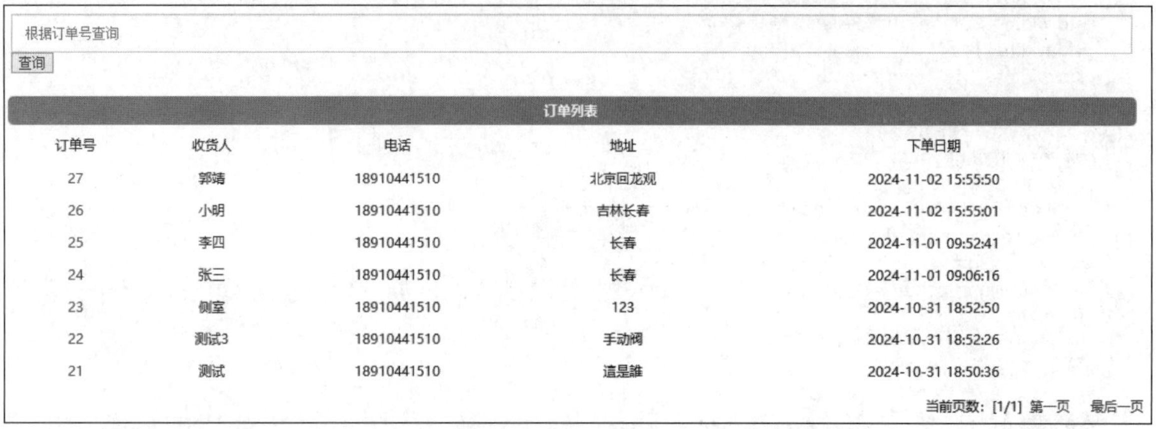

图 5.27　订单管理页面

订单管理模板页面 orders_list.html 的主要代码如下：

```html
<table width="1280" height="288"  border="0" align="center" cellpadding="0" cellspacing="0" bgcolor="#FFFFFF">
  <tr>
    <td align="center" valign="top">
      <table width="100%"  border="0" cellpadding="0" cellspacing="0">
        <!-- 省略部分代码 -->
        <tr>
          <td align="right"> </td>
          <td height="10" colspan="3">
            <form action="" method="get" >
              <input type="text" placeholder="根据订单号查询" name="keywords" id="orderId" />
              <input type="submit" value="查询" />
            </form>
          </td>
          <td> </td>
        </tr>
        <!-- 省略部分代码 -->
        <table width="96%" height="48"  border="1" cellpadding="0" cellspacing="0" bordercolor="#FFFFFF" bordercolordark="#CCCCCC" bordercolorlight="#FFFFFF">
          <tr align="center">
            <td width="8%" height="30">订单号</td>
            <td width="10%">收货人</td>
            <td width="15%">电话</td>
            <td width="15%">地址</td>
            <td width="26%">下单日期</td>
          </tr>
          {% for v in page_data.items %}
          <tr align="center">
            <td height="24">
              <a href="{{ url_for('admin.orders_detail',order_id=v.id)}}">{{ v.id }}</a>
```

```html
                    </td>
                    <td>{{ v.recevie_name}}</td>
                    <td>{{ v.recevie_tel}}</td>
                    <td>{{ v.recevie_address}}</td>
                    <td>{{ v.addtime}}</td>
                </tr>
            {% endfor %}
            </table>
            <table width="100%"  border="0" cellspacing="0" cellpadding="0">
                <tr>
                    <td height="30" align="right">当前页数：[{{page_data.page}}/{{page_data.pages}}] 
                        <a href="{{ url_for('admin.orders_list',page=1) }}">第一页</a>
                        {% if page_data.has_prev %}
                            <a href="{{ url_for('admin.orders_list',page=page_data.prev_num) }}">上一页</a>
                        {% endif %}
                        {% if page_data.has_next %}
                            <a href="{{ url_for('admin.orders_list',page=page_data.next_num) }}">下一页</a>
                        {% endif %}
                        <a href="{{ url_for('admin.orders_list',page=page_data.pages) }}">最后一页 </a>
                    </td>
                </tr>
            </table>
        </td>
    </tr>
</table>
```

5.10 项目运行

通过前述步骤，设计并完成了"乐购甄选在线商城"项目的开发。下面运行该项目，检验一下我们的开发成果。运行"乐购甄选在线商城"项目的步骤如下。

（1）打开 Shop\app__init__.py 文件，根据自己的数据库账号和密码修改如下代码：

```python
def create_app():
    app = Flask(__name__)
    app.config['SQLALCHEMY_DATABASE_URI']='mysql+pymysql://root:root@127.0.0.1:3306/shop'
    app.config['SQLALCHEMY_TRACK_MODIFICATIONS'] = True
    app.config['SECRET_KEY'] = 'mr'
    db.init_app(app)
```

（2）打开命令提示符对话框，进入 Shop 项目文件夹所在目录，在命令提示符对话框中输入如下命令来创建 venv 虚拟环境：

```
virtualenv venv
```

（3）在命令提示符对话框中输入如下命令来启动 venv 虚拟环境：

```
venv\Scripts\activate
```

（4）在命令提示符对话框中使用如下命令来安装 Flask 等依赖包：

```
pip install -r  requirements.txt
```

（5）在命令提示符对话框中使用如下命令创建数据表：

```
flask   db   init                    # 创建迁移仓库，首次使用
flask   db   migrate                 # 创建迁移脚本
flask   db   upgrade                 # 把迁移应用到数据库中
```

（6）新增的数据表中数据为空，所以需要导入数据。将 Shop\shop.sql 文件导入数据库中。

（7）在 PyCharm 的左侧项目结构中展开"乐购甄选在线商城"的项目文件夹 Shop，在其中选中 app.py 文件，单击鼠标右键，在弹出的快捷菜单中选择 Run 'app'命令，如图 5.28 所示。

（8）如果在 PyCharm 底部出现如图 5.29 所示的提示，说明程序运行成功。

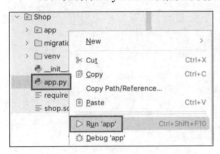

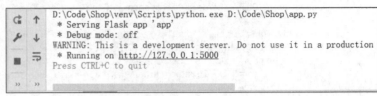

图 5.28　选择 Run 'app'　　　　　　　　图 5.29　程序运行成功提示

（9）在浏览器中输入网址 http://127.0.0.1:5000/即可进入乐购甄选在线商城的前台首页，效果如图 5.30 所示。在商城的前台可以登录注册、查看商品、推荐商品、加入购物车、提交订单等。

图 5.30　乐购甄选在线商城前台首页

（10）在浏览器中输入网址 http://127.0.0.1:5000/admin 即可进入后台登录页，在其中输入管理员账号和密码后，即可进入后台对商品、商品分类、会员、订单等信息进行管理，效果如图 5.31 所示。

图 5.31　乐购甄选在线商城后台页

本章主要介绍如何使用 Flask 框架实现乐购甄选在线商城项目。在本项目中，重点讲解了商城前后台功能的实现。在实现这些功能时，我们使用了 Flask 的流行模块，包括使用 Flask-SQLAlchemy 来操作数据库，使用 Flask-WTF 创建表单，以及使用模板渲染展示等。

5.11　源码下载

源码下载

本章虽然详细地讲解了如何编码实现"乐购甄选在线商城"的各个功能，但给出的代码都是代码片段，而非完整的源代码。为了方便读者学习，本书提供了该项目的完整源代码，读者可以通过扫描右侧的二维码进行下载。

第2篇 Django 框架项目

Django 是一个高级 Python Web 框架。它由经验丰富的开发者构建,旨在处理网站开发中的复杂部分,让开发者能够专注于编写应用程序的核心逻辑,而不是重复造轮子。它采用 MTV(Model-Template-View)设计模式,这是传统的 MVC(Model-View-Controller)模式的一种变体,其中控制器的责任被框架自动处理,从而简化了开发流程。Django 框架因其功能强大、安全性高和易维护性,成为许多开发者和团队优先选择的 Web 框架之一。

本篇将使用 Django 框架开发"心灵驿站聊天室""站内全局搜索引擎""综艺之家""智慧校园考试系统""吃了么外卖网"5 个流行的 Web 项目,带领读者全面体验使用 Django 框架开发 Web 项目的过程,并积累实战项目开发经验。

第 6 章
心灵驿站聊天室

——WebSocket + Django + Channels + Channels-Redis

随着网络技术的发展，人们开始寻求更快捷、更便捷的沟通方式，而在线聊天室是满足这一需求的重要形式之一。因此，设计一个适应时代需求、技术先进、注重用户体验的聊天室，是对社交趋势的一种前瞻布局，具有广阔的发展前景和社会价值。本章将使用 Django 框架及 Channels 模块开发一个聊天室项目。

项目微视频

本项目的核心功能及实现技术如下：

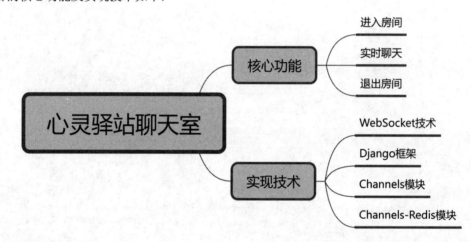

6.1 开发背景

在快节奏的现代社会里，即时通信已成为人们日常沟通不可或缺的一部分，它不仅缩短了人与人之间的距离，还极大地丰富了交流的形式和内容。在这个背景下，传统的沟通方式逐渐被更加灵活、便捷的在线聊天平台所取代，而聊天室作为其中的一个重要形态，其开发和优化显得尤为重要。

Django 框架是一个使用 Python 编写的 Web 框架，旨在帮助开发者快速构建高效、可扩展的 Web 应用。Channels 模块是一个强大的工具，它使得构建实时的 Web 应用变得更加容易。因此，将 Django 框架与 Channels 模块搭配使用，就可以方便地开发出支持异步的实时应用程序。这种组合搭配是开发聊天室项目的理想选择。

本项目的实现目标如下：
- ☑ 相同兴趣的用户进入相同的房间。
- ☑ 实现即时聊天。
- ☑ 支持一对一私聊和多人群聊。
- ☑ 界面简洁大方，适应不同年龄的用户。

6.2 系统设计

6.2.1 开发环境

本项目的开发及运行环境如下：
- ☑ 操作系统：推荐 Windows 10、Windows 11 或更高版本。
- ☑ 开发工具：PyCharm 2024（向下兼容）。
- ☑ 开发语言：Python 3.12。
- ☑ 数据库：Redis 5.0。
- ☑ Python Web 框架：Django 5.0。

6.2.2 业务流程

心灵驿站聊天室主要实现进入相同房间的人们之间进行即时聊天的功能。输入相同的房间号，即可进入相同的房间，在该房间内发送的消息，将会被同房间的其他人看到。本项目的业务流程如图 6.1 所示。

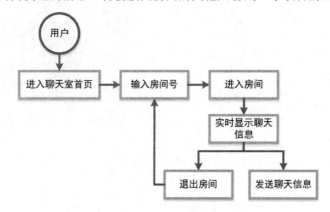

图 6.1 心灵驿站聊天室的业务流程

6.2.3 功能结构

本项目的功能结构已经在章首页中给出，其实现的具体功能如下：
- ☑ 输入房间号进入房间。如果输入的房间号相同，则进入相同的聊天室。
- ☑ 在相同聊天室的人们，可以进行实时聊天。
- ☑ 不想聊天时，可以退出房间。

6.3 技术准备

6.3.1 技术概览

在开发聊天室程序时，传统的实现方式是使用 Ajax 轮询，这种方式虽然简单，容易实现，但是频繁地请求会导致服务器压力增大和带宽消耗。现在推荐的实现方式是使用 WebSocket、Django 和 Channels 模块实现。WebSocket 是一种在客户端和服务器之间提供长期的全双工通信协议。它允许双方在建立一次连接后，能够实时地、双向地交换数据，无须像 HTTP 协议那样为每一个请求和响应建立新的连接。关于 WebSocket 的知识在《Python 从入门到精通（第3版）》中有详细的讲解，对该知识不太熟悉的读者可以参考该书对应的内容。

下面对实现本项目时用到的其他主要技术点进行必要介绍，如 Django 框架、Channels 模块、Channels-Redis 模块等，以确保读者可以顺利完成本项目。

6.3.2　Django 框架的基本使用

Django 是基于 Python 的开源 Web 框架，它拥有高度定制的 ORM、大量的 API、简单灵活的视图、优雅的 URL、适于快速开发的模板、强大的管理后台，这些使得它在 Python Web 开发领域占据不可动摇的地位。下面介绍 Django 框架的基本使用方法。

1．安装 Django Web 框架

安装 Django Web 框架非常简单，直接使用以下命令即可：

```
pip install django
```

2．创建并运行 Django 项目

创建及运行 Django 项目的步骤如下。

（1）在虚拟环境下创建一个名为 django_demo 的项目，命令如下：

```
django-admin startproject django_demo
```

（2）使用 Pycharm 打开 django_demo 项目，查看目录结构，如图 6.2 所示。

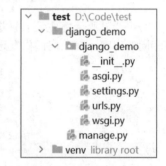

图 6.2　Django 项目目录结构

Django 项目中的文件及说明如表 6.1 所示。

表 6.1　Django 项目中的文件及说明

文件	说明
django_demo	Django 生成的和项目同名的配置文件夹
__init__.py	初始化包，在该文件中可以执行一些初始化操作
asgi.py	ASGI（异步服务器网关接口）的配置文件，用于配置异步 Web 服务器
settings.py	Django 总的配置文件，可以配置 App、数据库、中间件、模板等诸多选项
urls.py	Django 默认的路由配置文件，可以在其中引入其他路径下的 urls.py
wsgi.py	Django 实现的 WSGI 接口的文件，用来处理 Web 请求
manage.py	Django 程序执行的入口

（3）在虚拟环境中执行如下命令来运行项目：

```
python django_demo/manage.py runserver
```

运行结果如图 6.3 所示。

（4）从图 6.3 中可以看到，开发服务器已经开始监听 8000 端口的请求了，这时在浏览器中输入 http://127.0.0.1:8000 即可看到一个 Django 页面，如图 6.4 所示。

（5）接下来使用命令来创建后台应用，首先使用 Ctrl+C 快捷键关闭服务器，然后通过如下命令执行数据库的迁移操作，以生成数据表：

```
python django_demo/manage.py migrate
```

心灵驿站聊天室 第 6 章

图 6.3 启动项目

图 6.4 Django 页面

（6）执行如下命名创建超级管理员用户（这里需要输入用户名、密码、邮箱等内容）：

python django_demo/manage.py createsuperuser

（7）按照步骤（3）重新启动服务器，在浏览器中访问 http://127.0.0.1:8000/admin，即可进入后台登录页面，如图 6.5 所示。输入前面创建的用户名和密码，单击 LOG IN 按钮，即可进入后台的管理页面，如图 6.6 所示。

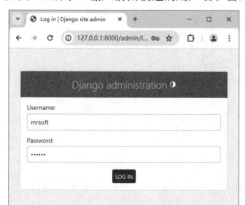

图 6.5 后台登录页面

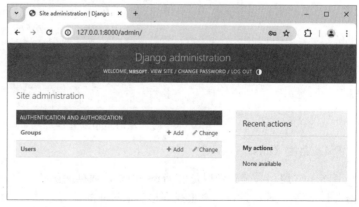

图 6.6 Django 项目后台管理页面

3. 创建一个 App

在 Django 项目中，推荐使用 App 来完成不同模块的任务。创建一个 App 非常简单，命令如下：

```
python django_demo/manage.py startapp app1
```

运行完成后，django_demo 目录下会多出一个名称为 app1 的目录，如图 6.7 所示。

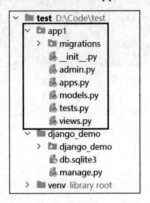

图 6.7 Django 项目的 App 目录结构

Django 项目中 App 目录的文件及说明如表 6.2 所示。

表 6.2 Django 项目中 App 目录的文件及说明

文件	说明
migrations	执行数据库迁移生成的脚本
__init__.py	初始化包，在该文件中可以执行一些初始化操作
admin.py	配置 Django 管理后台的文件
apps.py	单独配置添加的每个 App 的文件
models.py	创建数据库数据模型对象的文件
tests.py	用来编写测试脚本的文件
views.py	用来编写视图控制器的文件

接下来需要激活名为 app1 的 App，否则 app1 内的文件都不会生效。激活方式非常简单，在 django_demo/settings.py 配置文件中，找到 INSTALLED_APPS 列表，添加 app1，效果如图 6.8 所示。

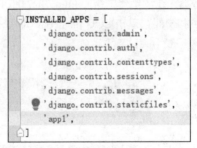

图 6.8 将创建的 App 名称添加到 settings.py 配置文件中

4. 路由（urls）

Django 路由系统的作用是使 views 中处理数据的函数与请求的 URL 建立映射关系，使请求到来后，根据 urls.py 里的关系条目，去查找到与请求对应的处理方法，从而返回给客户端 HTTP 页面数据。执行流程如图 6.9 所示。

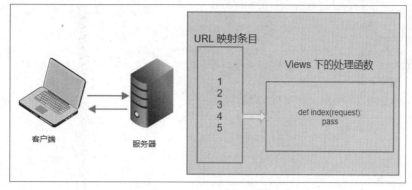

图 6.9 URL 映射流程

Django 项目中的 URL 规则定义放在 project 的 urls.py 目录下，默认如下：

```
from django.conf.urls import url
from django.contrib import admin

urlpatterns = [
    url(r'^admin/', admin.site.urls),
]
```

url()函数可以传递 4 个参数。其中，两个是必需的：regex 和 view；两个是可选的：kwargs 和 name。下面介绍每个参数的含义。

- ☑ regex：regex 是正则表达式的通用缩写，它是一种匹配字符串或 URL 地址的语法。Django 根据用户请求的 URL 地址，在 urls.py 文件中对 urlpatterns 列表中的每一项条目从头开始进行逐一对比，一旦遇到匹配项，立即执行该条目映射的视图函数或二级路由，其后的条目将不再继续匹配。因此，URL 路由的编写顺序至关重要！

> **说明**
> regex 不会去匹配 GET 或 POST 参数或域名，例如对于 https://www.example.com/myapp/，regex 只尝试匹配 myapp/。对于 https://www.example.com/myapp/?page=3，regex 也只尝试匹配 myapp/。

- ☑ view：当正则表达式匹配到某个条目时，自动将封装的 HttpRequest 对象作为第一个参数，将正则表达式"捕获"到的值作为第二个参数，传递给该条目指定的视图。如果是简单捕获，那么捕获值将作为一个位置参数进行传递；如果是命名捕获，那么将作为关键字参数进行传递。
- ☑ kwargs：任意数量的关键字参数可以作为一个字典传递给目标视图。
- ☑ name：对 URL 进行命名，可以在 Django 的任意处，尤其是模板内显式地引用它。相当于给 URL 取了个全局变量名。只需要修改这个全局变量的值，在整个 Django 中引用它的地方都将同样获得改变。

下面通过一个示例讲解 Django 路由的 URL 匹配方式，步骤如下：

（1）在项目 URL 配置文件 django_demo/urls.py 中添加如下代码：

```
urlpatterns = [
    path('admin/',admin.site.urls),
    path('app1/', include('app1.urls'))                # 引入 app1 模块下的一组路由
]
```

（2）在 app1 目录下创建 urls.py 文件，定义路由规则，代码如下：

```
from django.urls import path,re_path
from app1 import views as views
```

```python
urlpatterns = [
    path('index',views.index),                                                    # 精确匹配
    path('article/<int:id>', views.article),                                      # 匹配一个参数
    path('articles/<int:year>/<int:month>/<slug:slug>/', views.article_detail),   # 匹配两个参数和一个 slug
    re_path('articles/(?P<year>[0-9]{4})/', views.year_archive),                  # 正则匹配 4 个字符的年份
]
```

在上述代码中，列举了比较常见的几种 URL 匹配模式。其中，<类型: 变量名>是格式转换模式。例如，<int:id>将用户 URL 中的 id 参数自动转换为整型数据，否则默认为字符串型数据。

（3）在 app1/views.py 文件中编写视图函数，代码如下：

```python
from django.shortcuts import render
from django.http import HttpResponse


def index(request):
    return HttpResponse("Hello World")

def article(request,id):
    content = "This article's id is {}".format(id)
    return HttpResponse(content)

def article_detail(request,year,month,slug):
    content = 'the year is %s , the month is %s , the slug is %s.'.format(year,month,slug)
    return HttpResponse(content)

def year_archive(request,year):
    return HttpResponse(year)
```

完成以上步骤后，即可根据路由信息，在浏览器中输入相应 URL 来查看运行效果。例如，使用浏览器访问网址 http://127.0.0.1:8000/app1/articles/2024/04/python。

5．表单（forms）

在 app1 文件夹下创建一个 forms.py 文件，添加如下代码：

```python
from django import forms
class PersonForm(forms.Form):
    first_name = forms.CharField(label='你的名字', max_length=20)
    last_name = forms.CharField(label='你的姓氏', max_length=20)
```

上面代码定义了一个 PersonForm 表单类，两个字段类型为 forms.CharField，其对应 models.CharField，first_name 指字段的 label 为"你的名字"，并且指定该字段最大长度为 20 个字符。max_length 参数可以指定 forms.CharField 的验证长度。

PersonForm 类将呈现为下面的 html 代码：

```html
<label for="你的名字">你的名字: </label>
<input id="first_name" type="text" name="first_name" maxlength="20" required />
<label for="你的姓氏">你的姓氏: </label>
<input id="last_name" type="text" name="last_name" maxlength="20" required />
```

表单类 forms.Form 有一个 is_valid()方法，可以在 views.py 中验证提交的表单是否符合规则。

对于提交的内容，在 views.py 中编写如下代码进行 POST 或 GET 访问：

```python
from django.shortcuts import render
from django.http import HttpResponse, HttpResponseRedirect
from app1.forms import PersonForm

def get_name(request):
    # 判断请求方法是否为 POST
    if request.method == 'POST':
```

```python
        # 将请求数据填充到 PersonForm 实例中
        form = PersonForm(request.POST)
        # 判断 form 是否为有效表单
        if form.is_valid():
            # 使用 form.cleaned_data 获取请求的数据
            first_name = form.cleaned_data['first_name']
            last_name = form.cleaned_data['last_name']
            # 响应拼接后的字符串
            return HttpResponse(first_name + '' + last_name)
        else:
            return HttpResponseRedirect('/error/')
    # 请求方法为 GET
    else:
        return render(request, 'name.html', {'form': PersonForm()})
```

那么，在 html 文件中如何使用这个返回的表单呢？代码如下：

```html
<form action="/app1/get_name" method="post"> {% csrf_token %}
    {{ form }}
    <button type="submit">提交</button>
</form>
```

上面的代码中，{{form}}是 Django 模板的语法，用来获取页面返回的数据。该数据是一个 PersonForm 实例，Django 按照返回的数据渲染表单，但这里渲染的表单只是表单的字段，所以需要在 html 中手动添加 <form></form> 标签，并指出需要提交的路由 /app1/get_name 和请求的方法 post。另外，<form> 标签中需要加上 Django 的防止跨站请求伪造攻击的模板标签 {% csrf_token %}，这样可以避免提交 form 表单时，出现跨站请求伪造攻击的情况。

最后，添加 URL 到我们创建的 app1/urls.py 中，代码如下：

```python
path('get_name', app1_views.get_name)
```

此时访问页面 http://127.0.0.1:8000/app1/get_name，效果如图 6.10 所示。

图 6.10 在 Django 项目中创建表单

6．视图（views）

Django 中的视图类型有两种，分别是 FBV（Function-Based View，基于函数的视图）和 CBV（Class-Based View，基于类的视图），下面分别通过示例进行讲解。

1）FBV

下面通过一个示例讲解如何在 Django 项目中定义视图，代码如下：

```python
from django.http import HttpResponse            # 导入响应对象
import datetime                                  # 导入时间模块

def current_datetime(request):                   # 定义一个视图方法，必须带有请求对象作为参数
    now = datetime.datetime.now()                # 请求的时间
    html = "<html><body>It is now %s.</body></html>" % now   # 生成 html 代码
    return HttpResponse(html)                    # 将响应对象返回，数据为生成的 html 代码
```

上面的代码定义了一个函数，返回了一个 HttpResponse 对象，这就是 Django 中的 FBV，每个视图函数都要有一个 HttpRequest 对象作为参数，用来接收来自客户端的请求，并且必须返回一个 HttpResponse 对象，作为响应给客户端。

django.http 模块下有很多继承于 HttpReponse 的对象。例如，在查询不到数据时，给客户端一个 HTTP 404 的错误页面，可以利用 django.http 下面的 Http404 对象，代码如下：

```python
from django.shortcuts import render
from django.http import HttpResponse, HttpResponseRedirect, Http404
from app1.forms import PersonForm
```

```python
from app1.models import Person

def person_detail(request, pk):                              # url 参数 pk
    try:
        p = Person.objects.get(pk=pk)                        # 获取 Person 数据
    except Person.DoesNotExist:
        raise Http404('Person Does Not Exist')               # 获取不到时抛出 HTTP 404 错误页面
    return render(request, 'person_detail.html', {'person': p})   # 返回详细信息视图
```

这时，在浏览器中输入 http://127.0.0.1:8000/app1/person_detail/100/，会抛出异常，效果如图 6.11 所示。

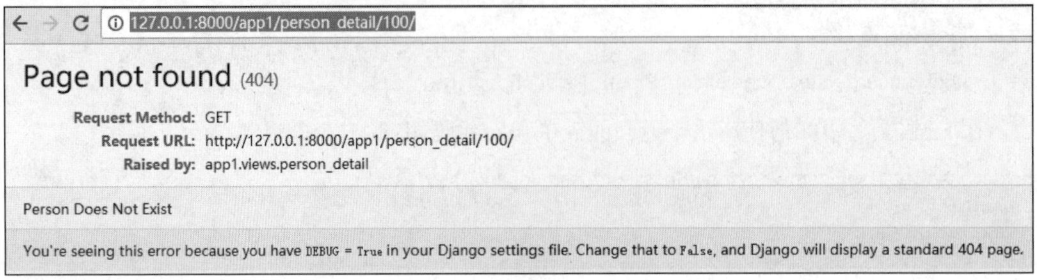

图 6.11　HTTP 404 错误页面

2）CBV

CBV 和 FBV 大同小异，下面通过示例进行讲解。

首先定义一个类视图，这个类视图需要继承一个基础的类视图，所有的类视图都继承自 views.View，类视图的初始化参数需要给出。将上面的 get_name() 方法改成 CBV，代码如下：

```python
from django.shortcuts import render
from django.http import HttpResponse, HttpResponseRedirect, Http404
from django.views import View
from app1.forms import PersonForm
from app1.models import Person

class PersonFormView(View):
    form_class = PersonForm                                  # 定义表单类
    initial = {'key': 'value'}                               # 定义表单初始化展示参数
    template_name = 'name.html'                              # 定义渲染的模板

    def get(self, request, *args, **kwargs):                 # 定义 GET 请求的方法
        return render(request, self.template_name, {'form': self.form_class(initial=self.initial)})  # 渲染表单

    def post(self, request, *args, **kwargs):                # 定义 POST 请求的方法
        form = self.form_class(request.POST)                 # 填充表单实例
        if form.is_valid():                                  # 判断请求是否有效
            # 使用 form.cleaned_data 获取请求的数据
            first_name = form.cleaned_data['first_name']
            last_name = form.cleaned_data['last_name']
            # 响应拼接后的字符串
            return HttpResponse(first_name + " " + last_name)    # 返回拼接的字符串
        return render(request, self.template_name, {'form': form})   # 如果表单无效，返回表单
```

接下来定义一个 URL，代码如下：

```python
from django.urls import path
from app1 import views as app1_views
urlpatterns = [
    path('get_name', app1_views.get_name),
    path('get_name1', app1_views.PersonFormView.as_view()),
    path('person_detail/<int:pk>/', app1_views.person_detail),
]
```

在浏览器中请求/app1/get_name1，会调用 PersonFormViews 视图的方法，如图 6.12 所示。

输入 hugo 和 zhang，并单击"提交"按钮，效果如图 6.13 所示。

图 6.12　请求定义的视图

图 6.13　请求视图结果

7. Django 模板

Django 指定的模板引擎在 settings.py 文件中定义，代码如下：

```
TEMPLATES = [
    {
        'BACKEND': 'django.template.backends.django.DjangoTemplates',    # 模板引擎，默认为 Django 模板
        'DIRS': [],                                                       # 模板所在的目录
        'APP_DIRS': True,                                                 # 是否启用 APP 目录
        'OPTIONS': {
        },
    },
]
```

Django 模板引擎使用{%%}来描述 Python 语句，使用{{}}来描述 Python 变量。Django 模板引擎中的标签及说明如表 6.3 所示。

表 6.3　Django 模板引擎中的标签及说明

标　　签	说　　明
{% extends 'base_generic.html'%}	扩展一个母模板
{%block title%}	指定母模板中的一段代码块，此处为 title，在母模板中定义 title 代码块，可以在子模板中重写该代码块。block 标签必须是封闭的，要由{% endblock %}结尾
{{section.title}}	获取变量的值
{% for story in story_list %}、{% endfor %}	和 Python 中的 for 循环用法相似，必须是封闭的

在 Django 模板中，过滤器非常实用，用来对返回的变量值做一些特殊处理，常用的过滤器如下：

- ☑　{{value|default:"nothing"}}：用来指定默认值。
- ☑　{{value|length}}：用来计算返回的列表的长度或字符串的长度。
- ☑　{{value|filesizeformat}}：用来将数字转换成人类可读的文件大小，如 13KB、128MB 等。
- ☑　{{value|truncatewords:30}}：用来对返回的字符串取固定的长度，此处为 30 个字符。
- ☑　{{value|lower}}：用来将返回的数据变为小写字母。

例如，下面是一个使用 Django 模板引擎的示例：

```
{% extends "base_generic.html" %}
{% block title %}{{ section.title }}{% endblock %}
{% block content %}
<h1>{{ section.title }}</h1>
{% for story in story_list %}
<h2>
  <a href="{{ story.get_absolute_url }}">
    {{ story.headline|upper }}
  </a>
</h2>
<p>{{ story.tease|truncatewords:"100" }}</p>
{% endfor %}
{% endblock %}
```

6.3.3 Channels 模块的基本使用

1. 什么是 Channels 模块

Channels 是 Django 框架的扩展，用于处理 Web 应用中的异步任务和即时通知。具体功能如下：

- ☑ 支持 WebSocket。Channels 模块允许开发者轻松处理 WebSocket 连接，实现实时双向通信。可以创建 WebSocket 消费者，处理来自客户端的 WebSocket 消息，并向客户端发送消息。
- ☑ 处理异步任务。Channels 模块允许在后台执行异步任务，而不会阻塞主线程。这对于处理长时间运行的任务非常有用。
- ☑ 即时通知。Channels 模块支持即时通知，可以将通知推送给客户端，实现实时提醒和更新。可以使用 CHANNEL_LAYERS 组件来实现通知的发送和接收。

2. 使用 Channels 模块的基本步骤

（1）要使用 Channels 模块，需先安装该模块，在虚拟环境中，通过执行下面的命令可以安装 Channels 模块。

```
pip install Channels
```

（2）修改 INSTALLED_APPS：在项目的 settings.py 文件中，需要把应用名添加到 INSTALLED_APPS 列表中。例如，把应用 chat 添加到 INSTALLED_APPS 列表中，关键代码如下：

```
INSTALLED_APPS = [
    'chat',
```

（3）在 settings.py 文件中，声明 ASGI_APPLICATION 变量，指向 ASGI 应用配置（通常位于 asgi.py 文件中）。例如，项目名称为 apps，对应的代码如下：

```
ASGI_APPLICATION = 'apps.asgi.application'
```

（4）在 asgi.py 文件中，导入必要的模块并定义路由。这包括指定如何处理 HTTP 和 WebSocket 连接。参考代码如下：

```
import os

from channels.auth import AuthMiddlewareStack
from channels.routing import ProtocolTypeRouter, URLRouter
from channels.security.websocket import AllowedHostsOriginValidator
from django.core.asgi import get_asgi_application

from chat.routing import websocket_urlpatterns

os.environ.setdefault("DJANGO_SETTINGS_MODULE", "apps.settings")
# Initialize Django ASGI application early to ensure the AppRegistry
# is populated before importing code that may import ORM models.
django_asgi_app = get_asgi_application()


application = ProtocolTypeRouter(
    {
        "http": django_asgi_app,
        "websocket": AllowedHostsOriginValidator(
            AuthMiddlewareStack(URLRouter(websocket_urlpatterns))
        ),
    }
)
```

（5）创建一个名为 routing.py 的文件，在该文件中定义 URL 路由和对应的消费者（Consumer）类。例如，定义一个 WebSocket 路由，该路由指向处理 WebSocket 连接的消费者类。关键代码如下：

```python
from django.urls import re_path
from . import consumers
websocket_urlpatterns = [
    re_path(r"ws/chat/(?P<room_name>\w+)/$", consumers.ChatConsumer.as_asgi()),
]
```

（6）在应用目录下创建一个名为 consumers.py 的文件，然后定义一个继承自 channels.generic.websocket.WebsocketConsumer 的消费者类。这个类将包含处理 WebSocket 连接、接收消息和断开连接的逻辑。参考代码如下：

```python
import json

from channels.generic.websocket import WebsocketConsumer
class ChatConsumer(WebsocketConsumer):
    def connect(self):
        self.accept()
    def disconnect(self, close_code):
        pass
    def receive(self, text_data):
        text_data_json = json.loads(text_data)
        message = text_data_json["message"]
        self.send(text_data=json.dumps({"message": message}))
```

（7）在前端代码中，使用 JavaScript 来创建与服务器的 WebSocket 连接，并处理消息的发送和接收。例如，创建 WebSocket 连接并在收到消息时获取并显示该消息，代码如下：

```html
<script>
    const roomName = JSON.parse(document.getElementById('room-name').textContent);
    const chatSocket = new WebSocket(
        'ws://' + window.location.host + '/ws/chat/' + roomName + '/'
    );

    chatSocket.onmessage = function(e) {
        const data = JSON.parse(e.data);
        document.querySelector('#chat-log').value += (data.message + '\n');
    };
</script>
```

6.3.4　在 Channels 项目中集成 Channels-Redis

在使用 Channels 的应用中，需要使用 Redis 作为 Channels 的内存数据库，为 Web 应用程序提供关键的实时通信和事件驱动。在 Channels 中，Redis 具有以下关键角色。

- ☑ 消息队列和路由。Redis 作为通道层（Channel Layer）的后端存储，负责接收、暂存及转发来自客户端的 WebSocket 消息以及 Django 应用服务器产生的响应。它充当了一个中间人的角色，确保消息能够在正确的通道（Channels）之间进行高效、有序的传输，这是实时通信的基础。
- ☑ 支持异步处理。Channels 使用 Redis 作为消息队列，以支持异步任务处理和事件驱动。当 Web 应用程序需要处理大量并发的请求，并且需要处理耗时的任务时，Redis 可以作为中间人来提供异步处理的能力。通过将任务放入 Redis 的消息队列中，可以降低主应用程序的负载，并提高响应速度。
- ☑ 缓存层。Redis 可以作为缓存层，用于存储经常使用的数据和结果。由于 Redis 将数据存储在内存中，它能提供高速的数据读写能力，这对于低延迟的实时消息传递至关重要。它作为临时存储，可以迅速缓存和分发消息，避免了直接与数据库交互可能带来的延迟。

- ☑ 订阅与发布。Channels 使用 Redis 的发布与订阅功能来实现实时通信。在 Web 应用程序中，如果需要实时更新数据或实时通知用户，可以使用 Redis 的发布与订阅功能。当数据发生变化时，可以将变化的数据发布到 Redis 的频道中，而订阅该频道的客户端可以及时接收到变化的通知。
- ☑ 会话存储。Redis 可以作为 Django 的会话存储后端，用于存储用户的会话数据。在 Web 应用程序中，用户需要登录并保持登录状态，而会话数据需要在多个请求之间进行共享。通过使用 Redis 来存储会话数据，可以提供分布式和高性能的会话管理功能。

在使用 Channels 模块的 Web 应用中，需要通过 Channels-Redis 模块来使用 Redis。使用 Channels-Redis 模块时，需要先安装 Channels 和 Channels-Redis 模块，在虚拟环境中，通过执行下面的命令可以安装 Channels-Redis 模块：

```
pip install Channels
pip install Channels-Redis
```

安装完成后，在 Django 项目的 settings.py 文件中，需要配置在通道层（Channel Layer）中使用 Redis。关键代码如下：

```
CHANNEL_LAYERS = {
    "default": {
        "BACKEND": "channels_redis.core.RedisChannelLayer",
        "CONFIG": {
            "hosts": [("127.0.0.1", 6379)],
        },
    },
}
```

确保 Redis 服务器正在运行，并根据实际情况调整 hosts 配置中的地址和端口。如果 Redis 没有运行，可以在命令行窗口中先进入 Redis 所在的目录，然后使用下面的命令运行。

```
redis-server.exe redis.windows.conf
```

运行结果如图 6.14 所示。

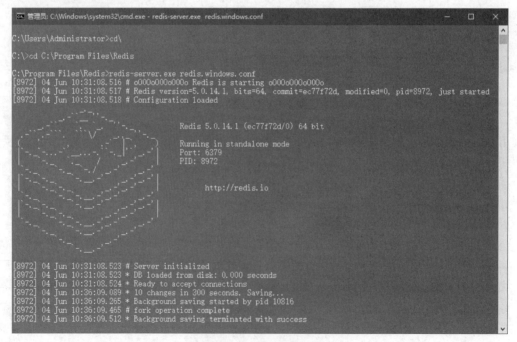

图 6.14　运行 Redis

接下来还需要配置 ASGI 服务器。Channels 模块推荐使用 Daphne 作为 ASGI 服务器，Daphne 也需要单独安装。在虚拟环境中，使用下面的命令进行安装。

```
pip install Daphne
```

6.4 创建项目

在实现心灵驿站聊天室时，需要先创建 Django 项目并安装所需的模块。在创建项目前，需要先安装项目中应用的模块，包括 Django、channels、channels-redis、daphne 和 redis。对应的命令如下：

```
pip install Django
pip install channels
pip install channels-redis
pip install daphne
pip install redis
```

接下来就可以创建项目了，这可以通过 Django 提供的命令实现。具体的命令如下：

```
django-admin startproject apps
```

创建项目后，还需要创建一个聊天室应用，这里为 chat。具体的命令如下：

```
python apps/manage.py startapp chat
```

项目创建完成后，将显示如图 6.15 所示的目录结构。

创建应用后，还需要注册该应用。打开自动生成的 apps/apps/settings.py 文件，找到 INSTALLED_APPS，在其中添加'chat'，添加后的代码如下：

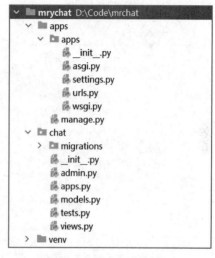

图 6.15 创建完成的项目

```
INSTALLED_APPS = [
    'django.contrib.admin',
    'django.contrib.auth',
    'django.contrib.contenttypes',
    'django.contrib.sessions',
    'django.contrib.messages',
    'django.contrib.staticfiles',
    'chat',
]
```

这样，聊天室的项目和应用就创建完成了。接下来就可以实现相应的功能了。

6.5 功能设计

心灵驿站聊天室主要有 3 部分功能，分别是进入房间、实时聊天和退出房间。在实现实时聊天时，需要完成异步任务。这里要通过 Channels 模块和 Channels-Redis 模块实现，下面分别进行介绍。

6.5.1 进入房间

在心灵驿站聊天室中，进入房间是在首页实现的。实现首页时，主要是创建首页对应的模板文件

index.html。在项目的 templates 目录下创建名称为 index.html 的模板文件。在该文件中，首先链接外部 CSS 样式，然后编写布局页面的 HTML 代码，并且添加用于输入房间号的文本框和"确认"按钮，最后编写 JavaScript 代码，实现按下 Enter 键或单击"确定"按钮时，跳转页面，这里为"/chat+房间号/"。index.html 模板文件的具体代码如下：

```html
<!-- chat/templates/chat/index.html -->
<!DOCTYPE html>
<html lang="en">
{% load static %}
<head>
<meta charset="UTF-8">
<meta name="viewport" content="width=device-width, initial-scale=1.0">
<link rel="stylesheet" href="{% static 'css/heart_css.css' %}">
<title>心灵驿站聊天室</title>
</head>
<body>
        <div class="chat_img"><img src="{% static 'images/chat_box.jpg' %}" alt="" /></div>
        <div class="chat_bottom">
            <div class="chat_input">
                请输入房间号：
                <input id="roomname" type="text" class="rounded-input" placeholder="房间号">
                <button  id="okbutton"   class="confirm-button">确认</button>
    </div>
  </div>
  <script>
        document.querySelector('#roomname').focus();                      // 让文本框获得焦点
        document.querySelector('#roomname').onkeyup = function(e) {
            if (e.key === 'Enter') {                                       // 如果按下 Enter 键
                document.querySelector('#okbutton').click();              // 执行按钮的单击事件
            }
        };
        document.querySelector('#okbutton').onclick = function(e) {
            var roomName = document.querySelector('#roomname').value;     // 获取输入的房间号
            window.location.pathname = '/chat/' + roomName + '/';         // 设置路径名
        };
    </script>
</body>
</html>
```

在上面的代码中，链接了外部样式表文件和显示了图片，这就涉及了包含静态资源，所以需要在全局范围内加载自定义的模板标签库。方法是打开 apps/apps/settings.py 文件，在 TEMPLATES 的 OPTIONS 字典中，为键 context_processors，添加值'django.template.context_processors.static'，然后将 DIRS 键的值设置为 [os.path.join(BASE_DIR, 'templates')]，添加后的代码如下：

```python
TEMPLATES = [
    {
        'BACKEND': 'django.template.backends.django.DjangoTemplates',
        'DIRS':    [os.path.join(BASE_DIR, 'templates')],
        'APP_DIRS': True,
        'OPTIONS': {
            'context_processors': [
                'django.template.context_processors.debug',
                'django.template.context_processors.request',
                'django.contrib.auth.context_processors.auth',
                'django.contrib.messages.context_processors.messages',
                'django.template.context_processors.static',
            ],
        },
    },
]
```

模板文件创建完成后，为了显示进入房间的 index.html 文件，还需要创建 chat/views.py 文件，并且在该文件中编写首页对应的视图函数，实际上这也是进入房间页面的视图函数，代码如下：

```
# chat/views.py
from django.shortcuts import render

def index(request):
    return render(request, "chat/index.html")
```

为了调用上面的视图函数，还需要为它配置路由。这需要创建 chat/urls.py 文件，并且在该文件配置首页对应的路由。代码如下：

```
# chat/urls.py
from django.urls import path

from . import views

urlpatterns = [
    path("", views.index, name="index"),
]
```

接下来将 URLconf 指向 chat.urls 模块。在 apps 目录下的 urls.py 文件中添加一个导入 chat.urls 的映射，具体代码如下：

```
# apps/urls.py
from django.contrib import admin
from django.urls import include, path

urlpatterns = [
    path("", include("chat.urls")),
    path("admin/", admin.site.urls),
]
```

至此，心灵驿站聊天室的首页已经完成，运行项目，在浏览器中输入 http://127.0.0.1:8000/，即可显示心灵驿站聊天室的首页，如图 6.16 所示。

图 6.16　心灵驿站聊天室首页

6.5.2　实时聊天

在心灵驿站聊天室的首页中，输入房间号（尽量使用字母或数字，并且不带特殊符号），单击"确认"按钮，将进入实时聊天页面，在该页面中可以与使用相同房间号的用户一起聊天。具体的实现过程如下。

（1）配置 Channels 路由，这实际上是一个 ASGI 应用程序，用于指定 Channels 服务器来接收 HTTP 请求。修改创建项目时自动生成的 apps/asgi.py 文件，在该文件中，通过创建 ProtocolTypeRouter 协议处理器来指定由 Channels 服务器接收 HTTP 请求，修改后的代码如下：

```
# apps/asgi.py
import os
from channels.auth import AuthMiddlewareStack
from channels.routing import ProtocolTypeRouter, URLRouter
from channels.security.websocket import AllowedHostsOriginValidator
from django.core.asgi import get_asgi_application
from chat.routing import websocket_urlpatterns
os.environ.setdefault("DJANGO_SETTINGS_MODULE", "apps.settings")
```

```python
# Initialize Django ASGI application early to ensure the AppRegistry
# is populated before importing code that may import ORM models.
application = ProtocolTypeRouter(
    {
        "http": get_asgi_application(),
    }
)
```

（2）在 apps/settings.py 文件中注册 Daphne 库，并配置根路由指向 Channels，具体代码如下：

```python
INSTALLED_APPS = [
    'daphne',
    'django.contrib.admin',
    'django.contrib.auth',
    'django.contrib.contenttypes',
    'django.contrib.sessions',
    'django.contrib.messages',
    'django.contrib.staticfiles',
    'chat',
]
# Daphne
ASGI_APPLICATION = "apps.asgi.application"
```

说明

Daphne 库是一个异步服务器，提供高效快速的 HTTP 和 WebSocket 服务，并且支持 WebSocket 和异步编程。

（3）在 mrchat\apps\chat\templates\chat\ 文件夹中创建实时聊天界面的模板文件 room.html，在该文件中布局聊天界面，在聊天界面中，主要包括用于输入发言内容的文本框、"发送"按钮和"退出房间"按钮，还包括显示聊天内容的<div>，具体代码如下：

```html
<!-- chat/templates/chat/room.html -->
<!DOCTYPE html>
<html>
{% load static %}
<head>
    <meta charset="utf-8"/>
    <title>心灵驿站聊天室</title>
    <meta name="viewport" content="width=device-width, initial-scale=1.0">
    <link rel="stylesheet" href="{% static 'css/heart_css.css' %}">
</head>
<body>
        <div class="chat_banner">
            <img src="{% static 'images/chat_banner.jpg' %}" alt="" />
        </div>
        <div class="chat_box">
            <div class="center"><div id="chat-log"></div><br></div>
        </div>
            <div class=" Chat_C">
        <div class="content">
            <img src="{% static 'images/smile.jpg' %}" align="center"    alt="" />
            <input id="chat-message-input" type="text" class="rounded-input" placeholder="">
            <button id="chat-message-submit" class="confirm-button">发送</button>
            <button id="chat-logout" class="confirm-button2">退出房间</button>
        </div>
            </div>
</body>
</html>
```

（4）在 room.html 文件中，编写 JavaScript 代码，创建与服务器的 WebSocket 连接，并处理消息的发送

和接收。代码如下:

```html
{{ room_name|json_script:"room-name" }}
<script>
    const roomName = JSON.parse(document.getElementById('room-name').textContent);
    const chatSocket = new WebSocket(
        'ws://'
        + window.location.host
        + '/ws/chat/'
        + roomName
        + '/'
    );                                                          // 发送 WebSocket 请求
    chatSocket.onmessage = function(e) {
        const data = JSON.parse(e.data);
        document.querySelector('#chat-log').innerHTML += (data.message + '<br>');
    };                                                          // 显示消息
    chatSocket.onclose = function(e) {
        console.error('Chat socket closed unexpectedly');
    };                                                          // 关闭 WebSocket 请求
    document.querySelector('#chat-message-input').focus();
    document.querySelector('#chat-message-input').onkeyup = function(e) {
        if (e.key === 'Enter') {                                // 按下 Enter 键
            document.querySelector('#chat-message-submit').click(); // 模拟单击"发送"按钮
        }
    };
    document.querySelector('#chat-message-submit').onclick = function(e) {
        const messageInputDom = document.querySelector('#chat-message-input');
        const message = messageInputDom.value;
        chatSocket.send(JSON.stringify({
            'message': message
        }));                                                    // 单击"发送"按钮发送消息
        messageInputDom.value = '';
    };
</script>
```

(5) 为了显示实时聊天界面的 room.html 文件,还需要在 chat/views.py 文件中添加实时聊天界面对应的视图函数 room(),在该函数中,返回实时聊天界面的渲染,代码如下:

```python
def room(request, room_name):
    return render(request, "chat/room.html", {"room_name": room_name})
```

(6) 为了调用上面的视图函数,还需要在 chat/urls.py 文件中为它配置路由。并且在该文件配置实时聊天界面对应的路由,该路由中需要加上输入的房间号。修改后的 chat/urls.py 文件的代码如下:

```python
# chat/urls.py
from django.urls import path

from . import views

urlpatterns = [
    path("", views.index, name="index"),
    path("chat/<str:room_name>/", views.room, name="room"),
]
```

(7) 当 Channels 接收到 WebSocket 连接时,会根据根路由的配置来查找消费者,然后调用各种函数,处理连接中的事件。这就需要配置接收路径上的 WebSocket 连接的消费者。在 apps\chat\目录下,创建消费者文件 consumers.py,具体代码如下:

```python
# chat/consumers.py
import json
from channels.generic.websocket import WebsocketConsumer
class ChatConsumer(WebsocketConsumer):
```

```python
    def connect(self):
        self.accept()
    def disconnect(self, close_code):
        pass
    def receive(self, text_data):
        text_data_json = json.loads(text_data)
        message = text_data_json["message"]
        # 发送消息到 WebSocket
        self.send(text_data=json.dumps({"message": message}))
```

上面创建的消费者文件会接收任何路径上的 WebSocket 连接，并且将其回显到同一个 WebSocket。

此时，这是一个同步的 WebSocket，它并不会向进入同一个房间内的其他客户端传递所发送的消息，只能显示自己发出的信息。下面还需要将其修改为异步形式。

（8）在 apps/chat 目录下，创建名称为 routing.py 的文件，在该文件中，代码如下：

```python
# chat/routing.py
from django.urls import re_path

from . import consumers

websocket_urlpatterns = [
    re_path(r"ws/chat/(?P<room_name>\w+)/$", consumers.ChatConsumer.as_asgi()),
]
```

（9）在 apps 下 asgi.py 的 ProtocolTypeRouter 列表中插入一个键，将根路由配置指向 chat.routing 模块，修改后的代码如下：

```python
# apps/asgi.py
import os
from channels.auth import AuthMiddlewareStack
from channels.routing import ProtocolTypeRouter, URLRouter
from channels.security.websocket import AllowedHostsOriginValidator
from django.core.asgi import get_asgi_application

from chat.routing import websocket_urlpatterns

os.environ.setdefault("DJANGO_SETTINGS_MODULE", "apps.settings")
# Initialize Django ASGI application early to ensure the AppRegistry
# is populated before importing code that may import ORM models.

application = ProtocolTypeRouter(
    {
        "http": get_asgi_application(),
        "websocket": AllowedHostsOriginValidator(
            AuthMiddlewareStack(URLRouter(websocket_urlpatterns))
        ),
    }
)
```

（10）为了验证路径的使用者是否有效，需要在本项目的虚拟环境中，使用下面的命令完成数据库迁移。

```
python apps/manage.py migrate
```

此项操作是因为 Django 框架本身需要数据库才能执行，而本项目涉及的数据并不会存储在数据库中。

执行数据库迁移命令的运行结果如图 6.17 所示。

图 6.17 执行数据库迁移命令的运行结果

（11）需要让多个 Consumer 能够相互通信，请确保已安装好 Redis 并启动，然后编辑 apps/settings.py 文件，添加以下代码配置与 Redis 的连接。

```
DEFAULT_AUTO_FIELD = 'django.db.models.BigAutoField'

# Channels 基本配置
ASGI_APPLICATION = "apps.asgi.application"
CHANNEL_LAYERS = {
    "default": {
        "BACKEND": "channels_redis.core.RedisChannelLayer",
        "CONFIG": {
            "hosts": [("127.0.0.1", 6379)],
        },
    },
}
```

（12）配置完成后，将 chat/consumers.py 文件替换成如下代码：

```
# chat/consumers.py
import json
from asgiref.sync import async_to_sync
from channels.generic.websocket import WebsocketConsumer
class ChatConsumer(WebsocketConsumer):
    def connect(self):
        self.room_name = self.scope["url_route"]["kwargs"]["room_name"]
        self.room_group_name = f"chat_{self.room_name}"
        # 进入聊天界面
        async_to_sync(self.channel_layer.group_add)(
            self.room_group_name, self.channel_name
        )
        self.accept()
    def disconnect(self, close_code):
        # 离开聊天界面
        async_to_sync(self.channel_layer.group_discard)(
            self.room_group_name, self.channel_name
```

```python
    )
    # 从WebSocket接收消息
    def receive(self, text_data):
        text_data_json = json.loads(text_data)
        message = text_data_json["message"]
        # 发送消息
        async_to_sync(self.channel_layer.group_send)(
            self.room_group_name, {"type": "chat.message", "message": message}
        )
    # 接收消息
    def chat_message(self, event):
        message = event["message"]
        # 发送消息到WebSocket
        self.send(text_data=json.dumps({"message": message}))
```

当用户发布消息时，JavaScript 函数将通过 WebSocket 将消息传输到 ChatConsumer。ChatConsumer 将接收该消息并将其转发到与房间名称对应的组。然后，同一组中的每个 ChatConsumer（因为在同一个房间中）将接收来自该组的消息，而且通过 WebSocket 将其转发回 JavaScript，并将其附加到聊天日志中。

（13）到目前为止，ChatConsumer 仍然是同步消费者，为了提供更高级别的性能，我们需要将其重写为异步，将下列代码替换到 chat/consumers.py 中：

```python
import json
from channels.generic.websocket import AsyncWebsocketConsumer
class ChatConsumer(AsyncWebsocketConsumer):
    async def connect(self):
        self.room_name = self.scope["url_route"]["kwargs"]["room_name"]
        # 直接从用户指定的房间名称构造Channels组名称，不进行任何引用或转义。
        # 组名只能包含字母、数字、连字符和句点。
        # 因此，使用包含其他字符的房间号将失败。
        self.room_group_name = f"chat_{self.room_name}"
        # 进入聊天室
        await self.channel_layer.group_add(self.room_group_name, self.channel_name)
        # 接受 WebSocket 连接。
        # 如果不在 connect()方法中调用 accept()，则拒绝并关闭连接。
        # 如果您选择接受连接，建议将 accept()作为 connect()中的最后一个操作来调用。
        await self.accept()
    async def disconnect(self, close_code):
        # 离开聊天室
        await self.channel_layer.group_discard(self.room_group_name, self.channel_name)
    # 从WebSocket接收消息
    async def receive(self, text_data):
        text_data_json = json.loads(text_data)
        message = text_data_json["message"]
        # 发送消息
        await self.channel_layer.group_send(
            self.room_group_name, {"type": "chat.message", "message": message}
        )
    # 从聊天室获取消息
    async def chat_message(self, event):
        message = event["message"]
        # 发送消息到WebSocket
        await self.send(text_data=json.dumps({"message": message}))
```

至此，您的聊天服务器是完全异步的了，开始启动项目，在浏览器中打开 http://127.0.0.1:8000/，将显示进入房间界面，输入房间号（如 w）后，将显示实时聊天界面。再开启一个浏览器窗口，仍然输入 http://127.0.0.1:8000/，并且房间号也输入 w，这时两个浏览器窗口就可以进行聊天了。效果如图 6.18 和图 6.19 所示。

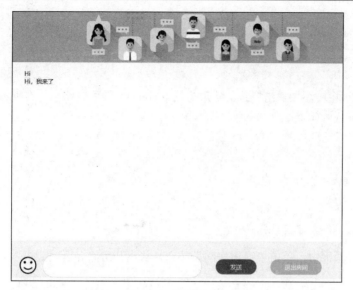

图 6.18 聊天界面 1

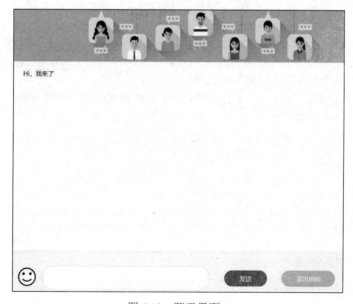

图 6.19 聊天界面 2

6.5.3 退出房间

退出房间功能十分简单,只要将页面重定向到首页即可。具体实现步骤如下。

(1) 在聊天页面的模板文件 room.html 中,添加 "退出房间" 按钮,并且为该按钮添加单击事件,实现将页面跳转到路由 "/logout",代码如下:

```
<input id="chat-logout" type="button" value="退出房间" onclick="window.location.href='/logout'">
```

(2) 在聊天室的 chat/urls.py 文件中,添加退出房间的路由,添加后的代码如下:

```
from django.urls import path
from . import views
```

```
urlpatterns = [
    path("", views.index, name="index"),
    path("chat/<str:room_name>/", views.room, name="room"),
    path("logout/",views.logout,name="logout"),
]
```

（3）在聊天室的视图中添加退出登录的实现方法 logout()，并配置路由。在 logout()方法中，将页面渲染到聊天室首页。实现代码如下：

```
def logout(request):
    """
    退出房间
    :param request: 请求对象
    :return: 重定向到首页
    """
    print("退出房间")
    return render(request, "chat/index.html")
```

运行程序，在聊天页面中，单击"退出房间"按钮，将离开房间并返回到进入房间页面。

6.6 项目运行

通过前述步骤，设计并完成了"心灵驿站聊天室"项目的开发。下面运行该项目，检验一下我们的开发成果。运行"心灵驿站聊天室"项目的步骤如下。

（1）打开 mrchat\apps\apps 文件，根据自己的 Redis 主机和端口号（通常不用修改端口号，除非安装或配置过程中进行了修改）修改如下代码：

```
# Channels 基本配置
ASGI_APPLICATION = "apps.asgi.application"
CHANNEL_LAYERS = {
    "default": {
        "BACKEND": "channels_redis.core.RedisChannelLayer",
        "CONFIG": {
            "hosts": [("127.0.0.1", 6379)],
        },
    },
}
```

（2）打开命令提示符对话框，进入 mrchat 项目文件夹所在目录，在命令提示符对话框中输入如下命令来创建 venv 虚拟环境：

```
virtualenv venv
```

（3）在命令提示符对话框中输入如下命令来启动 venv 虚拟环境：

```
venv\Scripts\activate
```

（4）在命令提示符对话框中使用如下命令来安装 Django 等依赖包：

```
pip install -r  requirements.txt
```

（5）启动 Redis。在命令提示符对话框中先进入 Redis 所在的目录，然后使用下面的命令运行。

```
redis-server.exe redis.windows.conf
```

说明

启动之前需要保证已经安装了 Redis 5.0 或以上版本。

Redis 启动过程如图 6.20 所示。

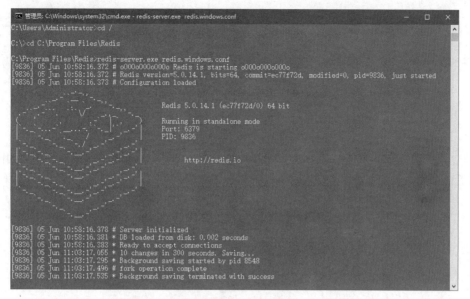

图 6.20　启动 Redis

（6）在虚拟环境下执行如下数据库迁移命令。

r python apps/manage.py migratef

（7）在 PyCharm 中打开项目文件夹 mrchat，在其中选中 apps\manage.py 文件，单击鼠标右键，在弹出的快捷菜单中选择 Modify Run Configuration…命令，如图 6.21 所示。

（8）在打开的对话框中的 Parameters 文本框中输入 runserver，并且单击 Apply 按钮，如图 6.22 所示。

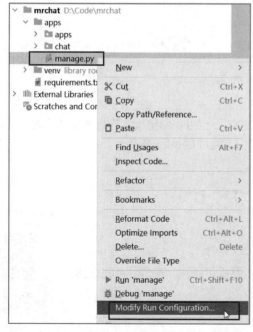

图 6.21　选择 Modify Run Configuration…命令

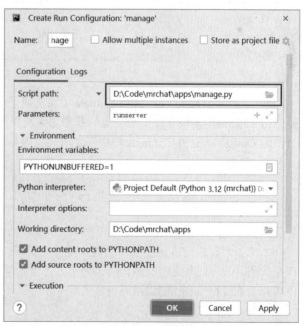

图 6.22　输入 runserver

（9）单击右上角的运行按钮 ，运行项目，如果在 PyCharm 底部出现如图 6.23 所示的提示，说明程序运行成功。

```
D:\Code\mrchat\venv\Scripts\python.exe D:\Code\mrchat\apps\manage.py runserver
Performing system checks...

Watching for file changes with StatReloader
System check identified no issues (0 silenced).
June 05, 2024 - 13:08:15
Django version 5.0.6, using settings 'apps.settings'
Starting ASGI/Daphne version 4.1.2 development server at http://127.0.0.1:8000/
Quit the server with CTRL-BREAK.
```

图 6.23　程序运行成功提示

（10）在浏览器中输入网址 http://127.0.0.1:5000/ 即可进入心灵驿站聊天室的首页，效果如图 6.24 所示。输入房间号后，单击"确定"按钮，即可进入聊天室，在该聊天室可以和进入该房间的用户聊天。

图 6.24　心灵驿站聊天室的首页

本项目使用 Django 框架和 Channels 模块实现了心灵驿站聊天室项目。通过项目的学习，能够使读者了解使用 Channels 模块开发异步任务的基本思路及实现技术的综合应用。希望读者通过本章的学习，可以学会使用 Django 框架和 Channels 模块开发聊天室等具有实时交互功能的 Web 应用。

6.7　源码下载

本章虽然详细地讲解了如何编码实现"心灵驿站聊天室"的各个功能，但给出的代码都是代码片段，而非完整的源代码。为了方便读者学习，本书提供了该项目的完整源代码，读者可以通过扫描右侧的二维码进行下载

源码下载

第 7 章 站内全局搜索引擎

——Django + Django-Haystack + Whoosh + Jieba

在信息爆炸的时代，网站内容的丰富性和多样性急剧增长，传统的导航式浏览方式难以满足用户快速定位目标信息的需求，特别是在内容庞大的网站中，用户往往会面临"信息迷航"的困境。因此，开发一个高效、智能的站内全局搜索引擎显得尤为重要。本章将使用 Django 框架及 Django-Haystack 模块开发一个站内全局搜索引擎。

项目微视频

本项目的核心功能及实现技术如下：

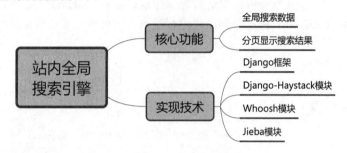

7.1 开发背景

在这个数据驱动的时代，用户对于快速、准确地获取所需信息的需求日益迫切。为了提高用户体验，满足用户对信息检索的高效性和精确性的要求，开发一个站内全局搜索引擎成为一项至关重要的任务。

Django-Haystack 是一个全文检索框架，支持多种全文检索引擎。而 Whoosh 则是一个纯 Python 编写的全文搜索引擎，虽然性能可能不及一些其他引擎，但其无二进制依赖，使得程序稳定性较高，适合小型站点使用。因此，采用 Django-Haystack 模块和 Whoosh 模块能够快速搭建起一个功能强大且易于维护的站内搜索引擎，提升用户体验和满意度。

本项目的实现目标如下：
- ☑ 不使用特定字段的模糊搜索，而采用全文检索，更高效地处理搜索请求。
- ☑ 支持中文分词，提高效率。
- ☑ 提供分页功能，控制每页显示的搜索结果数量，提高用户体验。
- ☑ 易于维护，降低运营成本。

7.2 系统设计

7.2.1 开发环境

本项目的开发及运行环境如下：

- ☑ 操作系统：推荐 Windows 10、Windows 11 或更高版本。
- ☑ 开发工具：PyCharm 2024（向下兼容）。
- ☑ 开发语言：Python 3.12。
- ☑ 数据库：MySQL 8.0+PyMySQL 驱动。
- ☑ Python Web 框架：Django 3.2。

> **说明**
> 到本书截稿为止，最新版本的 Django-Haystack 模块只支持 Django 3.2，所以本项目采用的 Django 版本为 3.2。

7.2.2 业务流程

站内全局搜索引擎主要实现在整个网站中进行分词搜索。用户在搜索输入框内输入搜索关键词，系统将自动对其进行中文分词处理，并且根据分词结果进行搜索，然后再对搜索结果进行分页显示。本项目的业务流程如图 7.1 所示。

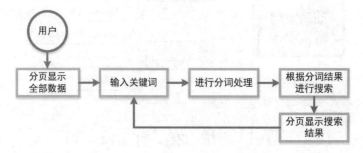

图 7.1　站内全局搜索引擎的业务流程

7.2.3 功能结构

本项目的功能结构已经在章首页中给出，其实现的具体功能如下：
- ☑ 对输入的关键词分词，并进行全局搜索数据。
- ☑ 分页显示搜索结果。

7.3 技术准备

7.3.1 技术概览

本项目采用 Django 框架作为主体 Web 框架，为了实现全局搜索，还结合了流行的 Django-Haystack 模块和 Whoosh 模块。另外，在实现搜索时，采用了 Jieba 框架对输入的关键词进行分词处理。有关 Django 框架的基本使用方法，在本书 6.3.2 节有详细的讲解，对该知识不太熟悉的读者可以参考该节的内容。

下面对实现本项目时用到的其他技术点进行必要介绍，如 Django 框架的模型与数据库、Django-Haystack 模块的基本使用方法、使用 Whoosh 模块，以及使用 jieba 模块进行中文分词，以确保读者可以顺利完成本项目。

7.3.2 Django 框架的模型与数据库

1. 定义数据模型（model）

在 app1 的 models.py 中定义一个 Person 数据模型，代码如下：

```python
from django.db import models

# Create your models here.
class Person(models.Model):
    """
    编写 Person 模型类，数据模型应该继承于 models.Model 或其子类
    """
    # 第一个字段使用 models.CharField 类型
    first_name = models.CharField(max_length=30)
    # 第二个字段使用 models.CharField 类型
    last_name = models.CharField(max_length=30)
```

上面的数据模型类在数据库中会创建如下数据表：

```sql
CREATE TABLE myapp_person (
    "id" serial NOT NULL PRIMARY KEY,
    "first_name" varchar(30) NOT NULL,
    "last_name" varchar(30) NOT NULL
);
```

另外，对于一些公有的字段，为了简化代码，可以使用如下的实现方式：

```python
from django.db import models

class CreateUpdate(models.Model):                    # 创建抽象数据模型，同样要继承于 models.Model
    # 创建时间，使用 models.DateTimeField
    created_at = models.DateTimeField(auto_now_add=True)
    # 修改时间，使用 models.DateTimeField
    updated_at = models.DateTimeField(auto_now=True)
    class Meta:                                      # 元数据，除了字段以外的所有属性
        # 设置模型为抽象类。指定该表不应该在数据库中创建
        abstract = True
```

上述代码创建了一个抽象数据模型，其中主要是定义模型的创建时间和修改时间，创建完成后，其他需要用到创建时间和修改时间的数据模型都可以继承该类，例如：

```python
class Person(CreateUpdate):                          # 继承 CreateUpdate 基类
    """
    编写 Person 模型类，数据模型应该继承于 models.Model 或其子类
    """
    # 第一个字段使用 models.CharField 类型
    first_name = models.CharField(max_length=30)
    # 第二个字段使用 models.CharField 类型
    last_name = models.CharField(max_length=30)

class Order(CreateUpdate):                           # 继承 CreateUpdate 基类
    """
    编写 Order 模型类，数据模型应该继承于 models.Model 或其子类
    """
    order_id = models.CharField(max_length=30, db_index=True)
    order_desc = models.CharField(max_length=120)
```

上面创建表时使用了两个字段类型：CharField 和 DateTimeField，它们分别表示字符串值类型和日期时间类型。此外，django.db.models 还提供了很多常见的字段类型，如表 7.1 所示。

表7.1 Django 数据模型中常见的字段类型

字 段 类 型	说　明
AutoField	id 自增的字段，但创建表过程中 Django 会自动添加一个自增的主键字段
BinaryField	保存二进制源数据的字段
BooleanField	布尔值字段，应该指明默认值，管理后台默认呈现为 CheckBox 形式
NullBooleanField	可以为 None 值的布尔值字段
CharField	字符串值字段，必须指明参数 max_length 值，管理后台默认呈现为 TextInput 形式
TextField	文本域字段，对于大量文本应该使用 TextField。管理后台默认呈现为 Textarea 形式
DateField	日期字段，代表 Python 中 datetime.date 的实例。管理后台默认呈现 TextInput 形式
DateTimeField	日期时间字段，代表 Python 中 datetime.datetime 实例。管理后台默认呈现 TextInput 形式
EmailField	邮件字段，是 CharField 的实现，用于检查该字段值是否符合邮件地址格式
FileField	上传文件字段，管理后台默认呈现 ClearableFileInput 形式
ImageField	图片上传字段，是 FileField 的实现。管理后台默认呈现 ClearableFileInput 形式
IntegerField	整数值字段，管理后台默认呈现 NumberInput 形式或者 TextInput 形式
FloatField	浮点数值字段，管理后台默认呈现 NumberInput 形式或者 TextInput 形式
SlugField	只保存字母、数字、下画线和连接符，用于生成 url 的短标签
UUIDField	保存一般统一标识符的字段，代表 Python 中 UUID 的实例，建议提供默认值 default
ForeignKey	外键关系字段，需提供外键的模型参数和 on_delete 参数（指定当该模型实例删除时，是否删除关联模型），如果要外键的模型出现在当前模型的后面，需要在第一个参数中使用单引号 'Manufacture'
ManyToManyField	多对多关系字段，与 ForeignKey 类似
OneToOneField	一对一关系字段，常用于扩展其他模型

2．数据库迁移

创建完数据模型后，需要进行数据库迁移，步骤如下。

（1）如果不使用 Django 默认自带的 SQLite 数据库，而是使用当下流行的 MySQL 数据库，则需要在 django_demo\settings.py 配置文件中进行一些修改。将如下代码：

```
DATABASES = {
    'default': {
        'ENGINE': 'django.db.backends.sqlite3',
        'NAME': os.path.join(BASE_DIR, 'db.sqlite3'),
    }
}
```

修改为：

```
DATABASES = {
    'default': {
        'ENGINE': 'django.db.backends.mysql',
        'NAME': 'demo',                       # 数据库名称
        'USER': 'root',                       # 数据库用户名
        'PASSWORD': 'root'                    # 数据库密码
    }
}
```

说明

这里需要安装 MySQL 数据库驱动（如 PyMySQL），并且需要在 MySQL 中创建相应的数据库。

（2）在 Django 项目中找到 django_demo__init__.py 文件，在行首添加如下代码：

```
import pymysql
pymysql.install_as_MySQLdb()                    # 为了使 PyMySQL 发挥最大数据库操作性能
```

（3）在终端命令窗口中执行以下命令创建数据表：

```
python manage.py makemigrations                 # 生成迁移文件
python manage.py migrate                        # 迁移数据库，创建新表
```

以上操作完成后，即可在数据库中查看这两张数据表了。Django 会默认按照 App 名称+下画线+模型类名称小写的形式创建数据表。对于上面两个模型，Django 创建了如下表：

- ☑ Person 类对应 app1_person 表。
- ☑ Order 类对应 app1_order 表。

在数据库管理软件中可以查看新创建的数据表，效果如图 7.2 所示。

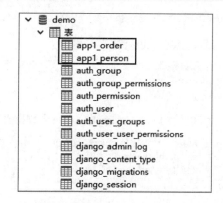

图 7.2　在数据库管理软件中查看新创建的数据表

3．Django 中的数据库操作命令

- ☑ 导入数据模型。命令如下：

```
# 导入 Person 和 Order 两个类
from app1.models import Person, Order
```

- ☑ 创建数据。添加数据有两种方法，分别如下：
  - ➢ 方法 1：

```
p = Person.objects.create(first_name="andy", last_name="feng")
```

  - ➢ 方法 2：

```
p=Person(first_name="andy", last_name="王")
p.save()                                        # 必须调用 save()才能写入数据库
```

- ☑ 查询数据：
  - ➢ 查询所有数据，命令如下：

```
Person.objects.all()
```

输出结果如下：

```
<QuerySet [<Person: Person object (1)>, <Person: Person object (2)>]>
```

  - ➢ 查询单个数据，命令如下：

```
Person.objects.get(id =1)                       # 括号内需要加入确定的条件，因为 get()方法只返回一个确定值
```

输出结果如下：

```
<Person: Person object (1)>
```

  - ➢ 查询指定条件的数据，命令如下：

```
Person.objects.filter(first_name__exact="andy") # 指定 first_name 字段值必须为 andy
```

输出结果如下：

```
<QuerySet [<Person: Person object (1)>, <Person: Person object (2)>]>
```

```
Person.objects.filter(id__gt=1)                    # 查找所有 id 值大于 1 的
Person.objects.filter(id__lt=100)                  # 查找所有 id 值小于 100 的
# 排除所有创建时间大于现在时间的，exclude 的用法是排除，和 filter 正好相反
Person.objects.exclude(created_at__gt=datetime.datetime.now(tz=datetime.timezone.utc))
# 过滤出所有 first_name 字段值包含 a 的，然后将之前的查询结果按照 id 进行排序
Person.objects.filter(first_name__contains="a").order_by("id")
Person.objects.exclude(first_name__icontains="a")  # 查询所有 first_name 值不包含 a 的记录
```

☑ 修改数据。修改之前需要查询到对应的数据或者数据集，代码如下：

```
p = Person.objects.get(id=1)
```

然后按照需求进行修改，例如：

```
p.first_name = "jack"
p.last_name = "ma"
p.save()
```

也可以使用 get_or_create()，如果数据存在就修改，不存在就创建，代码如下：

```
p, is_created = Person.objects.get_or_create(
    first_name="jackie",
    defaults={"last_name": "chan"}
)
```

get_or_create()返回一个元组、一个数据对象和一个布尔值，defaults 参数是一个字典。当获取数据时，defaults 参数里面的值不会被传入，也就是获取的对象只存在 defaults 之外的关键字参数的值。

☑ 删除数据。删除数据同样需要你先查找到对应的数据，然后进行删除，代码如下：

```
Person.objects.get(id=1).delete()
```

运行结果如下：

```
(1,({'app1.Person':1}))
```

7.3.3　Django-Haystack 模块的基本使用方法

Django-Haystack 模块是一个强大的全文检索框架，专为 Django 项目设计，可以快速实现站内搜索功能。它简化了全文搜索功能的实现流程。下面介绍 Django-Haystack 模块的基本使用方法。

（1）安装 Django-Haystack 模块。如果在当前环境下没有安装 Django-Haystack 模块，需要使用 pip install 命令进行安装。具体的命令如下：

```
pip install django_haystack
```

（2）添加 Django-Haystack 框架到 INSTALLED_APPS。例如，打开 Django 项目的 settings.py 文件，将 'haystack'添加到 INSTALLED_APPS 列表中，代码如下：

```
# Application definition
INSTALLED_APPS = [
    'django.contrib.admin',
    'django.contrib.auth',
    'django.contrib.contenttypes',
    'django.contrib.sessions',
    'django.contrib.messages',
    'django.contrib.staticfiles',
    'haystack',
]
```

（3）配置 HAYSTACK_CONNECTIONS 以指定搜索引擎和索引存储位置。例如，使用 Whoosh 搜索引

擎，可以参考以下配置。

```
# 配置 Django-Haystack
HAYSTACK_CONNECTIONS = {
    'default': {
        # 设置搜索引擎，文件是 apps 下的 search 的 whoosh_cn_backend.py
        # 如果 search 模块未在 apps 下，请自行替换或去掉 apps
        'ENGINE': 'apps.search.whoosh_cn_backend.WhooshEngine',
        'PATH': os.path.join(BASE_DIR, 'whoosh_index'),
        'INCLUDE_SPELLING': True,
    },
}
```

> **注意**
> 使用 Whoosh 作为搜索引擎时，考虑到性能因素，可能不适合处理非常大量的数据。

（4）创建数据模型。在应用的 models.py 文件中定义要搜索的数据模型，例如，Poetry 模型包含 name 和 detail 字段，代码如下：

```
from django.db import models
class Poetry(models.Model):
    name = models.CharField('名称', max_length=50)
    detail = models.CharField('内容', max_length=300)
    # 设置返回值
    def __str__(self):
        return self.name
```

（5）创建索引类。为每个数据模型创建一个相应的搜索索引类，该类继承自 indexes.SearchIndex 和 indexes.Indexable，并定义要索引的字段。参考代码如下：

```
from haystack import indexes
from apps.search.models import Poetry
class PoetryIndex(indexes.SearchIndex, indexes.Indexable):
    text = indexes.CharField(document=True, use_template=True)
    # 将索引类与模型 Poetry 进行绑定
    def get_model(self):
        return Poetry
    # 设置索引的查询范围
    def index_queryset(self, using=None):
        return self.get_model().objects.all()
```

（6）在 settings.py 文件中配置 HAYSTACK_SIGNAL_PROCESSOR，来指定何时应更新索引。例如，当数据库改变时自动更新索引，可以使用下面的代码：

```
# 当数据库改变时，自动更新索引
HAYSTACK_SIGNAL_PROCESSOR = 'haystack.signals.RealtimeSignalProcessor'
```

（7）编写视图函数处理搜索请求和创建模板文件展示搜索结果即可。

7.3.4 使用 Whoosh 模块

使用 Whoosh 模块，可以轻松地为 Python 项目添加全文搜索功能。基本使用步骤如下。

（1）安装 Whoosh 模块。如果在当前环境下没有安装 Whoosh 模块，需要使用 pip install 命令进行安装。具体的命令如下：

```
pip install Whoosh
```

（2）初始化环境。即创建一个目录来存储索引文件，并初始化 Whoosh 的 schema，定义文档的结构和可搜索字段。

（3）建立索引。创建一个索引目录并打开一个索引编写器（IndexWriter）来向索引中添加文档。

（4）配置分词器。Whoosh 默认支持英文分词，对中文的支持不佳，可以使用 jieba 等第三方模块替换 Whoosh 自带的分词组件以进行优化。

（5）构建查询。即使用 query 模块中的函数和方法构造查询语句。

（6）搜索处理。即使用 Searcher 类来执行搜索，可以指定评分算法、是否关闭 reader 等参数。

7.3.5 使用 jieba 模块进行分词

jieba 模块是一个强大的 Python 分词库，可以对简体中文、繁体中文进行分词。下面将详细介绍如何使用该模块。

1．特点

jieba 模块主要有以下 4 个特点。
- ☑ 支持 3 种分词模式（精确模式、全模式和搜索引擎模式）。
- ☑ 支持繁体分词。
- ☑ 支持自定义词典。
- ☑ 采用 MIT（开源软件许可协议）授权协议。

2．安装 jieba 模块

同其他第三方模块一样，jieba 模块也可以使用 pip 命令进行安装。安装 jieba 模块时，需要先进入命令提示符对话框中，然后在命令提示符对话框中执行如下命令代码。

```
pip install jieba
```

安装 jieba 模块以后，就可以在 Python 文件中使用 import jieba 语句引用该模块，之后就可以使用该模块对指定内容进行分词了。

3．使用 jieba 模块进行分词

jieba 模块提供了 3 种分词模式，分别是精确模式、全模式和搜索引擎模式。下面对这 3 种分词模式进行详细介绍。

1）精确模式

采用精确模式时，会对句子进行最精确的切割，从而让结果更适合文本分析。实现精确模式分词时使用 jieba 模块的 cut()方法或 lcut()方法实现。cut()方法的基本语法格式如下：

```
jieba.cut(sentence, cut_all=False, HMM=True)
```

参数说明如下：
- ☑ sentence：用于指定要进行分词的字符串。
- ☑ cut_all：用于指定模式类型。全模式为 True，精确模式为 False。默认值为 False。
- ☑ HMM：是否使用 HMM（Hidden Markov Model，隐马尔可夫模型）。

例如，采用精确模式对乔布斯的名言中的"每一件都要做得精彩绝伦"进行分词，可以使用下面的代码。

```
import jieba    # 导入中文分词模块

'''精确模式示例'''
print('='*30,'精确分词','='*30)
motto = '每一件都要做得精彩绝伦'
```

```
word_list = jieba.cut(motto)                    # 进行精确模式分词
print('精确模式分词的结果：',word_list)          # 直接输出分词结果
```

执行上面的代码将输出以下内容。

```
精确模式分词的结果：   <generator object Tokenizer.cut at 0x0000016B4FDD6930>
```

从上面的结果可以看出，jieba.cut()方法返回的结果是一个可迭代对象。想要查看它的具体内容可以通过循环遍历进行输出，也可以将其转换为列表或元组进行输入。例如，可以使用下面的代码将其转换为列表。

```
print('精确模式分词的结果：',list(word_list))    # 转换为列表进行输出
```

将显示如图 7.3 所示的运行结果。

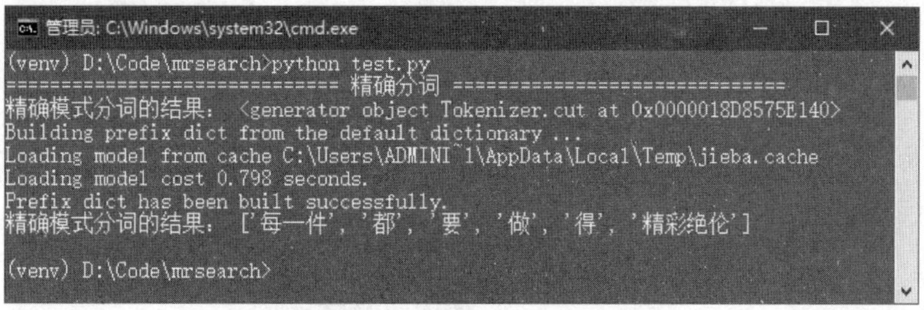

图 7.3　精确模式的分词结果

说明

lcut()方法的语法格式和 cut()方法一样，所不同的是，lcut()方法的输出结果为 list 列表对象。所以在使用该方法时，就不需要再将输出结果转换为列表对象了。

2）全模式

采用全模式时，会把文本中所有可能的词都扫描出来，可能会有冗余。实现全模式分词时也是使用 jieba 模块的 cut()方法或 lcut()方法来实现，只不过将其中的 cut_all 参数值设置为 Ture。

例如，采用全模式对乔布斯的名言中的"每一件都要做得精彩绝伦"进行分词，可以使用下面的代码。

```
import jieba                                    # 导入中文分词模块

'''全模式示例'''
print('='*30,'全分词','='*30)
motto = '每一件都要做得精彩绝伦'
word_list = jieba.cut(motto,cut_all=True)       # 进行全模式分词
print('全模式分词的结果：',list(word_list))     # 转换为列表进行输出
```

执行上面的代码将显示如图 7.4 所示的运行结果。

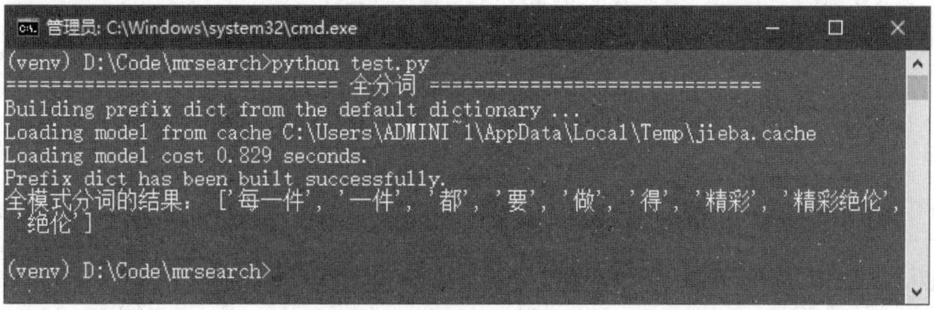

图 7.4　全模式的分词结果

> **说明**
> 从图 7.3 和图 7.4 中可以看出，在进行全模式分词时，把每一个可能是词的词都进行了分词。例如，"每一件"和"一件"都是一个词，还有"精彩"和"精彩绝伦"也都是一个词。在进行精确分词时，会把最大限度的词作为一个词进行切割。

3）搜索引擎模式

采用搜索引擎模式时，会在精确模式的基础上对长词再次切分。实现搜索引擎模式分词时使用 jieba 模块的 cut_for_search() 方法或 lcut_for_search() 方法来实现。cut_for_search() 方法的基本语法格式如下：

jieba.cut_for_search(sentence,HMM=True)

参数说明如下：

- ☑ sentence：用于指定要进行分词的字符串。
- ☑ HMM：是否使用 HMM。

例如，采用搜索引擎模式对乔布斯的名言中的"每一件都要做得精彩绝伦"进行分词，可以使用下面的代码。

```python
import jieba                                    # 导入中文分词模块

'''搜索引擎模式示例'''
print('='*30,'搜索引擎模式分词','='*30)
motto = '每一件都要做得精彩绝伦'
word_list = jieba.cut_for_search(motto)        # 进行搜索引擎模式分词
print('搜索引擎模式分词的结果：',list(word_list))    # 转换为列表进行输出
```

执行上面的代码将显示如图 7.5 所示的运行结果。

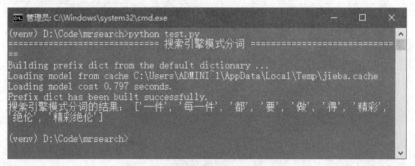

图 7.5 搜索引擎模式的分词结果

> **说明**
> lcut_for_search() 方法的语法格式和 cut_for_search() 方法一样，所不同的是，lcut_for_search() 方法的输出结果为 list 列表对象。所以在使用该方法时，就不需要再将输出结果转换为列表对象了。

7.4 数据库设计

7.4.1 数据库设计概要

站内全局搜索引擎使用 MySQL 数据库来存储数据，数据库名为 mrsearch，共包含 8 张数据表（包括

Django 默认的 7 张数据表），其数据库中的数据表如图 7.6 所示。

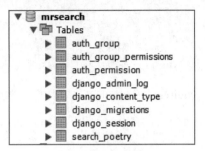

图 7.6 数据库中的数据表

mrsearch 数据库中的数据表对应的中文表名及主要作用如表 7.2 所示。

表 7.2 mrsearch 数据库中的数据表及作用

英 文 表 名	中 文 表 名	作 用
auth_group	授权组表	Django 默认的授权组
auth_group_permissions	授权组权限表	Django 默认的授权组权限信息
auth_permission	授权权限表	Django 默认的权限信息
django_admin_log	django 日志表	保存 Django 管理员登录日志
django_content_type	django contenttype 表	保存 Django 默认的 content type
django_migrations	django 迁移表	保存 Django 的数据库迁移记录
django_session	django session 表	保存 Django 默认的授权等 session 记录
search_poetry	搜索信息表	保存要搜索的数据

7.4.2 数据表模型

Django 框架自带的 ORM 可以满足绝大多数数据库开发的需求，在没有达到一定的数量级时，开发人员完全不需要担心 ORM 为项目带来的瓶颈。在本项目中，只有一张数据表为用户操作的数据表，即 search_poetry，该表中保存着要进行搜索的数据，对应的数据表模型代码如下：

```
from django.db import models
class Poetry(models.Model):
    id = models.AutoField('序号', primary_key=True)
    name = models.CharField('名称', max_length=50)
    author = models.CharField('作者', max_length=20)
    detail = models.CharField('内容', max_length=300)
    # 设置返回值
    def __str__(self):
        return self.name
```

7.5 创 建 项 目

在实现站内全局搜索引擎时，需要先创建 Django 项目并安装所需的模块。在创建项目前，需要先安装项目中应用的模块，本项目中使用的模块及对应版本如下：

```
Django==3.2
celery==5.4.0
```

```
eventlet==0.36.1
jieba==0.42.1
Whoosh==2.7.4
itsdangerous==2.2.0
django_haystack==3.2.1
mysqlclient==2.2.4
setuptools==70.0.0
```

将以上内容保存在一个名称为 requirements.txt 的文本文件中，然后在虚拟环境中通过下面的命令依次安装所需模块。

```
pip install -r requirements.txt
```

接下来按照如图 7.7 所示的目录结构创建项目。

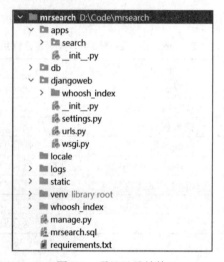

图 7.7　项目目录结构

项目创建完成后，就可以实现相应的功能了。

7.6　功能设计

开发站内全局搜索引擎时，需要实现全局搜索数据和分页显示搜索结果。针对这些需求，采用 Django 模块、Django-Haystack 模块和 Whoosh 模块进行开发比较高效，所以本项目采用这些模块。下面介绍具体的实现过程。

7.6.1　全局搜索数据

在站内全局搜索引擎中，一个最主要的功能就是实现全局搜索数据，即从数据库的指定表中按中文分词搜索。具体实现步骤如下。

（1）要实现站内搜索，就需要使用 Django-Haystack 模块。使用该模块时，需要在项目的 djangoweb\settings.py 文件中注册它。具体代码如下。

```
INSTALLED_APPS = [
    'django.contrib.admin',
    'django.contrib.auth',
    'django.contrib.contenttypes',
```

```
        'django.contrib.sessions',
        'django.contrib.messages',
        'django.contrib.staticfiles',
        'haystack',
        'apps.search',    # 全局搜索模块
]
# 配置 Django-HayStack
HAYSTACK_CONNECTIONS = {
    'default': {
        # 设置搜索引擎，文件是 apps 下的 search 的 whoosh_cn_backend.py
        # 如果 search 模块未在 apps 下，请自行替换或者去掉 apps
        'ENGINE': 'apps.search.whoosh_cn_backend.WhooshEngine',
        'PATH': os.path.join(BASE_DIR, 'whoosh_index'),
        'INCLUDE_SPELLING': True,
    },
}
# 设置每页显示的数据量
HAYSTACK_SEARCH_RESULTS_PER_PAGE = 2
# 当数据库改变时，自动更新索引
HAYSTACK_SIGNAL_PROCESSOR = 'haystack.signals.RealtimeSignalProcessor'
```

（2）在项目的 apps\search 目录下创建一个名称为 whoosh_cn_backend.py 的 Python 文件，然后打开 Python 安装目录下的 Lib\site-packages\haystack\backends\shoosh_backend.py 文件，将全文复制到 whoosh_cn_backend.py 文件中，修改如下内容：

```
# 顶部添加，引入 jieba 分词器模块
from jieba.analyse import ChineseAnalyzer
# 在 166 行左右找到如下代码
schema_fields[field_class.index_fieldname] = TEXT(stored=True, analyzer=StemmingAnalyzer (),
    field_boost=field_class.boost, sortable=True)
# 将其修改为如下格式
schema_fields[field_class.index_fieldname] = TEXT(stored=True, analyzer=ChineseAnalyzer(), field_boost=field_class.boost,
    sortable=True)
```

（3）在 search 模块下 models.py 文件中创建新的数据模型作为搜索引擎的搜索对象，具体代码如下：

```
from django.db import models
class Poetry(models.Model):
    id = models.AutoField('序号', primary_key=True)
    name = models.CharField('名称', max_length=50)
    author = models.CharField('作者', max_length=20)
    detail = models.CharField('内容', max_length=300)
    # 设置返回值
    def __str__(self):
        return self.name
```

（4）创建搜索引擎的索引，在数据量非常大的时候，我们就需要为指定的数据添加一个索引，来使搜索引擎快速找到符合条件的数据，所以在 search_indexes.py 中定义该模型的索引类，注意该文件名称不可改变，具体代码如下：

```
# 本文件名称不允许修改，否则将无法创建索引
from haystack import indexes
from apps.search.models import Poetry
# 类名必须为模型名+Index，比如模型 Poetry，则索引类为 PoetryIndex
# 其对应的索引模板路径应为/项目的应用模块名称/templates/search/indexes/项目的应用模块名称/模型（小写）_text.txt
class PoetryIndex(indexes.SearchIndex, indexes.Indexable):
    # doucument=True 代表搜索引擎将使用此字段的内容作为索引进行检索
    # use_template=True 代表使用索引模板建立索引文件
    text = indexes.CharField(document=True, use_template=True)
    # 将索引类与模型 Poetry 进行绑定
```

```
    def get_model(self):
        return Poetry
    # 设置索引的查询范围
    def index_queryset(self, using=None):
        return self.get_model().objects.all()
```

添加完成后，在索引模板 poetry_text.txt 文件中设置索引的检索字段，添加如下具体代码：

```
# templates/search/indexes/search/poetry_text.txt
{{ object.name }}
{{ object.author }}
{{ object.detail }}
```

（5）根据上面定义的索引类和索引模板创建索引文件。对应命令如下：

```
python manage.py rebuild_index
```

索引文件创建成功后会在项目文件夹下看到 whoosh_index 文件夹，该文件夹中包含索引文件。

（6）在 Django 中实现搜索功能，实现模型 Poetry 的全文检索。在 djangoweb\urls.py 文件中定义搜索引擎的 URL 地址，具体代码如下：

```
from django.urls import path
from apps.search import views
urlpatterns = [
    # 搜索引擎
    path('',views.MySearchView(), name='haystack'),
]
```

（7）在 URL 的视图文件 pps\search\views.py 中，编写实现全局搜索数据的业务逻辑，添加如下代码：

```
from django.shortcuts import render
from django.core.paginator import Paginator, EmptyPage, PageNotAnInteger
from django.conf import settings
from .models import *
from haystack.views import SearchView
# 视图以通用视图实现
class MySearchView(SearchView):
    # 模板文件
    template = 'search.html'
    # 重写响应方式，如果请求参数 q 为空，返回模型 Poetry 的全部数据，否则根据参数 q 搜索相关数据
    def create_response(self):
        if not self.request.GET.get('q', ''):
            show_all = True
            poetry = Poetry.objects.all()
            paginator = Paginator(poetry, settings.HAYSTACK_SEARCH_RESULTS_PER_PAGE)
            try:
                page = paginator.page(int(self.request.GET.get('page', 1)))
            except PageNotAnInteger:
                # 如果参数 page 的数据类型不是整型，则返回第一页数据
                page = paginator.page(1)
            except EmptyPage:
                # 用户访问的页数大于实际页数，则返回最后一页的数据
                page = paginator.page(paginator.num_pages)
            return render(self.request, self.template, locals())
        else:
            show_all = False
            qs = super(MySearchView, self).create_response()
            return qs
```

这样就完成了全局搜索数据，暂时还不能显示搜索到的数据，还需要实现分页显示搜索结果。

7.6.2 分页显示搜索结果

在 7.6.1 节中已经使用 Django-Haystack 模块和 Whoosh 模块实现了全局搜索数据。下面需要将搜索到的数据分页显示。具体的实现过程如下。

（1）将编写好的 CSS 代码文件复制到 static 文件夹下，以便于在模板文件中引用。
（2）编写 HTML 模板文件 apps\search\templates\search.html，用于展示数据。search.html 文件的代码如下：

```html
<!DOCTYPE html>
<html lang="en">
<head>
    <meta charset="UTF-8">
    <title>搜索引擎</title>
    {# 导入 CSS 样式文件 #}
    {% load static %}
    <link type="text/css" rel="stylesheet" href="{% static "css/common1.css" %}">
</head>
<body>
<div class="header">
    <div class="search-box">
        <form id="searchForm" action="" method="get">
            <div class="search-keyword">
                {# 搜索输入文本框必须命名为 q #}
                <input id="q" name="q" type="text" class="keyword" maxlength="120"/>
            </div>
            <input id="subSerch" type="submit" class="search-button" value="搜 索" />
        </form>
        <div id="suggest" class="search-suggest"></div>
    </div>
</div><!--end header-->

<div class="wrapper clearfix" id="wrapper">
<div class="mod_songlist">
    <ul class="songlist__header">
        <li class="songlist__header_name">名称</li>
        <li class="songlist__header_author">作者</li>
        <li class="songlist__header_album">内容</li>
    </ul>
    <ul class="songlist__list">
        {# 列出当前分页所对应的数据内容 #}
        {% if show_all %}
            {% for item in page.object_list %}
            <li class="js_songlist__child" mid="1425301" ix="6">
                <div class="songlist__item">
                    <div class="songlist__songname">{{ item.name }}</div>
                    <div class="songlist__artist">{{item.author}}</div>
                    <div class="songlist__album">{{ item.detail }}</div>
                </div>
            </li>
            {% endfor %}
        {% else %}
            {# 导入自带高亮功能 #}
            {% load highlight %}
            {% for item in page.object_list %}
            <li class="js_songlist__child" mid="1425301" ix="6">
                <div class="songlist__item">
                    <div class="songlist__songname">{% highlight item.object.name with query %}</div>
                    <div class="songlist__artist">{% highlight item.object.author with query %}</div>
                    <div class="songlist__album">{% highlight item.object.detail with query %}</div>
                </div>
            </li>
            {% endfor %}
        {% endif %}
```

```html
</ul>
{# 分页导航 #}
<div class="page-box">
<div class="pagebar" id="pageBar">
{# 上一页的 URL 地址 #}
{% if page.has_previous %}
    {% if query %}
        <a href="{% url 'haystack'%}?q={{ query }}&amp;page={{ page.previous_page_number }}"
        class="prev">上一页</a>
    {% else %}
        <a href="{% url 'haystack'%}?page={{ page.previous_page_number }}" class="prev">上一页</a>
    {% endif %}
{% endif %}
{# 列出所有的 URL 地址 #}
{% for num in page.paginator.page_range %}
    {% if num == page.number %}
        <span class="sel">{{ page.number }}</span>
    {% else %}
        {% if query %}
            <a href="{% url 'haystack' %}?q={{ query }}&amp;page={{ num }}" target="_self">{{num}}</a>
        {% else %}
            <a href="{% url 'haystack' %}?page={{ num }}" target="_self">{{num}}</a>
        {% endif %}
    {% endif %}
{% endfor %}
{# 下一页的 URL 地址 #}
{% if page.has_next %}
    {% if query %}
        <a href="{% url 'haystack' %}?q={{ query }}&amp;page={{ page.next_page_number }}"
        class="next">下一页</a>
    {% else %}
        <a href="{% url 'haystack' %}?page={{ page.next_page_number }}" class="next">
        下一页</a>
    {% endif %}
{% endif %}
</div>
</div>
</div><!--end mod_songlist-->
</div><!--end wrapper-->
</body>
</html>
```

添加完成后，启动项目，在浏览器中输入 http://127.0.0.1:8000/，将进入站内全局搜索引擎的首页，在该页面中将显示全部数据（默认为第 1 页），如图 7.8 所示。

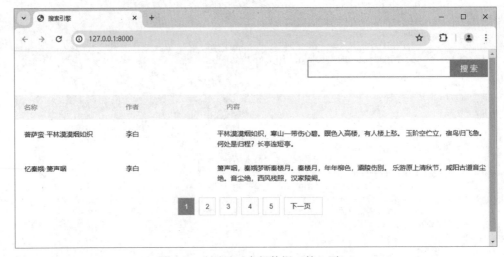

图 7.8　分页显示全部数据（第 1 页）

在右上角的搜索栏中输入要查询的关键字（如"月"），单击"搜索"按钮，将显示搜索到的数据，如图 7.9 所示。

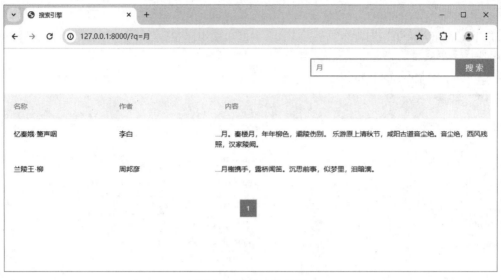

图 7.9 分词查询结果

7.7 项目运行

通过前述步骤，设计并完成了"站内全局搜索引擎"项目的开发。下面运行该项目，检验一下我们的开发成果。运行"站内全局搜索引擎"项目的步骤如下。

（1）打开 login\run.py 文件，根据自己的数据库账号和密码修改如下代码：

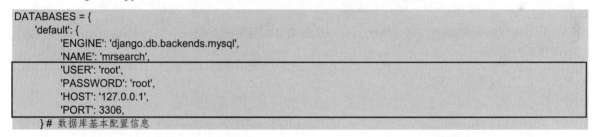

（2）打开命令提示符对话框，进入 mrsearch 项目文件夹所在目录，在命令提示符对话框中输入如下命令来创建 venv 虚拟环境：

virtualenv venv

（3）在命令提示符对话框中输入如下命令来启动 venv 虚拟环境：

venv\Scripts\activate

（4）在命令提示符对话框中使用如下命令来安装 Django 等依赖包：

pip install -r   requirements.txt

（5）创建数据库。可以使用 MySQL 命令行方式或 MySQL 可视化管理工具（如 Navicat）创建数据库。使用命令行方式时输入如下命令：

```
create database mrsearch default character set utf8;
```

（6）在命令提示符对话框中，执行 mrsearch.py 文件，用于创建数据表及添加默认数据。具体命令如下：

```
SOURCE mrsearch.sql
```

（7）在 PyCharm 中打开项目文件夹 mrsearch，在其中选中 manage.py 文件，单击鼠标右键，在弹出的快捷菜单中选择 Modify Run Configuration…命令，如图 7.10 所示。

（8）在打开的对话框中的 Parameters 文本框中输入 runserver，并且单击 Apply 按钮，如图 7.11 所示。

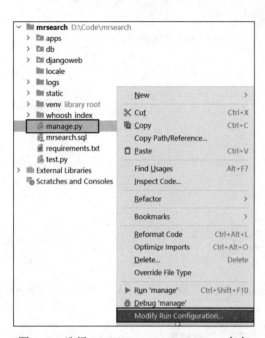

图 7.10　选择 Modify Run Configuration…命令　　　　图 7.11　输入 runserver

（9）单击右上角的运行按钮 ，运行项目，如果在 PyCharm 底部出现如图 7.12 所示的提示，说明程序运行成功。

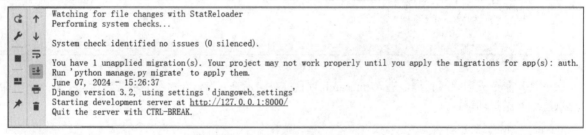

图 7.12　程序运行成功提示

（10）在浏览器中输入网址 http://127.0.0.1:8000/即可进入站内全局搜索引擎的首页，效果如图 7.13 所示。在该界面中，输入搜索关键字后，单击"搜索"按钮，可以将名称、作者和内容中包含该关键字的内容全部搜索出来。单击分页导航链接，可以按页查看搜索结果。

通过本章项目的学习，我们可以使用 Django-Haystack 模块和 Whoosh 模块为站点配置一个搜索引擎。其中，Django-Haystack 框架不仅支持 Whoosh，还支持 Solr 和 Elasticsearch 等搜索引擎，也可通过 Django-Haystack 直接切换引擎，且无须修改大量的搜索代码。另外，如果要使用中文搜索，还需为 Django-Haystack 配置 jieba 模块，实现中文分词。这些内容需要读者理解并能灵活应用。

图 7.13 站内全局搜索引擎的首页

7.8 源 码 下 载

本章虽然详细地讲解了如何编码实现"站内全局搜索引擎"的各个功能，但给出的代码都是代码片段，而非完整的源代码。为了方便读者学习，本书提供了该项目的完整源代码，读者可以通过扫描右侧的二维码进行下载。

源码下载

第 8 章 综艺之家

——Django-Spirit + ECharts

项目微视频

互联网技术的不断发展与人们精神文化水平的不断提高，使得电影电视领域与互联网的联系愈发紧密，网络上的各类综艺节目信息数量出现显著上升。为实现综艺节目信息的整理聚合，本章将主要使用 Django-Spirit 模块和 ECharts 模块实现一个综艺节目信息可视化展示与交流系统——综艺之家。

本项目的核心功能及实现技术如下：

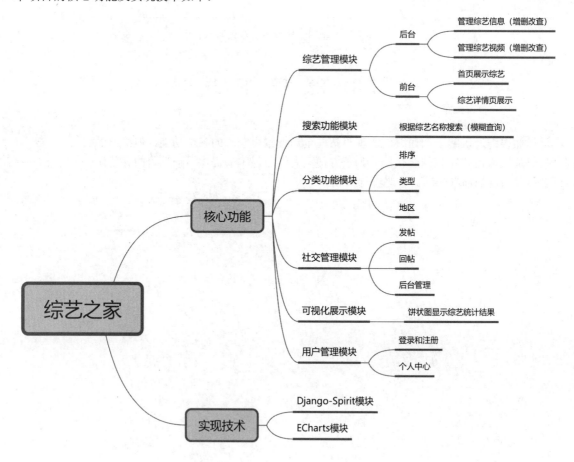

8.1 开发背景

伴随着科技的飞速发展与经济水平的不断提升，多种多样的电视综艺节目成为丰富人们娱乐生活的重要方式之一，综艺节目的数量出现了明显上升。然而，海量的节目数据也增加了观众获取信息的难度。同时由于商业利益、版权归属等问题，各视频平台之间难以实现资源共享，降低了观众搜索并观看节目内容的便捷程度。除此之外，现有的各类网站功能大多只限于视频的播放，面对海量的节目数据，很难从中直观总结出节目规律、流行趋势等。基于以上问题及思考，本章将实现一个综艺之家网站，用于电视节目的可视化展示与交流。

Django 框架是一个使用 Python 编写的 Web 框架，旨在帮助开发者快速构建高效、可扩展的 Web 应用。Django-Spirit 模块是一个基于 Django 框架的开源论坛系统，可以快速构建一个功能丰富、易于维护的论坛。ECharts 模块是一款基于 JavaScript 的数据可视化图表库，提供直观、生动、可交互、可个性化定制的数据可视化图表。因此，将 Django-Spirit 模块和 ECharts 模块搭配使用，就可以方便地开发出带可视化信息展示与交流功能的网站。

本项目的实现目标如下：
- ☑ 提供全面的综艺节目资源。
- ☑ 强大的搜索引擎，帮助用户快速找到感兴趣的节目。
- ☑ 提供节目排行榜功能，方便用户了解当下热点。
- ☑ 提供分页功能，提升页面性能。
- ☑ 提供多维统计分析饼状图。
- ☑ 建立论坛，让用户可以讨论节目内容，分享观点。
- ☑ 确保网站加载和播放速度快，减少缓冲时间，提升用户体验。
- ☑ 具备清晰、友好的用户界面，方便新用户快速上手并浏览内容。

8.2 系统设计

8.2.1 开发环境

本项目的开发及运行环境如下：
- ☑ 操作系统：推荐 Windows 10、Windows 11 或更高版本。
- ☑ 开发工具：PyCharm 2024（向下兼容）。
- ☑ 开发语言：Python 3.12。
- ☑ 数据库：MySQL 8.0+PyMySQL 驱动。
- ☑ Python Web 框架：Django-Spirit 0.13。

Django-Spirit 模块集成了 Django 框架，所以安装 Django-Spirit 模块时，会自动安装匹配版本的 Django 框架。

8.2.2 业务流程

综艺之家网站主要实现一个综艺节目信息可视化展示与交流系统。本系统分为前后台：前台为用户操作，在前台中，用户可以查看、搜索和交流综艺信息和视频；后台为管理员操作，在后台中，管理员可以管理综艺信息和视频（包括添加、修改、删除和查询操作）。本项目的业务流程如图 8.1 所示。

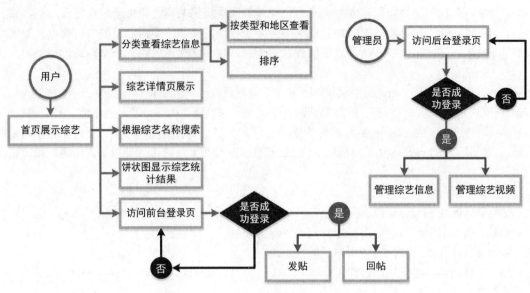

图 8.1 综艺之家网站的业务流程

8.2.3 功能结构

本项目的功能结构已经在章首页中给出，其实现的具体功能如下：

- ☑ 综艺管理模块：包括后台管理综艺信息和视频、前台首页展示和综艺详情页展示等功能。
- ☑ 搜索功能模块：用户可直接按节目名进行搜索。
- ☑ 分类功能模块：提供分类联合筛选功能，用户可根据个人喜好按综艺类型和地区进行筛选，同时也可以查看最近热播和热门排行（排序）。
- ☑ 社交管理功能：用户可在独立的论坛版块发贴和回帖等。
- ☑ 可视化展示功能：以图表形式，呈现综艺节目相关数据占比饼状图等。
- ☑ 用户管理模块：包括为用户提供登录、注册、查看个人主页，以及编辑管理个人信息等。

8.3 技术准备

8.3.1 技术概览

本项目主体采用 Django-Spirit 模块实现，同时，为了实现统计结果饼状图显示还使用了 ECharts 模块。下面对实现本项目时用到的这两个主要技术点进行必要介绍，以确保读者可以顺利完成本项目。

8.3.2 Django-Spirit 模块的基本使用方法

Django-Spirit 是一个基于 Django 框架的开源论坛系统，使用它可以快速搭建功能丰富、易于维护的论坛，因为它集成了许多论坛必备的功能，如用户认证、主题分类、帖子管理、回复、用户权限控制等。

在使用该模块时，需要先安装 Django 和 Django-Spirit 模块，由于 Django 框架是使用 Django-Spirit 时必备的框架，所以安装 Django-Spirit 模块时，会自动安装其所必需的框架，这里直接通过 pip 命令来安装

Django-Spirit 即可。对应的命令如下：

```
pip install django-spirit
```

另外，还需要手动安装 Django-Spirit 模块必需的支持模块 setuptools，对应的命令如下：

```
pip install setuptools==70.0.0
```

接下来就可以创建 Django 项目、集成 Django-Spirit 并创建论坛了，基本步骤如下。

（1）创建 Django 项目，具体代码如下：

```
django-admin startproject wforum
cd wforum
```

（2）在项目的 settings.py 文件中，把'spirit.core'添加到 INSTALLED_APPS 列表中，并且确保相关依赖已添加。完成后的 INSTALLED_APPS 列表的代码如下：

```
INSTALLED_APPS = [
    'django.contrib.admin',
    'django.contrib.auth',
    'django.contrib.contenttypes',
    'django.contrib.sessions',
    'django.contrib.messages',
    'django.contrib.staticfiles',
    'django.contrib.humanize',
    'spirit.core',
    'spirit.admin',
    'spirit.user',
    'spirit.user.admin',
    'spirit.user.auth',
    'spirit.category',
    'spirit.topic',
    'spirit.comment',
    'djconfig',
    'haystack',
]
```

（3）配置 Django 使用的数据库（默认使用 SQLite3），通常情况下，在创建 Django 项目时，会自动在 settings.py 中配置完成。对应的代码如下：

```
DATABASES = {
    'default': {
        'ENGINE': 'django.db.backends.sqlite3',
        'NAME': BASE_DIR / 'db.sqlite3',
    }
}
```

（4）在 settings.py 中配置网站根目录的 URL 地址，例如，在 settings.py 中添加下面的代码。

```
ST_SITE_URL = 'http://127.0.0.1:8000/'
```

（5）在 HAYSTACK_CONNECTIONS 列表中，指定全文搜索的基本配置信息，代码如下：

```
import os
HAYSTACK_CONNECTIONS = {
    'default': {
        'ENGINE': 'haystack.backends.whoosh_backend.WhooshEngine',
        'PATH': os.path.join(BASE_DIR, 'st_search'),
    },
}
HAYSTACK_SIGNAL_PROCESSOR = 'spirit.search.signals.RealtimeSignalProcessor'
```

（6）在 MIDDLEWARE 列表中，添加处理请求和响应的中间件，这里添加 djconfig 对应的中间件。添加后的代码如下：

```
MIDDLEWARE = [
    'django.middleware.security.SecurityMiddleware',
    'django.contrib.sessions.middleware.SessionMiddleware',
    'django.middleware.locale.LocaleMiddleware',
    'django.middleware.common.CommonMiddleware',
    'django.middleware.csrf.CsrfViewMiddleware',
    'django.contrib.auth.middleware.AuthenticationMiddleware',
    'django.contrib.messages.middleware.MessageMiddleware',
    'django.middleware.clickjacking.XFrameOptionsMiddleware',
    'spirit.user.middleware.TimezoneMiddleware',
    'djconfig.middleware.DjConfigMiddleware',
]
```

（7）在 TEMPLATES 列表中配置 djconfig 对应的模板引擎，对应的代码如下：

```
'djconfig.context_processors.config',
```

（8）配置 Django 项目的缓存设置和更新缓存配置，对应的代码如下：

```
CACHES = {
    'default': {
        'BACKEND': 'django.core.cache.backends.db.DatabaseCache',
        'LOCATION': 'spirit_cache',
    },
    'st_rate_limit': {
        'BACKEND': 'django.core.cache.backends.db.DatabaseCache',
        'LOCATION': 'spirit_rl_cache',
        'TIMEOUT': None
    }
}
CACHES.update({
    'default': {
        'BACKEND': 'django.core.cache.backends.locmem.LocMemCache',
    },
    'st_rate_limit': {
        'BACKEND': 'django.core.cache.backends.locmem.LocMemCache',
        'LOCATION': 'spirit_rl_cache',
        'TIMEOUT': None
    }
})
```

（9）配置用户登入登出相关的配置项，代码如下：

```
LOGIN_URL = 'spirit:user:auth:login'
LOGIN_REDIRECT_URL = 'spirit:user:update'
LOGOUT_REDIRECT_URL = 'spirit:index'
```

（10）应用数据库迁移命令创建 Django-Spirit 所需的数据表。在虚拟环境中，输入如下命令完成。

```
python manage.py migrate
```

（11）打开 wforum\wforum\urls.py 文件，在该文件中配置论坛前台首页和后台管理首页的路由，代码如下：

```
from django.contrib import admin
from django.conf.urls import include
from django.urls import re_path
urlpatterns = [
    re_path(r'^forums/', include('spirit.urls')),
    re_path(r'^admin/', admin.site.urls),
]
```

启动项目，访问 http://127.0.0.1:8000/forums/，将打开论坛的前台首页，如图 8.2 所示。

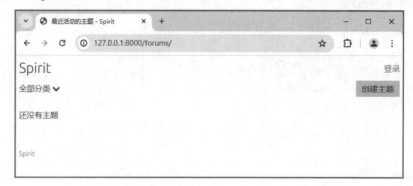

图 8.2　论坛的前台首页

8.3.3　使用 ECharts 模块显示图表

ECharts 是一个基于 JavaScript 的开源可视化图表库，可以流畅地运行在 PC 和移动设备上，兼容当前绝大部分浏览器（IE6/7/8/9/10/11、chrome、firefox、Safari 等），底层依赖轻量级的 Canvas 类库 ZRender，提供直观、生动、可交互、可高度个性化定制的数据可视化图表。创新的拖拽重计算、数据视图、值域漫游等特性大大增强了用户体验，赋予了用户对数据进行挖掘、整合的能力。

它支持折线图（区域图）、柱状图（条状图）、散点图（气泡图）、K 线图、饼图（环形图）、雷达图（填充雷达图）、和弦图、力导向布局图、地图、仪表盘、漏斗图、事件河流图等 12 类图表，同时提供标题、详情气泡、图例、值域、数据区域、时间轴、工具箱等 7 个可交互组件，支持多图表、组件的联动和混搭展现。

1．获取 ECharts

可以通过以下几种方式获取 Apache ECharts。
- ☑　从 Apache ECharts 官网下载界面获取官方源码包后构建。
- ☑　在 ECharts 的 GitHub 获取。
- ☑　通过 npm 获取 ECharts：npm install echarts -save。
- ☑　通过 CDN 方式引入。

为简单起见，本项目中直接通过 CDN 的方式来引入 ECharts。这种方式读者需要保证计算机可以正常访问 ECharts 的 CDN。实例代码如下：

```
<script src="https://cdn.bootcdn.net/ajax/libs/echarts/4.7.0/echarts-en.common.js">
</script>
```

2．引入 ECharts

通过标签方式直接引入构建好的 ECharts 文件，关键代码如下：

```
<!DOCTYPE html>
<html>
<head>
    <meta charset="utf-8">
    <title>ECharts</title>
    <!-- 引入 echarts.js -->
    <script src="e https://cdn.bootcdn.net/ajax/libs/echarts/4.7.0/echarts-en.common.js "></script>
</head>
</ html>
```

3. 绘制一个简单的图表

在绘图前我们需要为 ECharts 准备一个具备宽高的 DOM 容器。

```html
<body>
    <!-- 为 ECharts 准备一个具备大小（宽高）的 DOM -->
    <div id="main" style="width: 600px;height:400px;"></div>
</body>
```

然后就可以通过 echarts.init() 方法初始化一个 echarts 实例并通过 setOption() 方法生成一个简单的柱状图，完整代码如下：

```html
<!DOCTYPE html>
<html>
<head>
    <meta charset="utf-8">
    <title>ECharts</title>
    <!-- 引入 echarts.js -->
    <script src="e https://cdn.bootcdn.net/ajax/libs/echarts/4.7.0/echarts-en.common.js "></script>
</head>
<body>
    <!-- 为 ECharts 准备一个具备大小（宽高）的 Dom -->
    <div id="main" style="width: 600px;height:400px;"></div>
    <script type="text/javascript">
        // 基于准备好的 dom，初始化 echarts 实例
        var myChart = echarts.init(document.getElementById('main'));

        // 指定图表的配置项和数据
        var option = {
            title: {
                text: 'ECharts 入门示例'
            },
            tooltip: {},
            legend: {
                data:['销量']
            },
            xAxis: {
                data: ["衬衫","羊毛衫","雪纺衫","裤子","高跟鞋","袜子"]
            },
            yAxis: {},
            series: [{
                name: '销量',
                type: 'bar',
                data: [5, 20, 36, 10, 10, 20]
            }]
        };

        // 使用刚指定的配置项和数据显示图表
        myChart.setOption(option);
    </script>
</body>
</html>
```

运行结果如图 8.3 所示。

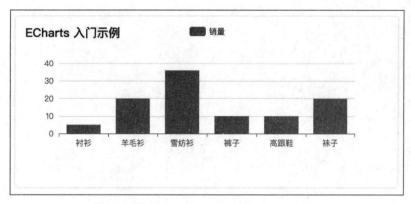

图 8.3　ECharts 柱状图示例

8.4　数据库设计

8.4.1　数据库设计概要

本项目采用 MySQL 数据库，数据库名称为 variety，主要数据表名称及作用如表 8.1 所示。

表 8.1　数据库中的数据表及作用

表　名	含　义	作　用
variety	综艺节目表	用于存储综艺节目信息
video	综艺的每期视频表	用于存储综艺节目下的每期视频
slide	幻灯片表	用于存储幻灯片信息
auth_user	用户表	用于存储用户信息
spirit_user_userprofile	论坛用户表	用于存储论坛用户信息
spirit_category_category	论坛分类表	用于存储论坛分类信息
spirit_topic_topic	论坛主题表	用于存储论坛主题信息

8.4.2　数据表模型

Django 框架自带的 ORM 可以满足绝大多数数据库开发的需求，在没有达到一定的数量级时，完全不需要担心 ORM 为项目带来的瓶颈。下面是综艺之家网站中使用的 ORM 数据模型，由于篇幅有限，这里只给出 models.py 模型文件中比较重要的代码。关键代码如下：

```
from django.db import models

# 地区
Region = [
    (0,'内地'),
    (3,'欧美'),
    (6,'其他')
]
# 综艺类型
Type = [
    (0,'脱口秀'),
```

```
    (1,'真人秀'),
    (2,'搞笑'),
    (3,'选秀'),
    (4,'情感'),
    (5,'访谈'),
    (6,'音乐'),
    (7,'职场'),
    (8,'体育'),
    (9,'其他')
]
# 年份
Year = [
    ('2015','2015'),
    ('2016','2016'),
    ('2017','2017'),
    ('2018','2018'),
    ('2019','2019'),
    ('2020','2020'),
    ('2021','2021'),
    ('2022', '2022'),
    ('2023', '2023'),
    ('2024','2024')
]
Hot = [
    (False,'否'),
    (True,'是')
]
Recommend = [
    (False,'否'),
    (True,'是')
]

class Variety(models.Model):
    # 综艺表信息
    id = models.AutoField(primary_key=True)
    variety_name = models.CharField(max_length=100,verbose_name='综艺名')
    type = models.SmallIntegerField(choices=Type,blank=False,verbose_name='类型')
    year = models.CharField(choices=Year,max_length=4,verbose_name='年代')
    region = models.SmallIntegerField(choices=Region,blank=False,verbose_name='地区')
    ranking = models.IntegerField(verbose_name='全网排名')
    platform = models.CharField(max_length=100,default='',verbose_name='播出平台')
    star = models.CharField(max_length=200,verbose_name='明星')
    review = models.TextField(max_length=500,null=True,verbose_name='简介')
    is_hot = models.BooleanField(choices=Hot,default=False,verbose_name='是否热门')
    is_recommended = models.BooleanField(choices=Recommend,default=False,verbose_name='是否推荐')
    image = models.ImageField(upload_to='variety', verbose_name='图片', null=True)

    class Meta:
        db_table = 'variety'
        verbose_name = '综艺管理'
        verbose_name_plural = '综艺管理'

    def __str__(self):
        return self.variety_name

class Video(models.Model):
    # 视频信息
    id = models.AutoField(primary_key=True)
    title = models.CharField(max_length=100,verbose_name='标题')
    desc = models.CharField(max_length=255,verbose_name='描述',default='')
```

```
image = models.ImageField(upload_to='video', verbose_name='图片', null=True)
video_url = models.CharField(max_length=300,verbose_name='视频链接')
release_date = models.DateField(verbose_name='上映日期')
# 关联 Variety 表
variety = models.ForeignKey(Variety,on_delete=models.CASCADE,related_name='video',verbose_name='所属综艺')

    class Meta:
        db_table = 'video'
        verbose_name = '视频管理'
        verbose_name_plural = '视频管理'

    def __str__(self):
        return self.title

class Slide(models.Model):
    # 幻灯片
    id = models.AutoField(primary_key=True)
    title = models.CharField(max_length=100,verbose_name='名称')
    desc = models.CharField(max_length=100,verbose_name='描述',default='')
    ranking = models.IntegerField(verbose_name='排序')
    image = models.ImageField(upload_to='slide', verbose_name='图片', null=True)
    jump_url = models.CharField(max_length=255,verbose_name='链接地址',default='')
    created_date = models.DateTimeField(auto_now_add=True, verbose_name='创建时间')
    modified_date = models.DateTimeField(auto_now=True, null=True, blank=True, verbose_name='更新时间')

    class Meta:
        db_table = 'slide'
        verbose_name = '幻灯片管理'
        verbose_name_plural = '幻灯片管理'

    def __str__(self):
        return self.title
```

8.4.3 数据表关系

本项目中有一组主要的数据表关系，一个综艺节目（variety 表）对应多个综艺视频（video 表），它们之间是一对多的关系，每个 video 表中的 variety_id 字段都对应着 variety 表中的 id 字段。我们使用 ER 图来直观地展现数据表之间的关系，如图 8.4 所示。

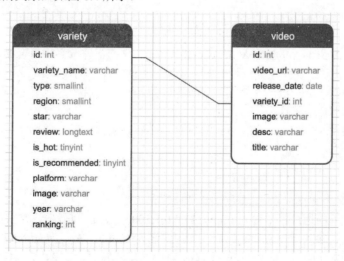

图 8.4 主要数据表关系

8.5 综艺管理模块设计

综艺管理模块是本项目的核心模块，用于管理和展示综艺信息和视频，主要包括后台录入综艺信息和视频以及前台展示综艺视频等内容，下面分别进行介绍。

8.5.1 实现后台录入综艺信息和视频的功能

Django 框架自带后台管理系统 Admin，使用 Admin 管理后台只需要通过配置几行简单的代码就可以实现一个完整的后台数据管理控制平台。

在 8.4.2 小节中，已经创建了数据表模型，Admin 管理后台通过读取模型数据，快速构造出一个可以对实际数据进行管理的 Web 站点。关键代码如下：

```python
from django.contrib import admin
from variety.models import Variety,Video,Slide,Star,HotWord
class VarietyAdmin(admin.ModelAdmin):
    # 显示列表
    list_display = ('variety_name','type','region','year')
    # 右侧筛选条件
    list_filter = ('region','type')
    # 查询字段
    search_fields = ('variety_name', 'type')


class VideoAdmin(admin.ModelAdmin):
    # 显示列表
    list_display = ('title','release_date')
    # 查询字段
    search_fields = ('variety_name',)
    # 获取视频所属的综艺名
    def get_variety_name(self, obj):
        return obj.variety.variety_name
    # 列名的描述信息
    get_variety_name.short_description = '综艺名'

admin.site.register(Variety,VarietyAdmin)
admin.site.register(Video,VideoAdmin)
```

上述代码中创建了 VarietyAdmin 和 VideoAdmin 两个类，它们都继承系统后台模块的 admin.ModelAdmin 类。在这两个类中，只需要定义对应的属性，即可实现相应的功能。例如，list_display 属性用于设置后台列表页显示的字段名和数据，list_filter 属性用于设置筛选的条件。最后，再使用 admin.site.register() 方法将模型和定义的两个类注册到后台模块。

此外，本项目使用了 Simpleui 主题，在配置文件中安装该主题应用，代码如下：

```python
INSTALLED_APPS = [
    'simpleui',  # 使用 Simpleui 主题
    'django.contrib.admin',
    'django.contrib.auth',
```

启动项目，在浏览器的地址栏中输入 http://127.0.0.1:8000/admin，将进入综艺管理的后台。如果没有登录，则先显示登录页面，输入用户名和密码后（如用户名为 admin，密码为 admin），将进入综艺之家网站的后台，在左侧的列表中选择相应的列表项，将显示对应的功能界面。其中，综艺信息列表页如图 8.5 所示。编辑综艺信息页如图 8.6 所示。

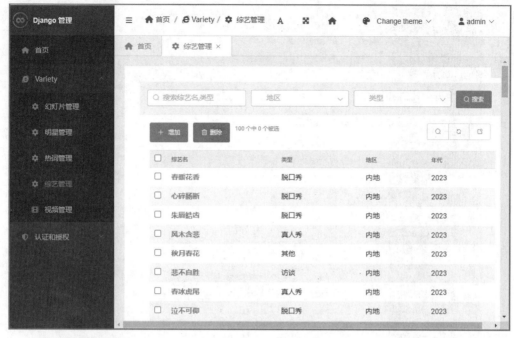

图 8.5　综艺信息列表页运行效果

图 8.6　编辑综艺信息页运行效果

后台综艺视频列表页如图 8.7 所示。编辑综艺视频页如图 8.8 所示。

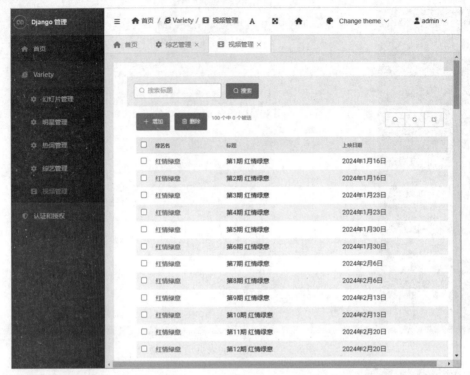

图 8.7 综艺视频列表页运行效果

图 8.8 编辑综艺视频页运行效果

8.5.2 实现前台首页展示功能

网站前台首页是网站的门面,页面要设计简洁,并且要展示重要信息。在本项目中,首页内容包括头部信息、导航栏分类信息、幻灯片信息、正在热播综艺信息和重磅推荐信息等内容。下面重点介绍正在热播的综艺信息和重磅推荐信息功能的实现。

由于网站的综艺信息和视频信息较多，不可能全部展示在首页，所以，设置了热播综艺和重磅推荐两个栏位。这两个栏位显示的内容是由管理员在后台设置的。在 variety 表中，is_hot 字段对应热播综艺栏位，is_recommended 对应重磅推荐栏位，取值内容如下：

- ☑ is_hot：值为 0，表示非热播综艺；值为 1，表示热播综艺。
- ☑ is_recommended：值为 0，表示不推荐；值为 1，表示推荐。

在 variety\views.py 文件中编写实现首页的热播综艺和重磅推荐功能，关键代码如下：

```python
def index(request):
    """
    首页
    """
    # 获取幻灯片
    slide = Slide.objects.order_by('ranking')[:10]
    # 热播综艺
    hot_variety = Variety.objects.filter(is_hot=True).order_by("-year")[:12]
    hot = []
    for item in hot_variety:
        last = item.video.all().last()
        if last:
            hot.append(last)
    # 推荐综艺
    recommend_variety = Variety.objects.filter(
                         is_recommended=True).order_by("-year")[:12]
    recommend = []
    for item in recommend_variety:
        last = item.video.all().last()
        if last:
            recommend.append(last)
    return render(request, 'index.html',
                  {'slide':slide,'hot':hot,'recommend':recommend,
                   'type':Type[:8],'region': Region, })
```

上述代码中，根据 filter(is_hot =True)筛选条件获取最多 12 条热播综艺，并且使用 order_by()依据时间进行降序排列。接下来，通过 variety 表和 video 表的一对多关系，获取该综艺下的所有视频，然后选择最后一个视频信息。重磅推荐的代码实现与热播综艺类似，这里不再赘述。

获取到数据后，使用 render()方法渲染前台首页模板 variety\templates\index.html，关键代码如下：

```html
<div class="m-rebo p-mod" id="js-rebo">
    <div class="p-mod-title">
        <span class="p-mod-label">正在热播</span>
    </div>
    <div class="content">
        <ul class="rebo-list w-newfigure-list g-clear js-list">
            {% for item in hot %}
                <li title='{{item.variety.variety_name}}' >
                    <a href="{% url 'detail' id=item.variety.id %}"
                       data-url='{{item.video_url}}' data-specialurl='' class='js-link'>
                        <div class='w-newfigure-imglink g-playicon js-playicon'>
                            <img src='/media/{{item.image}}'
                                 alt='{{item.variety.variety_name}}'/>
                            <span class='w-newfigure-hint'>{{item.release_date}}期</span>
                        </div>
                        <div class='w-newfigure-detail'>
                            <p class='title g-clear'>
                                <span class='s1' style="padding-left:8px">
                                    {{item.variety.variety_name}}</span></p>
                            <p class='w-newfigure-desc' style="padding-left:8px">
```

```html
                                    {{item.title}}</p>
                            </div>
                        </a></li>
                {% endfor %}
            </ul>
        </div>
</div>

<div class="p-mod-title">
        <span class="p-mod-label">重磅推荐！</span>
</div>
<div class="content">
        <ul class="rebo-list w-newfigure-list g-clear js-list">
                {% for item in recommend %}
                        <li title='{{item.variety.variety_name}}' >
                                <a href="{% url 'detail' id=item.variety.id %}"
                                        data-url='{{item.video_url}}' data-specialurl='' class='js-link'>
                                        <div class='w-newfigure-imglink g-playicon js-playicon'>
                                                <img src='/media/{{item.image}}'
                                                        alt='{{item.variety.variety_name}}'/>
                                                <span class='w-newfigure-hint'>{{item.release_date}}期</span></div>
                                        <div class='w-newfigure-detail'>
                                                <p class='title g-clear'>
                                                        <span class='s1' style="padding-left:8px">
                                                                {{item.variety.variety_name}}</span></p>
                                                <p class='w-newfigure-desc' style="padding-left:8px">
                                                        {{item.title}}</p>
                                        </div>
                                </a></li>
                {% endfor %}
        </ul>
</div>
```

上述代码中，item 对象就是获取到的每一个视频信息，此外，还需要获取该视频所属的综艺，由于在数据模型中设置了一对多的关系，所以，可以通过 item.variey 属性来获取到 variety 对象，然后再获取对应的综艺信息。

启动项目，在浏览器的地址栏中输入 http://127.0.0.1:8000/，即可进入前台首页，该页面的热播综艺和重磅推荐的运行效果如图 8.9 所示。

图 8.9　热播综艺和重磅推荐的运行效果

8.5.3 实现综艺详情页展示功能

在综艺之家网站的首页，单击综艺名称或综艺图片即可进入综艺详情页。综艺详情页是对综艺信息的详细描述，包括了综艺名称、图片、上映时间、发布平台、综艺介绍，以及该综艺下的所有视频信息。此外，当用户单击了综艺详情页后，还需要记录一下用户的浏览信息。

在 variety\views.py 文件实现综艺详情页功能，关键代码如下：

```python
def detail(request, id):
    """
    详情页
    :param request:
    :param id: 综艺 id
    :return:
    """
    try:
        variety = Variety.objects.get(pk=id)  # 根据 id 获取对象
        # 实现浏览记录功能
        cookies = request.COOKIES.get('variety_cookies','')
        if cookies == '':
            # 第一次浏览综艺详情，本地还没有生成综艺的 cookie 信息，
            # 那么直接将这个综艺的 id 存到 cookie。
            cookies = str(id)+';'                    # '1;2;3;'
        elif cookies != '':
            # 说明不是第一次浏览综艺详情，本地已经存在综艺的 cookie 信息；
            # 从 '1;2;3;'这个 cookie 字符串中，取出每一个综艺的 id
            variety_id_list = cookies.split(';')     # ['1','2','3']
            if str(id) in variety_id_list:
                # 说明当前这个综艺记录已经存在了，将这个记录从 cookie 中删除
                variety_id_list.remove(str(id))

            variety_id_list.insert(0,str(id))
            if len(variety_id_list) >= 6:
                variety_id_list = variety_id_list[:5]
            cookies = ';'.join(variety_id_list)
    except Variety.DoesNotExist:                     # 如果不存在，返回 404 页面
        return render(request, '404.html')
    response = render(request,'detail.html',
                      {'variety': variety,'region':Region,'year':Year,'type':Type})
    response.set_cookie('variety_cookies', cookies)
    return response
```

上述代码中，首先接收一个参数综艺 id，该参数是唯一的，通过它获取综艺节目信息。如果接收的综艺 id 在 variety 综艺表中不存在，则返回 404 页面。接下来，为了记录浏览信息，设置了一个名为 variety_cookies 的 cookie。首次访问时，该值为空字符串，将综艺 id 写入 variety_cookies 中。再次访问时，将原有综艺 id 删除，把该综艺 id 写入第一个位置。最后渲染 detail.html 模板。

variety\templates\detail.html 模板文件的关键代码如下：

```html
<div data-block="tj-info" class="top-info">
    <div class="top-info-title g-clear">
        <div class="title-left g-clear">
            <h1>{{variety.variety_name}}</h1>
            <p class="tag">更新至{{variety.video.all.last.release_date}}期</p>
            <a href="#" class="rank" data-block="tj-排行"
               monitor-shortpv-c-sub="tab_排名">全网综艺排名第{{variety.ranking}}名</a>
            <img src="https://p4.ssl.qhimg.com/t01460566f2d9f59a1b.png" />
        </div>
        <div class="s-title-right">
        </div>
```

```html
        </div>
        <div id="js-desc-switch" class="top-info-detail g-clear">
            <div class="base-item-wrap g-clear">
                <p class="item item42"><span class="cat-title">类型：</span>
                    {% for item in type %}
                        {% if item.0 == variety.type %}
                            {{item.1}}
                        {% endif %}
                    {% endfor %}
                </p>
                <p class="item item41"><span>年代：</span>{{variety.year}}年</p>
                <p class="item item41"><span>地区：</span>
                    {% for item in region %}
                        {% if item.0 == variety.region %}
                            {{item.1}}
                        {% endif %}
                    {% endfor %}
                </p>
                <p style='clear:both'></p>
                <p class="item item41"><span>播出：</span>{{variety.platform}}</p>
                <p class="item item44 item-actor">
                    <span>明星：</span>
                    {{variety.star}}
                </p>
            </div>
            <div class="item-desc-wrap g-clear js-open-wrap"><span>简介：</span>
                <p class="item-desc">{{variety.review}}
                </p></div>
        </div>
</div>

<div data-block="tj-juji" class="juji-main-wrap">
    <ul class="list w-newfigure-list g-clear">
        {% for item in variety.video.all.values %}
            <li title="{{item.title}}">
                <a href="{{item.video_url}}" data-url="{{item.video_url}}"
                    data-specialurl="" data-daochu="to=qiyi" class="js-link">
                    <div class="w-newfigure-imglink g-playicon js-playicon">
                        <img src="/media/{{item.image}}"
                            data-src="/media/{{item.image}}" alt="{{item.title}}">
                        <span class="w-newfigure-hint">{{ item.release_date }}期</span>
                    </div>
                    <div class="w-newfigure-detail">
                        <p class="title g-clear">
                            <span class="s1">
                                {% if item.desc %}
                                    {{ item.desc }}
                                {% else %}
                                    {{ item.title }}
                                {% endif %}
                            </span>
                        </p>
                    </div>
                </a>
            </li>
        {% endfor %}
    </ul>
</div>
```

上述代码中，第一部分是获取综艺信息，直接通过 vareity 对象的属性就可以获取到基本信息，但是对于更新时间这个栏位，需要使用 variety.video.all 对象获取全部视频，然后再来获取最后一个视频对象的发布日期属性，即 variety.video.all.last.release_date。

第二部分是获取该综艺下的所有视频，由于 variety 表和 video 表的一对多关系，可以使用 variety.video.all.values 获取所有的 video 对象，然后再遍历每一个 video 对象，获取相应的视频属性。

综艺详情页的运行效果如图 8.10 所示。

图 8.10　综艺详情页运行效果

8.6　搜索功能模块设计

为了快速查找到想要观看的综艺信息，可以使用顶部导航栏的搜索功能。输入关键字，然后单击"全网搜"按钮，即可搜索到所有包含该关键字的综艺信息。本项目使用的是模糊查询，即通过 MySQL 中的 like 关键字结合%来匹配所有综艺名称。如果匹配成功，获取搜索的综艺信息，否则，提示搜索内容不存在。

在 variety\views.py 文件中添加 search()方法实现搜索功能，关键代码如下：

```
def search(request):
    keyword = request.GET.get('keyword', '')
    variety_list = Variety.objects.filter(variety_name__contains=keyword)
    # 分页效果
    paginator = Paginator(variety_list, 8)
    page_number = request.GET.get('page')
    page_obj = paginator.get_page(page_number)
    page_range = paginator.page_range
    return render(request, 'search.html',
                  {'keyword': keyword, 'page_obj': page_obj,
                   'page_range': page_range,'region':Region,'type':Type})
```

上述代码中，接收关键字 keyword，然后使用 filter()方法中的"字段名+_contains"参数来查询所有综艺名字中包含关键字的 variety 对象。由于查询结果可能很多，所有使用 Paginator 对象实现分页。最后渲染

search.html 模板。

variety\templates\search.html 模板文件的关键代码如下：

```html
<div class="p-body g-clear js-logger">
    {% if not page_obj %}
    <span style="font-size:20px">
        您搜索的名字不存在，换一个名字试试！
    </span>
    {% else %}
    {% for variety in page_obj %}
        <div >
        <div class="m-mainpic">
            <a href="{% url 'detail' id=variety.id %}"
                title="{{variety.variety_name}}">
                <img src="/media/{{ variety.image }}" />
                <span>{{variety.video.all.values.last.release_date}}期</span>
            </a>
        </div>
        <div class="cont" style="width:80%">
            <h3 class="title">
                <a href="{% url 'detail' id=variety.id %}" >
                    <b>{{ variety.variety_name }}</b>
                </a>
                <span class="playtype">
                    {% for item in type %}
                        {% if item.0 == variety.type %}
                            {{item.1}}
                        {% endif %}
                    {% endfor %}
                    · {{variety.year}}</span>
                <div class="m-score"></div>
            </h3>
            <ul class="index-zongyi-ul g-clear" style="padding:0px">
                <li class='area'><b>地  区 :  </b>
                    {% for item in region %}
                        {% if item.0 == variety.region %}
                            <span>{{item.1}}</span>
                        {% endif %}
                    {% endfor %}
                </li>
                <li class='director'><b>明  星 :  </b>
                    {{ variety.star }}
                </li>
            </ul>
            <div class="m-description">
                <p><i>简  介 :  </i>
                    {{ variety.review }}
                </p>
            </div>
            <div class="index-zongyi-tabview js-zongyi-tabview">
                <div class="views js-zongyi-views">
                    <div>
                        {% for item in variety.video.all.values %}
                            <a href="{{item.video_url}}" title="{{ item.title }}">
                                <span class="data" style="width:105px">
                                    {{ item.release_date }}期</span>
                                <span class="name">{{ item.title }}</span>
                            </a>
                        {% endfor %}
                    </div>
                </div>
            </div>
```

```
        </div>
      </div>
    {% endfor %}
  {% endif %}
</div>
```

上述代码中,使用{% if %}标签来判断搜索内容是否存在,如果存在,再使用{% for %}标签来遍历每一个综艺,并展示综艺信息和该综艺下的视频信息。

启动项目,在前台首页的顶部输入搜索关键字,如"明眸皓齿",单击"全网搜"按钮将显示如图8.11所示的效果。否则,提示综艺不存在,运行效果如图8.12所示。

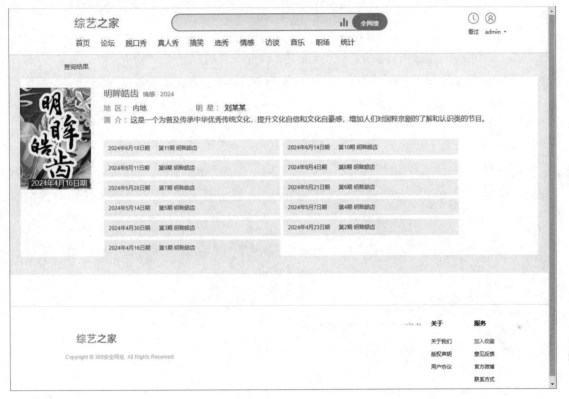

图8.11 显示搜索结果

图8.12 搜索结果不存在时的运行效果

8.7 分类功能模块设计

综艺节目根据类型可以划分为"脱口秀""真人秀""搞笑""选秀""情感""音乐"等。为了方便用户查找同一类型的综艺，本项目提供了分类筛选工能。筛选条件可以分为如下 3 类：
- ☑ 排序：最近热映、热门排行。
- ☑ 类型：全部、脱口秀、真人秀、搞笑等。
- ☑ 地区：内地、欧美、其他等。

在筛选时这 3 个分类属于"并且"关系，即综艺节目需要同时满足 3 个筛选条件，才会被筛出来，可以通过 URL 中的参数来设置分类的条件。分类页面中，一个完整的 URL 示例如下：

```
http://127.0.0.1:8000/lists/?tag=2&page=1&region=0&ranking=rank_order
```

重点关注"?"后的参数，tag 表示类型，page 表示页码，region 表示地区，ranking 表示排序。通过获取这几个参数，就能确定最终的筛选条件。

在 variety\views.py 文件中添加 lists()方法，实现分类功能，关键代码如下：

```python
def lists(request):
    # 获取参数
    tag = request.GET.get('tag', '全部')
    region = request.GET.get('region', '全部')
    ranking = request.GET.get('ranking', '最近热映')
    condition_dict = {}                              # 筛选条件字典
    if tag != '全部':                                # 筛选类型
        condition_dict['type'] = tag
    if region != '全部':                             # 筛选地区
        condition_dict['region'] = region
    if ranking == 'rank_hot':                        # 筛选热门综艺，后台设置是否热门
        condition_dict['is_hot'] = True
        variety_list = Variety.objects.filter(**condition_dict)
    else:                                            # 根据排名进行排序
        variety_list = Variety.objects.filter(**condition_dict).order_by(
            'ranking')
    # 分页功能实现
    paginator = Paginator(variety_list, 14)          # 设置每页显示条数
    page_number = request.GET.get('page')            # 获取当前页面
    page_obj = paginator.get_page(page_number)       # 获取分页对象
    page_range = paginator.page_range                # 分页迭代对象

    return render(request, 'lists.html', {'type':Type,'region': Region,
                    'page_obj': page_obj, 'page_range': page_range})
```

上述代码中，先来获取 3 个分类变量，然后加入 condition_dict 字典中，接下来使用 filter(**condition_dict) 进行多条件筛选。此外，还要结合分页功能。最后渲染 list.html 模板。

variety\templates\list.html 模板文件中的关键代码如下：

```html
<div data-channel="zongyi">
    <div class="filter-container" >
        <div class="s-filter">
            <dl class="s-filter-item g-clear">
                <dt class="type">排序</dt>
                <dd class="item g-clear js-filter-content">
                    <a class="ranking" href="javascript:;"
                        data-ranking='rank_hot' >最近热映</a>
                    <a class="ranking"    href="javascript:;"
                        data-ranking='rank_order'> 热门排行</a>
```

```
                </dd>
            </dl>
            <dl class="s-filter-item js-s-filter">
                <dt class="type">类型</dt>
                <dd class="item g-clear js-filter-content">
                    <a class="tag" href="javascript:;" data-tag="全部">全部</a>
                    {% for item in type %}
                        <a class="tag" href="javascript:;" data-tag="{{ item.0 }}">
                            {{ item.1 }}
                        </a>
                    {% endfor %}
                </dd>
            </dl>
            <dl class="s-filter-item js-s-filter">
                <dt class="type">地区</dt>
                <dd class="item g-clear js-filter-content">
                    <a class="region" href="javascript:;" data-region="全部">全部</a>
                    {% for item in region %}
                        <a class="region" href="javascript:;"
                            data-region="{{ item.0 }}">
                            {{ item.1 }}
                        </a>
                    {% endfor %}
                </dd>
            </dl>
        </div>
    </div>
    <div class="js-tab-container" data-block="tj-list" >
        <div class="s-tab">
            <div class="s-tab-main">
                <ul class="list g-clear js-list">
                    {% for variety in page_obj %}
                        <li class="item">
                            <a class="js-tongjic"
                                href="{% url 'detail' id=variety.id %}" >
                                <div class="cover g-playicon">
                                    <img src="/media/{{variety.image}}">
                                    <div class="mask-wrap">
                                        <span class="hint">
                                            {% if   variety.video.all.last.release_date %}
                                                {{variety.video.all.last.release_date}}期
                                            {% endif %}
                                        </span>
                                    </div>
                                </div>
                                <div class="detail">
                                    <p class="title g-clear">
                                        <span class="s1">{{variety.variety_name}}</span>
                                    </p>
                                    <p class="star">{{variety.video.all.last.title}}</p>
                                </div>
                            </a>
                        </li>
                    {% endfor %}
                </ul>
            </div>
        </div>
    </div>
</div>
```

上述代码中，使用{% for %}标签遍历获取每一个variety对象，然后获取对应的属性。以上只是根据分类条件获取对象。当单击分类右侧的名称时，对应的筛选条件发生改变，还需要保证其余的条件不变。例如，当前的筛选条件为"排序：热门排行；类型：脱口秀；地区：内地；页码：2"，当单击类型"真人秀"

时，只有类型发生变化，而其他条件保持不变，即筛选条件为"排序：热门排行；类型：真人秀；地区：内地；页码：2"。

为了实现以上功能，在 variety\templates\list.html 文件中使用了如下 JavaScript 代码：

```javascript
<script>
$(".tag , .region , .ranking").each(function () {
    $(this).click(function () {
        class_name = $(this).attr('class');
        var data_tag = $(this).data(class_name);
        matchUrl(class_name,data_tag);
    });
});

// 添加选中样式
$(document).ready(function(){
    // 清除原来选中的选项
    $(".on").removeClass("on");
    // 获取 tag 值，默认为"all"
    var tag =    getUrlParam('tag') ? getUrlParam('tag') : '全部';
    var region = getUrlParam("region") ? getUrlParam("region") : '全部';
    var ranking = getUrlParam("ranking") ? getUrlParam("ranking") : 'rank_hot';
    // 为 tag 添加选中样式
    console.log(tag)
    $(".tag , .region , .ranking").each(function(){
        if($(this).data('tag') == tag){
            $(this).addClass("on");
        }
        if($(this).data('region') == region){
            $(this).addClass("on");
        }
        if($(this).data('ranking') == ranking){
            $(this).addClass("on");
        }
    });
});
</script>
```

启动项目，在综艺之家网站的前台首页中，单击"类型"中的"情感"超链接，将进入分类功能页面，运行效果如图 8.13 所示。

图 8.13 分类筛选页面的效果

8.8 社交管理模块设计

社交管理模块，也可以称为社区或论坛模块，为用户提供相互交流和评论点赞的平台。社交模块的主要功能就是发帖、回帖和收藏等。为提供更好的服务，用户在发帖或回帖前，需要先登录网站，而未登录的游客只能浏览帖子，无法进行互动。

本项目使用开源模块 Django-Spirit 实现社交管理功能，将 Django-Spirit 作为一个应用整合到项目中。把 Django-Spirit 的配置文件作为项目的配置文件，将 variety 综艺应用添加到 INSTALLED_APPS 配置列表中，这里在 config\settings\base.py 文件中进行配置，关键配置代码如下：

```
INSTALLED_APPS = [
    'simpleui',
    'django.contrib.admin',
    'django.contrib.auth',
    'django.contrib.contenttypes',
    'django.contrib.sessions',
    'django.contrib.messages',
    'django.contrib.staticfiles',
    'django.contrib.humanize',

    'spirit.core',
    'spirit.admin',
    'spirit.search',

    'spirit.user',
    'spirit.user.admin',
    'spirit.user.auth',

    'spirit.category',
    'spirit.category.admin',

    'spirit.topic',
    'spirit.topic.admin',
    'spirit.topic.favorite',
    'spirit.topic.moderate',
    'spirit.topic.notification',
    'spirit.topic.private',
    'spirit.topic.unread',

    'spirit.comment',
    'spirit.comment.bookmark',
    'spirit.comment.flag',
    'spirit.comment.flag.admin',
    'spirit.comment.history',
    'spirit.comment.like',
    'spirit.comment.poll',

    'djconfig',
    'haystack',
    'variety'
]
```

接下来，配置路由文件，关键代码如下：

```
urlpatterns = [
    url(r'^', include('variety.urls')),
    url(r'^forum/', include('spirit.urls')),
    url(r'^admin/', admin.site.urls),
]
```

配置完成后，当访问 http://127.0.0.1:8000/forum/时，页面会跳转至论坛首页。运行效果如图 8.14 所示。

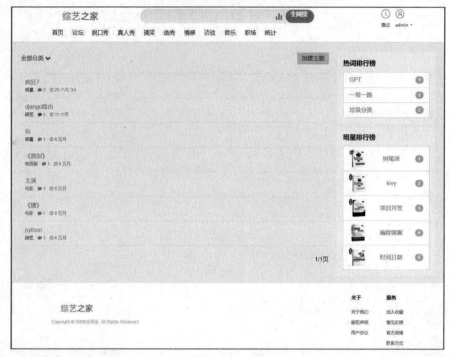

图 8.14　论坛首页运行效果

8.8.1　实现发帖和回帖功能

单击某个帖子，会进入帖子的详情页。在详情页，会展示帖子的标题、发布时间、发布人、发布的内容等，运行效果如图 8.15 所示。

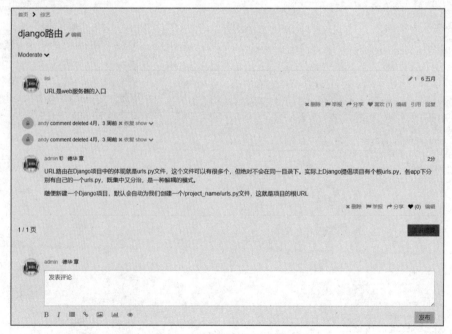

图 8.15　论坛详情页运行效果

如果用户没有登录，单击"回复"按钮，页面会跳转至登录页，运行效果如图 8.16 所示。如果没有账号，则需要单击"没有账号，去创建"超链接，页面会跳转至注册页，运行效果如图 8.17 所示。

图 8.16 登录页面运行效果　　　　　　　　图 8.17 注册页面运行效果

如果用户已经登录，则可以正常回帖，回帖页面使用了富文本编辑器，可以设置样式，或从本地插入图片，运行效果如图 8.18 所示。回复完成后的运行效果如图 8.19 所示。

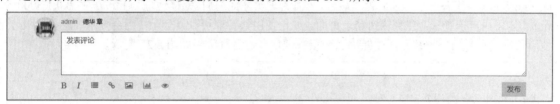

图 8.18 回复帖子运行效果

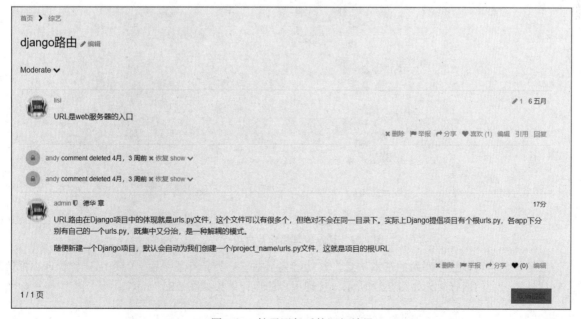

图 8.19 帖子回复后的运行效果

用户可以在论坛首页单击"创建主题"进行发帖，在发帖页面，需要填写帖子的标题，选择帖子的分类，用富文本编辑器编写帖子的内容，发布帖子的运行效果如图 8.20 所示。

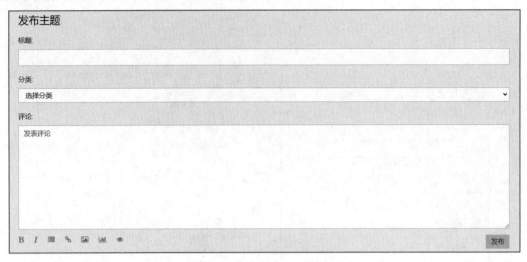

图 8.20　发布帖子的运行效果

8.8.2　实现论坛后台管理功能

　　Django-Spirit 模块有一个单独的管理后台，只有管理员才能访问。链接地址为 http://127.0.0.1:8000/forum/st/admin/。如果用户未登录，会跳转到登录页面。如果使用非管理员账号登录，则会提示"没有权限访问"。只有使用管理员账号访问该链接，才能进入论坛管理后台，运行效果如图 8.21 所示。

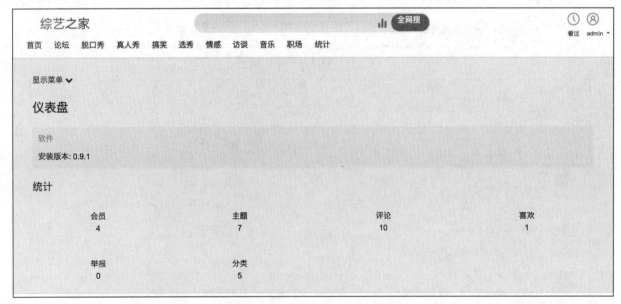

图 8.21　论坛管理后台首页运行效果

　　论坛管理后台可以设置网站的基本信息，其功能包括帖子分类、管理主题、管理用户、管理举报信息等。管理帖子分类的运行效果如图 8.22 所示。管理用户的运行效果如图 8.23 所示。

图 8.22 管理帖子分类运行效果

图 8.23 管理用户运行效果

8.9 可视化展示模块设计

正所谓一图胜千言，使用图表可以更直观地展示项目中的综艺信息数据。例如，每个类型的综艺占比，所有平台的综艺节目数量占比等。本项目中，选择使用比较流行的开源可视化图表库 ECharts 和 Ajax 来更直观地展示这些数据信息。

在本项目中，使用 ECharts 生成饼状图来展示数据所占比例。创建模板文件 variety\templates\statistics.html，关键代码如下：

```
<div class="col-9">
    <div class="dropdown" style="padding-bottom:20px">
        <a href="#" role="button" id="dropdownMenuLink" >
            类型
```

```html
                </a>
                <ul class="dropdown-menu" aria-labelledby="dropdownMenuLink">
                    <li><a class="dropdown-item type" href="#" id="all-categories">
                        所有类型节目数量占比饼状图</a></li>
                    <li><a class="dropdown-item type" href="#" id="all-platforms">
                        所有平台播出节目数量占比</a></li>
                    <li><a class="dropdown-item type" href="#" id="MRTV2">
                        MRTV2播出各类节目占比</a></li>
                </ul>
            </div>

            <!-- 展示图表 -->
            <div id="main" style="width: 1000px;height:400px;"></div>

</div>

<script src="/static/variety/js/jquery.js"></script>
<script src="https://cdn.bootcdn.net/ajax/libs/echarts/4.7.0/
            echarts-en.common.js"></script>
<script>
    // 自动加载时，执行单击事件
    $(document).ready(function(){
        $('#all-categories').click();
    });
    // 单击事件
    $('.type').click(function(){
        words = $(this).text()
        id = $(this).attr('id')
        $('#dropdownMenuLink').html(words)
        $(".shows").hide()
        $("."+id).show()
        var myChart = echarts.init(document.getElementById('main'));
        // 显示标题，图例和空的坐标轴
        myChart.setOption({
            title: {
                text: words,
                left: 'center'
            },
            tooltip: {
                trigger: 'item'
            },
            legend: {
                orient: 'vertical',
                left: 'left',
            },
            series : [
                {
                    type: 'pie',
                    radius: '55%',
                    data:[]
                }
            ]
        })

        url = '/chart/'+id
        console.log(id)
        // 异步加载数据
        $.get(url).done(function (data) {
            console.log(data)
            // 填入数据
            myChart.setOption({
                series: [{
                    // 根据名字对应到相应的系列
```

```
                    name: '销量',
                    data: data.data
                }]
            });
        });
    })
</script>
```

上述代码中，主要通过 JavaScript 的 click 单击事件实现图表切换，然后使用 Ajax 发送 get 请求，请求地址的 URL 为 http://127.0.0.1:8000/chart/，最后再将返回的 JSON 数据填充到 ECharts 中 series 对象的 data 属性。返回的 JSON 格式示例如下：

```
# 示例数据
data['data'] = [
    {'value':235, 'name':'视频广告'},
    {'value':274, 'name':'联盟广告'},
    {'value':310, 'name':'邮件营销'},
    {'value':335, 'name':'直接访问'},
    {'value':400, 'name':'搜索引擎'}
]
```

接下来，在 variety\urls.py 文件中设置路由，代码如下：

```
path('chart/<type>',views.chart,name='chart'),
```

然后在 variety\views.py 文件中创建视图，关键代码如下：

```
def dictfetchall(cursor):
    "将获取到的行数据以字典方式展示"
    desc = cursor.description
    return [
        dict(zip([col[0] for col in desc], row))
        for row in cursor.fetchall()
    ]

def transfor_type(data):
    "将类型由数字转化为名称"
    l = []
    for i in data:
        for j in Type:
            if i['name'] == j[0]:
                l.append({'name':j[1],'value':i['value']})
    return l
def chart(request,type):
    "生成图表数据"
    data = {}
    cursor = connection.cursor()
    if type == 'all-platforms':        # 所有平台综艺占比
        cursor.execute("select platform as name,count(*) as value from variety
                        where platform != '' group by platform")
        variety = dictfetchall(cursor)
        data['data'] = variety
    elif type == 'all-categories':     # 所有类型综艺占比
        cursor.execute("select type as name,count(*) as value from variety
                        group by type")
        variety = dictfetchall(cursor)
        data['data'] = transfor_type(variety)
    # 返回 json 格式数据
    return JsonResponse(data)
```

在上述代码中，首先接收 type 参数，通过 type 参数的值，判定要显示的图表内容。接下来，使用

cursor.execute()函数执行 SQL 语句。在 SQL 语句中，主要使用 group by 进行分组统计，并使用 as 关键字为返回的字段设置别名，方便后面整合数据。

此外，使用自定义函数 dictfetchall()将输出的列表类型数据转换为字典类型数据。使用 transfor_type()函数将数字转换为对应的文字。例如 type 为 1，转换为"真人秀"。

最后，使用 JsonResponse()函数将获取到的数据转换为 JSON 格式数据返回。

> **说明**
> 在获取数据时并没有使用 Django 自带的 ORM，而是使用了原生的 SQL 语句，因为当筛选的条件比较复杂时，使用 ORM 编写 SQL 比较麻烦，而且可读性不好。

启动项目，在前台首页的导航栏上单击"统计"超链接，将显示所有类型节目占比饼状图，运行效果如图 8.24 所示，在下拉列表中选择"所有平台播出节目数量占比"选项，将显示所有平台播出节目数量占比的饼状图，运行效果如图 8.25 所示。

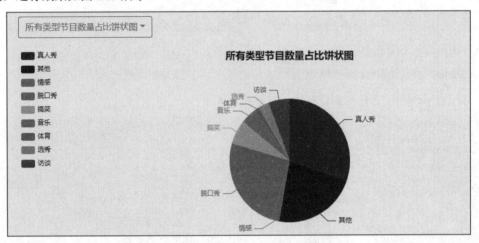

图 8.24 所有类型节目占比饼状图运行效果

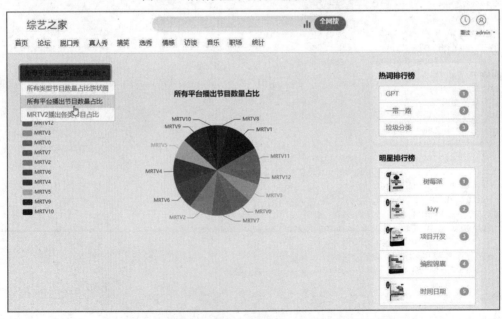

图 8.25 所有平台播出节目数量占比饼状图运行效果

8.10 项目运行

通过前述步骤，设计并完成了"综艺之家"项目的开发。下面运行该项目，检验一下我们的开发成果。运行"综艺之家"项目的步骤如下：

（1）打开 config\settings\dev.py 文件，根据自己的数据库账号和密码修改如下代码：

```
DATABASES = {
    'default': {
        'ENGINE': 'django.db.backends.mysql',
        'NAME': 'variety',
        'USER': 'root',
        'PASSWORD': 'root'
    }
} # 数据库基本配置信息
```

（2）打开命令提示符对话框，进入 variety 项目文件夹所在目录，在命令提示符对话框中输入如下命令来创建 venv 虚拟环境：

```
virtualenv venv
```

（3）在命令提示符对话框中输入如下命令来启动 venv 虚拟环境：

```
venv\Scripts\activate
```

（4）在命令提示符对话框中使用如下命令来安装所需的模块：

```
pip install -r requirements.txt
```

（5）创建数据库。可以使用 MySQL 命令行方式或 MySQL 可视化管理工具（如 Navicat）创建数据库。使用命令行方式时输入如下命令：

```
create database variety default character set utf8mb4;
```

（6）在命令提示符对话框中，执行 variety.py 文件，用于创建数据表及添加默认数据。具体命令如下：

```
SOURCE variety.sql
```

（7）在 PyCharm 中打开项目文件夹 variety，在其中选中 manage.py 文件，单击鼠标右键，在弹出的快捷菜单中选择 Modify Run Configuration…命令，如图 8.26 所示。

（8）在打开的对话框中的 Parameters 文本框中输入 runserver，并且单击 Apply 按钮，如图 8.27 所示。

（9）单击右上角的运行按钮 ，运行项目，如果在 PyCharm 底部出现如图 8.28 所示的提示，说明程序运行成功。

（10）在浏览器中输入网址 http://127.0.0.1:8000/即可进入综艺之家网站的首页，效果如图 8.29 所示。在该界面中，可以按类型、地区查看综艺信息和视频，也可以进入论坛讨论节目内容，分享观点等。

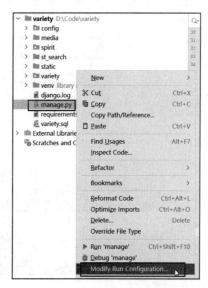

图 8.26 选择 Modify Run Configuration…命令

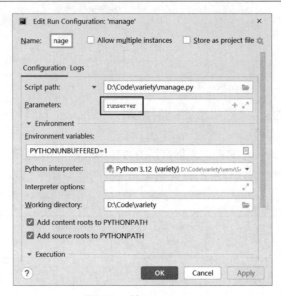

图 8.27　输入 runserver

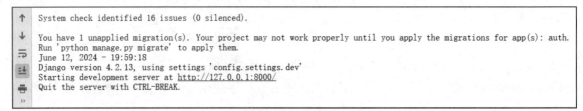

图 8.28　程序运行成功提示

图 8.29　综艺之家网站的首页

本章重点介绍了综艺节目信息可视化交互系统的设计与实现流程。其中，本章项目采用 Django-Spirit 模块，使用 MVC 模式进行开发的。因此，每部分功能都是模块化的，更加灵活方便。希望通过本章的学习，读者可以理解 Django 框架的模块化思想，以及掌握 Django-Spirit 模块的基本配置技术。

8.11 源码下载

本章虽然详细地讲解了如何编码实现"综艺之家"的各个功能，但给出的代码都是代码片段，而非完整的源代码。为了方便读者学习，本书提供了该项目的完整源代码，读者可以通过扫描右侧的二维码进行下载。

第 9 章 智慧校园考试系统

——Django + MySQL + Redis + 文件上传技术 + xlrd

智慧校园指的是以互联网为基础的智慧化的校园工作、学习、生活一体化环境,这个一体化环境以各种应用服务系统为载体,将教学、科研、管理和校园生活进行充分融合。在智慧校园体系中,考试系统则是一个不可或缺的重要环节。使用考试系统,通过简单配置,即可创建出一份精美的考卷,考生可以在计算机上进行考试或练习。本章将使用 Python Web 框架 Django,结合 MySQL 数据库、Redis 数据库等技术开发一个智慧校园考试系统。

本项目的核心功能及实现技术如下:

项目微视频

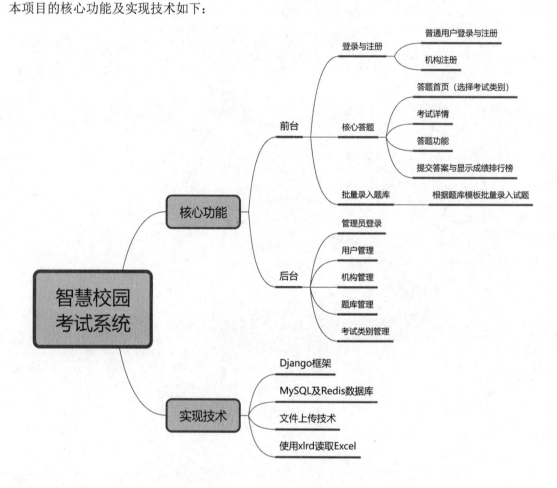

9.1 开发背景

随着信息技术的快速发展,智慧校园已成为高校信息化建设的重要方向。考试是教学活动的重要组成部分，传统的纸质考试方式已无法满足现代教育的需求。因此，开发一款智慧校园考试系统，实现考试流程的自动化、信息化和智能化，对于提高考试效率、减轻教师负担、提升学生体验具有重要意义。本项目选用 Django 框架，结合 MySQL 数据库、Redis 数据库等技术进行开发，主要基于以下考虑：

- ☑ Django 框架采用模型（model）、视图（view）和控制器（controller）的架构模式，使得代码结构清晰，易于维护和扩展，同时提供了大量的内置功能和工具，可以极大地加快开发速度，减少重复工作。
- ☑ MySQL 是一个成熟且稳定的数据库管理系统，适用于大规模的数据存储和查询，同时支持事务处理，可以保证数据的一致性和完整性。
- ☑ Redis 是一个内存数据库，读写速度非常快，适合用于缓存和实时数据处理。另外，它支持数据持久化，即使服务器重启，数据也不会丢失。而且，Redis 支持多种数据类型，如字符串、列表、哈希等，方便存储和操作考试相关的各种数据。

本项目的主要实现目标如下：

- ☑ 通过自动化和智能化的考试流程，减少人工操作，提高考试组织、实施的效率。
- ☑ 通过技术手段防止作弊行为，确保考试的公平性和公正性。
- ☑ 提供友好的用户界面和便捷的操作方式，降低用户使用门槛，提升用户体验。
- ☑ 通过收集和分析考试数据，为教学管理提供有力支持。

9.2 系统设计

9.2.1 开发环境

本项目的开发及运行环境如下：

- ☑ 操作系统：推荐 Windows 10、Windows 11 或更高版本。
- ☑ 开发工具：PyCharm 2024（向下兼容）。
- ☑ 开发语言：Python 3.12。
- ☑ 数据库：MySQL 8.0+PyMySQL 驱动、Redis。
- ☑ Python Web 框架：Django 5.0。

9.2.2 业务流程

在启动项目后，首先进入系统前台首页，这里将列出各类考试。如果想要进行考试，则需要判断用户是否登录，如果没有登录，则提示进行登录。登录时有 3 种身份，分别是普通用户、机构和管理员。其中，普通用户身份需要进行注册，登录后可以答题；机构同样需要注册，登录后可以录入题库和配置考试；管理员身份需要通过系统后台入口进行登录，登录后，可以对本系统的用户、机构、题库、考试类别等信息进行管理。

本项目的业务流程如图 9.1 所示。

说明

本章主要讲解智慧校园考试系统中主要功能模块的实现逻辑，其完整功能实现可以参考资源包中的源代码。

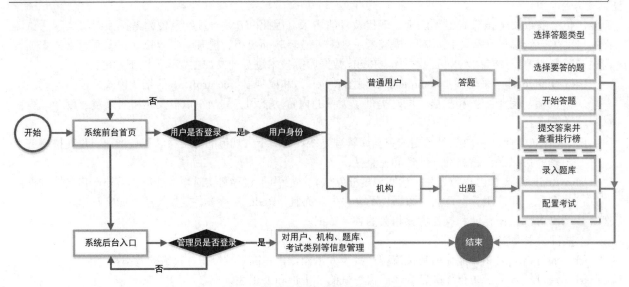

图 9.1 智慧校园考试系统业务流程

9.2.3 功能结构

本项目的功能结构已经在章首页中给出。作为在线考试方面的热门应用，本项目实现的具体功能如下：
- 登录与注册：包括普通用户登录与注册和机构注册等功能。其中，机构注册成功后可自主出题并配置考试信息。
- 邮件激活功能：用户注册完成后，需要登录邮箱激活。
- 快速出题功能：机构用户可下载题库模板，根据模板创建题目，上传题库。
- 分类功能：用户选择某类知识进行答题。
- 答题功能：用户参与考试后，可以选择上一题和下一题进行答题。
- 提交答案与显示成绩排行榜功能：用户答完所有题目并提交答案后，可以通过排行榜，查看考试成绩。
- 后台管理功能：管理员对用户、机构、题库、考试类别等信息进行管理。

9.3 技术准备

9.3.1 技术概览

本项目的主体框架采用 Django 框架实现，关于 Django 框架的基本用法请参考本书的 6.3.2 节和 7.3.2 节。另外，还用到了数据库存储技术、文件上传技术、使用 xlrd 模块读取 Excel 等。下面对这些技术点进行必要介绍，以确保读者可以顺利完成本项目。

9.3.2 数据存储技术

本项目中的数据存储主要使用了 MySQL 数据库及 Redis 数据库，其中操作 MySQL 数据库时使用了 PyMySQL 模块，关于该知识在《Python 从入门到精通（第 3 版）》中有详细的讲解，对该知识不太熟悉的读者可以参考该书对应的内容。下面对Python中如何操作Redis数据库进行详细介绍。

在Python中操作Redis数据库主要通过redis-py库实现，在使用该库时，需要先使用pip install命令安装它，具体的命令如下：

```
pip install redis
```

安装完成后，就可以使用redis-py库操作Redis数据库了。下面是一些常用的操作。

1．连接Redis

在Python中，可以使用redis.StrictRedis()方法连接Redis数据库。例如，创建一个Redis数据库连接对象，连接本地Redis数据库中的第一个数据库，端口为默认的6379，代码如下：

```
import redis
# 创建一个StrictRedis实例连接到本地Redis服务器
redis_client = redis.StrictRedis(host='localhost', port=6379, db=0)
```

说明

在Redis中，db=0表示使用第一个数据库。Redis默认提供了16个数据库，编号从0到15，你可以通过配置databases参数来调整这个数量。每个数据库都是独立的，它们之间不会共享数据。

2．设置键值对

使用Redis数据库连接对象提供的set()方法设置键值对，语法格式如下：

```
redis_client.set('key', 'value')
```

例如，将mykey键设置为"mr"，可以使用下面的代码：

```
redis_client.set('mykey', 'mr')
```

3．获取键值

使用Redis数据库连接对象提供的get()方法获取键对应的值，语法格式如下：

```
value = redis_client.get('key')
```

例如，获取并输出键mykey的值，可以使用下面的代码：

```
print('key 的值为：',redis_client.get('mykey'))
```

4．判断键是否存在

可以通过使用Redis数据库连接对象提供的exists()方法判断键是否存在，语法格式如下：

```
exists = r.exists('key')
```

例如，判断键mykey是否存在，可以使用下面的代码：

```
exists = r.exists('mykey')
print(exists)
```

输出值为1时，表示存在；否则输出值为0，表示不存在。

5．删除键

使用Redis数据库连接对象提供的delete()方法删除键，语法格式如下：

```
value = redis_client.delete('key')
```

例如，删除键 mykey 的值，可以使用下面的代码：

```
redis_client.delete('mykey')
```

6．增加计数器

Redis 提供了一个自动计数的功能，通过它可以轻松实现数据计数功能。实现增加计数的计数器，可以使用 Redis 数据库连接对象提供的 incr()方法，其语法格式如下：

```
redis_client.incr('counter')
```

例如，创建一个增加计数的计数器 mycounter，可以使用下面的代码：

```
redis_client.incr('mycounter')
```

7．减少计数器

使用 Redis 还可以实现减少计数的计数器，可以使用 Redis 数据库连接对象提供的 decr()方法，其语法格式如下：

```
redis_client.decr('counter')
```

例如，创建一个减少计数的计数器 mycounter，可以使用下面的代码：

```
redis_client.decr('mycounter')
```

8．向集合中添加元素

使用 Redis 数据库连接对象提供的 sadd()方法，可以实现向集合中添加元素，其语法格式如下：

```
redis_client.sadd('setname', 'member1')
```

例如，向 myset 集合中添加两个元素，分别是 mr 和 qq，可以使用下面的代码：

```
redis_client.sadd('myset', 'mr')
redis_client.sadd('myset', 'qq')
```

9．获取集合的所有成员

使用 Redis 数据库连接对象提供的 smembers()方法，可以获取集合中的所有成员，其语法格式如下：

```
redis_client.smembers('setname')
```

例如，获取并输出 myset 集合中所有的成员，可以使用下面的代码：

```
print(redis_client.smembers('myset'))
```

10．使用字典 API 操作哈希表

使用 Redis 数据库连接对象提供的 hset()和 hget()方法，设置或获取哈希表的值。例如，先设置哈希表 myhash 中字段 fiels1 的值，再获取并输入该字段的值，可以使用下面的代码：

```
redis_client.hset('myhash', 'field1', 'value1')
print(redis_client.hget('myhash', 'field1'))        # 输出：b'value1'
```

11．设置键的过期时间

使用 Redis 数据库连接对象提供的 setex()方法，设置指定键的过期时间。例如，设置键 mykey 的过期时间为 10 秒，可以使用下面的代码：

```
redis_client.setex('mykey', 10, 'value')        # 10 秒后过期
```

12. 关闭连接

使用 Redis 数据库连接对象提供的 close()方法，关闭 Redis 数据库的连接。具体代码如下：

```
redis_client.close()
```

9.3.3 Django 中的文件上传技术

当在 Django 应用中上传一个文件时，文件数据被放在 request.FILES 中。视图将在 request.FILES 中接收文件数据，request.FILES 是一个字典，它对每个 FileField（或者是 ImageField、FileField 的子类等）都包含一个 key。所以从表单传输的数据可以通过 request.FILES.get("key")或 request.FILES['key']键来访问。

例如，本项目中在上传题库时选择的是 Excel 文件，关键步骤如下：

（1）创建 Form 表单，通常使用 POST 方式提交上传文件，示例代码如下：

```html
<form method="post" action="" enctype="multipart/form-data" >
    {% csrf_token %}
    <input type="file" name="template" />
    <input type="submit" value="提交"/>
</form>
```

> **说明**
>
> 在上传文件时，需要将 Form 表单的 enctype 属性值设置为 multipart/form-data，request.FILES 中才包含文件数据，否则 request.FILES 为空。

（2）创建视图函数。在视图函数中，需要设置文件上传路径，判断上传文件后缀是否为 xls。然后，读取文件内容，最后将其写入指定的路径。关键代码如下：

```python
def upload_bank(request):
    """
    上传文件
    """
    template = request.FILES.get('template', None)           # 获取模板文件
    if not template:
        return render(request, 'err.html', FileNotFound)     # 模板不存在
    if template.name.split('.')[-1] not in ['xls']:
        return render(request, 'err.html', FileTypeError)    # 模板格式为 xls
    if not os.path.exists(settings.BANK_REPO):
        os.mkdir(settings.BANK_REPO)                         # 不存在该目录则创建
    final_path = settings.BANK_REPO + '.xls'                 # 生成文件名
    with open(final_path, 'wb+') as f:                       # 保存到目录
        f.write(template.read())
```

9.3.4 使用 xlrd 读取 Excel

使用 xlrd 能够很方便地读取 Excel 文件内容，而且 xlrd 是一个跨平台的库，它能够在 Windows、Linux、UNIX 等多个平台上面使用。下面介绍其使用步骤。

（1）安装 xlrd。使用 pip 安装 xlrd 的命令如下：

```
pip  install xlrd
```

（2）xlrd 的基本使用。xlrd 模块的 API 非常语言化，常用的 API 如下：

```python
data = xlrd.open_workbook('excelFile.xls')           # 打开一个 Excel 文件
# 获取工作表相关
```

```
table = data.sheets()[0]                          # 通过索引顺序获取
table = data.sheet_by_index(0)                    # 通过索引顺序获取
table = data.sheet_by_name(u'Sheet1')             # 通过名称获取
# 获取整行和整列的值（数组）
table.row_values(i)
table.col_values(i)
# 行数和列数
nrows = table.nrows
ncols = table.ncols
# 行列表数据
for i in range(nrows ):
        print table.row_values(i)
# 单元格相关
cell_A1 = table.cell(0,0).value
cell_C4 = table.cell(2,3).value
# 使用行列索引
cell_A1 = table.row(0)[0].value
cell_A2 = table.col(1)[0].value
```

下面通过一个示例讲解如何从 Excel 表格中读取数据。

例如，有一个名为 myfile.xls 的 Excel 表，该表中包含一个"学生信息表"Sheet 页，如图 9.2 所示。使用 xlrd 读取 Excel 表中数据的代码如下：

```
import xlrd
book = xlrd.open_workbook("myfile.xls")
print("一共有{}个 worksheets".format(book.nsheets))
print("Worksheet 的名字是：{}".format(book.sheet_names()))
sh = book.sheet_by_index(0)
print("{0} 有{1}行{2}列".format(sh.name, sh.nrows, sh.ncols))
for rx in range(1,sh.nrows):
    name = sh.row(rx)[0].value
    age  = int(sh.row(rx)[1].value)
    print("姓名：{} 年龄：{}".format(name,age))
```

运行结果如图 9.3 所示。

图 9.2　学生信息表数据

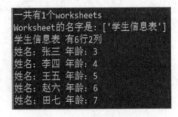

图 9.3　读取出的数据

9.4　数据库设计

9.4.1　数据库设计概要

智慧校园考试系统使用 MySQL 数据库来存储数据，数据库名为 exam，共包含 22 张数据表（包括 Django 默认的 10 张数据表），其数据库中的数据表如图 9.4 所示。

exam 数据库中的数据表对应的中文表名及主要作用如表 9.1 所示。

图 9.4 数据库中的数据表

表 9.1 exam 数据库中的数据表及作用

英 文 表 名	中 文 表 名	作 用
account_profile	用户信息表	保存授权后的账户信息
account_userinfo	用户填写信息表	保存用户填写的表单信息
auth_group	授权组表	Django 默认的授权组
auth_group_permissions	授权组权限表	Django 默认的授权组权限信息
auth_permission	授权权限表	Django 默认的权限信息
auth_user	授权用户表	Django 默认的用户授权信息
auth_user_groups	授权用户组表	Django 默认的用户组信息
auth_user_user_permissions	授权用户权限表	Django 默认的用户权限信息
business_appconfiginfo	机构 app 配置表	保存机构 app 配置信息
business_businessaccountinfo	机构账户表	保存机构账户信息
business_businessappinfo	机构 app 表	保存机构 app 信息，与配置信息关联
business_userinfoimage	表单图片链接表	保存每个表单字段的图片链接
business_userinforegex	表单验证正则表	保存每个表单字段的正则表达式信息
competition_bankinfo	题库信息表	保存题库信息
competition_choiceinfo	选择题表	保存选择题信息
competition_competitionkindinfo	考试信息表	保存考试信息和考试配置信息
competition_competitionqainfo	答题记录表	保存答题记录
competition_fillinblankinfo	填空题表	保存填空题信息
django_admin_log	Django 日志表	保存 Django 管理员登录日志
django_content_type	Django contenttype 表	保存 Django 默认的 content type
django_migrations	Django 迁移表	保存 Django 的数据库迁移记录
django_session	Django session 表	保存 Django 默认的授权等 session 记录

9.4.2 数据表模型

Django 框架自带的 ORM 可以满足绝大多数数据库开发的需求，在没有达到一定的数量级时，开发人员完全不需要担心 ORM 为项目带来的瓶颈。下面是智慧校园考试系统中使用 ORM 来管理一个考试信息的数据模型，关键代码如下：

```python
class CompetitionKindInfo(CreateUpdateMixin):
    """考试类别信息类"""
    IT_ISSUE = 0
    EDUCATION = 1
    CULTURE = 2
    GENERAL = 3
    INTERVIEW = 4
    REAR = 5
    GEO = 6
    SPORT = 7

    KIND_TYPES = (
        (IT_ISSUE, u'技术类'),
        (EDUCATION, u'教育类'),
        (CULTURE, u'文化类'),
        (GENERAL, u'常识类'),
        (GEO, u'地理类'),
        (SPORT, u'体育类'),
        (INTERVIEW, u'面试题')
    )

    kind_id = ShortUUIDField(_(u'考试id'), max_length=32, blank=True, null=True,
                             help_text=u'考试类别唯一标识', db_index=True)
    account_id = models.CharField(_(u'出题账户id'), max_length=32, blank=True, null=True,
                                   help_text=u'商家账户唯一标识', db_index=True)
    app_id = models.CharField(_(u'应用id'), max_length=32, blank=True, null=True,
                               help_text=u'应用唯一标识', db_index=True)
    bank_id = models.CharField(_(u'题库id'), max_length=32, blank=True, null=True,
                                help_text=u'题库唯一标识', db_index=True)
    kind_type = models.IntegerField(_(u'考试类型'), default=IT_ISSUE, choices=KIND_TYPES,
                                     help_text=u'考试类型')
    kind_name = models.CharField(_(u'考试名称'), max_length=32, blank=True, null=True,
                                  help_text=u'竞赛类别名称')
    sponsor_name = models.CharField(_(u'赞助商名称'), max_length=60, blank=True, null=True,
                                     help_text=u'赞助商名称')
    total_score = models.IntegerField(_(u'总分数'), default=0, help_text=u'总分数')
    question_num = models.IntegerField(_(u'题目个数'), default=0, help_text=u'出题数量')
    # 周期相关
    cop_startat = models.DateTimeField(_(u'考试开始时间'), default=timezone.now,
                                        help_text=_(u'考试开始时间'))
    period_time = models.IntegerField(_(u'答题时间'), default=60, help_text=u'答题时间(min)')
    cop_finishat = models.DateTimeField(_(u'考试结束时间'), blank=True, null=True,
                                         help_text=_(u'考试结束时间'))

    # 参与相关
    total_partin_num = models.IntegerField(_(u'total_partin_num'), default=0,
                                            help_text=u'总参与人数')
    class Meta:
        verbose_name = _(u'考试类别信息')
        verbose_name_plural = _(u'考试类别信息')

    def __unicode__(self):
        return str(self.pk)

    @property
    def data(self):
        return {
            'account_id': self.account_id,
            'app_id': self.app_id,
```

```
            'kind_id': self.kind_id,
            'kind_type': self.kind_type,
            'kind_name': self.kind_name,
            'total_score': self.total_score,
            'question_num': self.question_num,
            'total_partin_num': self.total_partin_num,
            'cop_startat': self.cop_startat,
            'cop_finishat': self.cop_finishat,
            'period_time': self.period_time,
            'sponsor_name': self.sponsor_name,
        }
```

与 CompetitionKindInfo 类相似，本项目中的其他类也继承基类 CreateUpdateMixin，该类中主要定义一些通用的信息，关键代码如下：

```
from django.db import models                              # 基础模型
from django.utils.translation import ugettext_lazy as _   # 引入延迟加载方法，只有在视图渲染时该字段才会呈现出翻译值
from TimeConvert import TimeConvert as tc

class CreateUpdateMixin(models.Model):
    """模型创建和更新时间戳 Mixin"""
    status = models.BooleanField(_(u'状态'), default=True, help_text=u'状态', db_index=True)   # 状态值，True 和 False
    # 创建时间
    created_at = models.DateTimeField(_(u'创建时间'), auto_now_add=True, editable=True, help_text=_(u'创建时间'))
    # 更新时间
    updated_at = models.DateTimeField(_(u'更新时间'), auto_now=True, editable=True, help_text=_(u'更新时间'))

    class Meta:
        abstract = True                                   # 抽象类，只为继承用，不会生成表
```

9.5 登录与注册模块设计

9.5.1 普通用户登录与注册模块概述

普通用户登录与注册模块主要对进入智慧校园考试系统的用户信息进行验证，本项目中使用邮箱和密码的方式进行登录，用户登录页面运行效果如图 9.5 所示。

图 9.5 用户登录页面

9.5.2 使用 Django 默认授权机制实现普通登录

Django 默认的用户授权机制可以提供绝大多数场景的登录功能,为了更加适应智慧校园考试系统的登录需求,这里对其进行简单修改。

1. 用户登录接口

在 account app 下创建一个 login_views.py 文件,用来作为接口视图,该文件中编写一个 normal_login() 函数,用来实现用户正常的用户名和密码登录功能,代码如下:

```python
@csrf_exempt
@transaction.atomic
def normal_login(request):
    """
    普通登录视图
    :param request: 请求对象
    :return: 返回 json 数据: user_info: 用户信息;has_login: 用户是否已登录
    """
    email = request.POST.get('email', '')                  # 获取 email
    password = request.POST.get('password', '')            # 获取 password
    sign = request.POST.get('sign', '')                    # 获取登录验证码的 sign
    vcode = request.POST.get('vcode', '')                  # 获取用户输入的验证码
    result = get_vcode(sign)                               # 从 Redis 中校验 sign 和 vcode
    if not (result and (result.decode('utf-8') == vcode.lower())):
        return json_response(*UserError.VeriCodeError)     # 校验失败则返回错误码 300003
    try:
        user = User.objects.get(email=email)               # 使用 email 获取 Django 用户
    except User.DoesNotExist:
        return json_response(*UserError.UserNotFound)      # 获取失败则返回错误码 300001
    user = authenticate(request, username=user.username, password=password) # 授权校验
    if user is not None:                                   # 校验成功,获得返回用户信息
        login(request, user)                               # 登录用户,设置登录 session
        # 获取或创建 Profile 数据
        profile, created = Profile.objects.select_for_update().get_or_create(
            email=user.email,
        )
        if profile.user_src != Profile.COMPANY_USER:
            profile.name = user.username
            profile.user_src = Profile.NORMAL_USER
            profile.save()
        request.session['uid'] = profile.uid               # 设置 Profile uid 的 session
        request.session['username'] = profile.name         # 设置用户名的 session
        set_profile(profile.data)                          # 将用户信息保存到 Redis 中,用户信息从 Redis 中查询
    else:
        return json_response(*UserError.PasswordError)     # 校验失败,返回错误码 300002
    return json_response(200, 'OK', {                      # 返回 JSON 格式数据
        'user_info': profile.data,
        'has_login': bool(profile),
    })
```

以上实现的是用户登录的接口,编写完上面代码后,需要在 api 模块下的 urls.py 中添加路由,代码如下:

```python
path('login_normal', login_views.normal_login, name='normal_login'),
```

在 web 目录下的 base.html 文件中,定义一个 Ajax 异步请求方法,用来处理用户登录的表单,代码如下:

```javascript
$('#signInNormal').click(function () {              // 单击"登录"按钮
    refreshVcode('signin');                          // 刷新验证码
    $('#signInModalNormal').modal('show');           // 显示弹窗
```

```javascript
$('#signInVcodeImg').click(function () {                // 单击验证码图片，刷新验证码
    refreshVcode('signin');
});
$('#signInPost').click(function () {                    // 单击"登录"按钮
    // 获取表单数据
    var email = $('#signInId').val();
    var password = $('#signInPassword').val();
    var vcode = $('#signInVcode').val();
    // 验证 email
    if(!checkEmail(email)){
        $('#signInId').val('');
        $('#signInId').attr('placeholder', '邮件格式错误');
        $('#signInId').css('border', '1px solid red');
        return false;
    }else{
        $('#signInId').css('border', '1px solid #C1FFC1');
    }
    // 验证密码
    if(!password){
        $('#signInPassword').attr('placeholder', '请填写密码');
        $('#signInPassword').css('border', '1px solid red');
    }else{
        $('#signInPassword').css('border', '1px solid #C1FFC1');
    }
    // Ajax 异步提交
    $.ajax({
        url: '/api/login_normal',                       // 提交地址
        data: {                                         // 提交数据
            'email': email,
            'password': password,
            'sign': loginSign,
            'vcode': vcode
        },
        type: 'post',                                   // 提交类型
        dataType: 'json',                               // 返回数据类型
        success: function(res){                         // 回调函数
            if (res.status === 200){                    // 登录成功
                $('#signInModalNormal').modal('hide');  // 隐藏弹窗
                window.location.href = '/';             // 跳转到首页
            }
            else if(res.status === 300001) {
                alert('用户名错误');
            }
            else if(res.status === 300002) {
                alert('密码错误');
            }
            else if(res.status === 300003) {
                alert('验证码错误');
            }
            else {
                alert('登录错误');
            }
        }
    })
});
```

  登录使用异步方式实现，当用户单击页面上的"登录"按钮时，使用 Bootstrap 框架的 modal 插件弹出登录框，用户输入邮箱账户、密码和验证码时，会根据不同的错误信息给用户一个友好的提示。

  当前端验证全部通过时，Ajax 发起请求，后台会校验用户输入的数据是否合理有效，如果验证全部通过，将在用户单击"登录"按钮时，显示出存储在 session 中的用户名。

> **说明** 在登录过程中，还需要添加刷新验证码功能。这可以通过在项目中创建 utils/codegen.py/CodeGen 类来实现，具体代码请查看资源包中的源码文件。

2. 用户注册接口

用户注册同样是使用 Ajax 异步请求的方式，在弹出的 modal 框中输入表单内容，然后通过正则表达式规则进行校验，如果校验成功，会将输入提交到后台进行校验，如果校验通过，将会返回一个新渲染的视图，并提示用户发送邮件去验证邮箱。

发送邮件需要通过异步请求的接口实现，用户注册的视图函数代码如下：

```python
@csrf_exempt
@transaction.atomic
def signup(request):
    email = request.POST.get('email', '')                              # 邮箱
    password = request.POST.get('password', '')                        # 密码
    password_again = request.POST.get('password_again', '')            # 确认密码
    vcode = request.POST.get('vcode', '')                              # 注册验证码
    sign = request.POST.get('sign')                                    # 注册验证码检验位
    if password != password_again:                                     # 两次密码不一样，返回错误码 300002
        return json_response(*UserError.PasswordError)
    result = get_vcode(sign)                                           # 校验 vcode，逻辑和登录视图相同
    if not (result and (result.decode('utf-8') == vcode.lower())):
        return json_response(*UserError.VeriCodeError)
    if User.objects.filter(email__exact=email).exists():               # 检查数据库是否存在该用户
        return json_response(*UserError.UserHasExists)                 # 返回错误码 300004
    username = email.split('@')[0]                                     # 生成一个默认的用户名
    if User.objects.filter(username__exact=username).exists():
        username = email                                               # 默认用户名已存在，使用邮箱作为用户名
    User.objects.create_user(                                          # 创建用户，并设置为不可登录
        is_active=False,
        is_staff=False,
        username=username,
        email=email,
        password=password,
    )
    Profile.objects.create(                                            # 创建用户信息
        name=username,
        email=email
    )
    sign = str(uuid.uuid1())                                           # 生成邮箱校验码
    set_signcode(sign, email)                                          # 在 Redis 设置 30 分钟时限的验证周期
    return json_response(200, 'OK', {                                  # 返回 JSON 数据
        'email': email,
        'sign': sign
    })
```

在 api 的 urls.py 中加入如下路由：

```python
path('signup', login_views.signup, name='signup'),
```

响应接口数据后，注册过程并未完成，需要用户手动触发邮箱验证。当用户单击"发送邮件"按钮时，Ajax 将会提交数据到以下接口路由：

```python
path('sendmail', login_views.sendmail, name='sendmail'),
```

上面路由对应的视图函数为 sendmail()，该函数用来完成一个使用 django.core.sendmail 发送邮件的过程，其实现代码如下：

```python
def sendmail(request):
    to_email = request.GET.get('email', '')              # 在 url 中获取的注册邮箱地址
    sign = request.GET.get('sign', '')                   # 在 url 中获取的 sign 标识
    if not get_has_sentregemail(to_email):               # 检查用户是否在同一时间多次单击"发送邮件"
        title = '[Quizz.cn 用户激活邮件]'                  # 定义邮件标题
        sender = settings.EMAIL_HOST_USER                # 获取发送邮件的邮箱地址
        # 回调函数
        url = settings.DOMAIN + '/auth/email_notify?email=' + to_email + '&sign=' + sign
        # 邮件内容
        msg = '您好，Quizz.cn 管理员想邀请您激活您的用户，单击链接激活。{}'.format(url)
        # 发送邮件并获取发送结果
        ret = send_mail(title, msg, sender, [to_email], fail_silently=True)
        if not ret:
            return json_response(*UserError.UserSendEmailFailed)  # 发送出错，返回错误码 300006
        set_has_sentregemail(to_email)                   # 正常发送，设置 3 分钟的继续发送限制
        return json_response(200, 'OK', {})              # 返回空 JSON 数据
    else:
        # 如果用户同一时间多次单击发送，返回错误码 300005
        return json_response(*UserError.UserHasSentEmail)
```

> **说明**
>
> 在上面发送邮件的视图函数 sendmail()中添加了一个回调函数，用来检查用户是否确认邮件。回调函数是个普通的视图渲染函数。

在 config 模块的 urls.py 中添加总的授权路由，代码如下：

```python
urlpatterns += [
        path('auth/', include(('account.urls','account'), namespace='auth')),
]
```

然后在 account 的 urls.py 中添加授权回调函数的路由：

```python
path('email_notify', login_render.email_notify, name='email_notify'),
```

授权回调函数 email_notify()主要用来验证是否在邮箱中对用户信息进行了激活，如果激活，则对用户信息进行配置，实现代码如下：

```python
@transaction.atomic
def email_notify(request):
    email = request.GET.get('email', '')                 # 获取要验证的邮箱
    sign = request.GET.get('sign', '')                   # 获取校验码
    signcode = get_signcode(sign)                        # 在 Redis 中校验邮箱
    if not signcode:
        return render(request, 'err.html', VeriCodeTimeOut)   # 校验失败则返回错误视图
    if not (email == signcode.decode('utf-8')):
        return render(request, 'err.html', VeriCodeError)     # 校验失败则返回错误视图
    try:
        user = User.objects.get(email=email)             # 获取用户
    except User.DoesNotExist:
        user = None
    if user is not None:                                 # 激活用户
        user.is_active = True
        user.is_staff = True
        user.save()
        login(request, user)                             # 登录用户
        profile, created = Profile.objects.select_for_update().get_or_create(  # 配置用户信息
            name=user.username,
            email=user.email,
        )
        profile.user_src = Profile.NORMAL_USER           # 配置用户为普通登录用户
```

```python
        profile.save()

        request.session['uid'] = profile.uid                    # 配置 session
        request.session['username'] = profile.name
        return render(request, 'web/index.html', {              # 渲染视图，并返回已登录信息
            'user_info': profile.data,
            'has_login': True,
            'msg': "激活成功",
        })
    else:
        return render(request, 'err.html', VerifyFailed)        # 校验失败则返回错误视图
```

前端单击"注册"按钮时链接的 Ajax 请求如下：

```javascript
$('#signUpPost').click(function () {                            // 单击"注册"按钮
    // 获取表单数据
    var email = $('#signUpId').val();
    var password = $('#signUpPassword').val();
    var passwordAgain = $('#signUpPasswordAgain').val();
    var vcode = $('#signUpVcode').val();
    // 验证邮箱
    if(!checkEmail(email)) {
        $('#signUpId').val('');
        $('#signUpId').attr('placeholder', '邮箱格式错误');
        $('#signUpId').css('border', '1px solid red');
        return false;
    }else{
        $('#signUpId').css('border', '1px solid #C1FFC1');}
    // 验证2次密码是否一致
    if(!(password === passwordAgain)) {
        $('#signUpPasswordAgain').val('');
        $('#signUpPasswordAgain').attr('placeholder', '两次密码输入不一致');
        $('#signUpPassword').css('border', '1px solid red');
        $('#signUpPasswordAgain').css('border', '1px solid red');
        return false;
    }else{
        $('#signUpPassword').css('border', '1px solid #C1FFC1');
        $('#signUpPasswordAgain').css('border', '1px solid #C1FFC1');}
    // Ajax 异步请求
    $.ajax({
        url: '/api/signup',                                     // 请求 URL
        type: 'post',                                           // 请求方式
        data: {                                                 // 请求数据
            'email': email,
            'password': password,
            'password_again': passwordAgain,
            'sign': loginSign,
            'vcode': vcode},
        dataType: 'json',                                       // 返回数据类型
        success: function (res) {                               // 回调函数
            if(res.status === 200) {                            // 注册成功
                sign = res.data.sign;
                email = res.data.email;
                // 拼接验证邮箱 URL
                window.location.href = '/auth/signup_redirect?email=' + email +
                '&sign=' + sign;
            }else if(res.status === 300002) {
                alert('两次输入密码不一致');
            }else if(res.status === 300003) {
                alert('验证码错误');
            }else if(res.status === 300004) {
                alert('用户名已存在');
```

```
            }
        }
    })
});
```

发送邮件的 Ajax 请求代码如下：

```
$('#sendMail').click(function () {                              // 单击"发送邮件"
    $('#sendMailLoading').modal('show');                        // 显示弹窗
    // Ajax 异步请求
    $.ajax({
        url: '/api/sendmail',                                   // 请求 URL
        type: 'get',                                            // 请求方式
        data: {                                                 // 请求数据
            'email': '{{ email|safe }}',
            'sign': '{{ sign|safe }}'
        },
        dataType: 'json',                                       // 返回数据类型
        success: function (res) {                               // 回调函数
            if(res.status === 200) {                            // 请求成功
                $('#sendMailLoading').modal('hide');
                alert('发送成功，快去登录邮箱激活账户吧');
            }
            else if(res.status === 300005) {
                $('#sendMailLoading').modal('hide');
                alert('您已经发送过邮件，请稍等再试');
            }
            else if(res.status === 300006) {
                $('#sendMailLoading').modal('hide');
                alert('验证邮件发送失败!');
            }
        }
    })
});
```

> **说明**
> 修改密码和重置密码的实现方式与用户注册的实现方式类似，这里不再赘述。

用户注册页面效果如图 9.6 所示。

图 9.6 用户注册页面

9.5.3 机构注册功能的实现

在智慧校园考试系统中还提供了机构注册的功能，当单击"成为机构"导航按钮时，需要根据用户的 uid 来判断用户是否已经注册过机构账户。如果没有注册过，渲染一个表单，该表单使用 Ajax 来异步请求；如果已经注册过，则返回一个信息提示，引导用户重定向到出题页面。下面讲解机构注册功能的实现过程。

在 config/urls.py 中添加机构 app 的路由，代码如下：

```
path('biz/', include(('business.urls','business'), namespace='biz')),      # 机构
```

在 business app 的 urls.py 中添加渲染机构页面的路由，代码如下：

```
path('^$', biz_render.home, name='index'),
```

上面的代码中用到了页面渲染视图函数，函数名称为 index()，其具体实现代码如下：

```python
def home(request):
    uid = request.GET.get('uid', '')                              # 获取 uid
    try:
        profile = Profile.objects.get(uid=uid)                    # 根据 uid 获取用户信息
    except Profile.DoesNotExist:
        profile = None                                            # 未获取到用户信息则 profile 变量置空
    types = dict(BusinessAccountInfo.TYPE_CHOICES)                # 所有的机构类型
    # 渲染视图，返回机构类型和是否存在该账户绑定过的机构账户
    return render(request, 'bussiness/index.html', {
        'types': types,
        'is_company_user': bool(profile) and (profile.user_src == Profile.COMPANY_USER)
    })
```

在 web/business/index.html 页面中添加一个 Bootstrap 框架的 panel 控件，用来存放机构注册表单，代码如下：

```html
<div class="panel panel-info">
    <div class="panel-heading"><h3 class="panel-title">注册成为机构</h3></div>
    <div class="panel-body">
        <form id="bizRegistry" class="form-group">
            <label for="bizEmail">邮箱</label>
            <input type="text" class="form-control" id="bizEmail"
                placeholder="填写机构邮箱" />
            <label for="bizCompanyName">名称</label>
            <input type="text" class="form-control" id="bizCompanyName"
                placeholder="填写机构名称" />
            <label for="bizCompanyType">类型</label>
            <select id="bizCompanyType" class="form-control">
                {% for k, v in types.items %}
                    <option value="{{ k }}">{{ v }}</option>
                {% endfor %}
            </select>
            <label for="bizUsername">联系人</label>
            <input type="text" class="form-control" id="bizUsername"
                placeholder="填写机构联系人" />
            <label for="bizPhone">手机号</label>
            <input type="text" class="form-control" id="bizPhone"
                placeholder="填写联系人手机" />
            <input type="submit" id="bizSubmit" class="btn btn-primary"
                value="注册机构" style="float: right;margin-top: 20px" />
        </form>
    </div>
</div>
```

在JavaScript脚本中添加申请成为机构的Ajax请求方法，代码如下：

```javascript
$('#bizSubmit').click(function () {                                    // 单击"注册机构"
    // 获取表单信息
    var email = $('#bizEmail').val();
    var name = $('#bizCompanyName').val();
    var type = $('#bizCompanyType').val();
    var username = $('#bizUsername').val();
    var phone = $('#bizPhone').val();
    // 正则表达式验证邮箱
    if(!email.match('^\\w+([-+.]\\w+)*@\\w+([-.]\\w+)*\\.\\w+([-.]\\w+)*$')) {
        $('#bizEmail').val('');
        $('#bizEmail').attr('placeholder', '邮箱格式错误');
        $('#bizEmail').css('border', '1px solid red');
        return false;
    }else{
        $('#bizEmail').css('border', '1px solid #C1FFC1');
    }
    // 正则表达式验证机构名称
    if(!(name.match('^[a-zA-Z0-9_\\u4e00-\\u9fa5]{4,20}$'))) {
        $('#bizCompanyName').val('');
        $('#bizCompanyName').attr('placeholder', '请填写4-20个中文、字母、数字或者下画线组成的机构名称');
        $('#bizCompanyName').css('border', '1px solid red');
        return false;
    }else{
        $('#bizCompanyName').css('border', '1px solid #C1FFC1');
    }
    // 正则表达式验证用户名
    if(!(username.match('^[\u4E00-\u9FA5A-Za-z]+$'))){
        $('#bizUsername').val('');
        $('#bizUsername').attr('placeholder', '联系人姓名应该为汉字或大小写字母');
        $('#bizUsername').css('border', '1px solid red');
        return false;
    }else{
        $('#bizUsername').css('border', '1px solid #C1FFC1');
    }
    // 正则表达式验证手机
    if(!(phone.match('^1[3|4|5|8][0-9]\\d{4,8}$'))){
        $('#bizPhone').val('');
        $('#bizPhone').attr('placeholder', '手机号不符合规则');
        $('#bizPhone').css('border', '1px solid red');
        return false;
    }else{
        $('#bizPhone').css('border', '1px solid #C1FFC1');
    }
    // Ajax 异步请求
    $.ajax({
        url: '/api/checkbiz',                                          // 请求URL
        type: 'get',                                                   // 请求方式
        data: {                                                        // 请求数据
            'email': email
        },
        dataType: 'json',                                              // 返回数据类型
        success: function (res) {                                      // 回调函数
            if(res.status === 200) {                                   // 注册成功
                if(res.data.bizaccountexists) {
                    alert('您的账户已存在，请直接登录');
                    window.location.href = '/';
                }
                else if(res.data.userexists && !res.data.bizaccountexists) {
                    if(confirm('您的邮箱已被注册为普通用户，我们将会为您绑定该用户。')){
                        bizPost(email, name, type, username, phone, 1);
```

```javascript
                    window.location.href = '/biz/notify?email=' + email + '&bind=1';
                }else {
                    window.location.href = '/{% if request.session.uid %}?
                        uid={{ request.session.uid }}{% else %}{% endif %}';
                }
            }
            else{
                bizPost(email, name, type, username, phone, 2);
                window.location.href = '/biz/notify?email=' + email;
            }
        }
    }
});
// 验证邮箱方法
function bizPost(email, name, type, username, phone, flag) {
    // Ajax 异步请求
    $.ajax({
        url: '/api/regbiz',                                    // 请求 URL
        data: {                                                // 请求数据
            'email': email,
            'name': name,
            'type': type,
            'username': username,
            'phone': phone,
            'flag': flag
        },
        type: 'post',                                          // 请求类型
        dataType: 'json'                                       // 返回数据类型
    })
}
});
```

单击"注册"按钮时，首先验证表单是否符合正则表达式，当这些验证都通过时，先请求一个/api/check_biz 接口，其对应的路由和接口函数如下：

```python
def check_biz(request):
    email = request.GET.get('email', '')                       # 获取邮箱
    try:                                                       # 检查数据库中是否有该邮箱注册过的数据
        biz = BusinessAccountInfo.objects.get(email=email)
    except BusinessAccountInfo.DoesNotExist:
        biz = None
    return json_response(200, 'OK', {                          # 返回是否已经被注册过和是否已经有此用户
        'userexists': User.objects.filter(email=email).exists(),
        'bizaccountexists': bool(biz)
    })
```

上面的接口用来检查用户输入的邮箱是否存在对应的普通用户账户和机构账户。如果普通用户账户存在，但是机构账户不存在，那么会提示用户绑定已有账户，注册成为机构账户；如果普通用户账户不存在，并且机构账户也不存在，则会为该邮箱创建一个未激活的普通用户账户和一个机构账户。这时，要注册的用户必须去自己的邮箱里面验证并激活，这需要将表单信息提交到/api/regbiz 接口。因此，需要在 api 模块的 urls.py 中添加如下路由：

```python
# bussiness
urlpatterns += [
    path('regbiz', biz_views.registry_biz, name='registry biz'),
    path('checkbiz', biz_views.check_biz, name='check_biz'),
]
```

然后，在 business 的 biz_views.py 中添加如下函数进行注册：

```python
@csrf_exempt
```

```python
@transaction.atomic
def registry_biz(request):
    email = request.POST.get('email', '')                              # 获取填写的邮箱
    name = request.POST.get('name', '')                                # 获取填写的机构名
    username = request.POST.get('username', '')                        # 获取填写的机构联系人
    phone = request.POST.get('phone', '')                              # 获取填写的手机号
    ctype = request.POST.get('type', BusinessAccountInfo.INTERNET)     # 获取机构类型
    # 获取一个标记位，代表用户是创建新用户还是使用绑定老用户的方式
    flag = int(request.POST.get('flag', 2))
    uname = email.split('@')[0]                                        # 创建一个账户名
    if not User.objects.filter(username__exact=name).exists():
        final_name = username
    elif not User.objects.filter(username__exact=uname).exists():
        final_name = uname
    else:
        final_name = email
    if flag == 2:                                                      # 如果标记位是2，那么将为其创建新用户
        user = User.objects.create_user(
            username=final_name,
            email=email,
            password=settings.INIT_PASSWORD,
            is_active=False,
            is_staff=False
        )
    if flag == 1:                                                      # 如果标记位是1，那么为其绑定老用户
        try:
            user = User.objects.get(email=email)
        except User.DoesNotExist:
            return json_response(*UserError.UserNotFound)
    pvalues = {
        'phone': phone,
        'name': final_name,
        'user_src': Profile.COMPANY_USER,
    }
    # 获取或创建用户信息
    profile, _ = Profile.objects.select_for_update().get_or_create(email=email)
    for k, v in pvalues.items():
        setattr(profile, k, v)
    profile.save()
    bizvalues = {
        'company_name': name,
        'company_username': username,
        'company_phone': phone,
        'company_type': ctype,
    }
    # 获取或创建机构账户信息
    biz, _ = BusinessAccountInfo.objects.select_for_update().get_or_create(
        email=email,
        defaults=bizvalues
    )
    return json_response(200, 'OK', {              # 响应JSON格式数据，这个标记位在发送验证邮件时还有用
        'name': final_name,
        'email': email,
        'flag': flag
    })
```

表单提交后，如果是新创建的用户，验证用户的邮件，该步骤和之前的注册普通用户步骤类似。整个过程完成，如果普通用户注册成为机构用户，那么他就可以在快速出题的导航页中录入题库，并且配置考试了。机构注册页面效果如图 9.7 所示。

图 9.7 机构注册页面效果

9.6 核心答题功能设计

9.6.1 答题首页设计

答题首页运行效果如图 9.8 所示。该页主要呈现考试的分类，我们将所有考试划分为 6 个类别和 1 个热门考试，对应的参数及说明如下。

- ☑ hot：代表所有热门考试前 10 位。
- ☑ tech：代表技术类热门考试前 10 位。
- ☑ edu：代表教育类考试前 10 位。
- ☑ culture：代表文化类考试前 10 位。
- ☑ sport：代表体育类考试前 10 位。
- ☑ general：代表常识类考试前 10 位。
- ☑ interview：代表面试题考试前 10 位。

图 9.8 答题首页

在答题首页单击某一个类别后,将进入该类别下的考试列表,其对应的路由如下:

re_path('games/s/(\w+)', cop_render.games, name='query_games'),

这里使用 re_path()函数来进行正则匹配,比如选择单击"热门考试",则进入如下 URL:

/bs/games/s/hot

在 re_path()函数中进行正则匹配时用到了 games()函数,该函数可以根据 URL 中最后一个参数的值来判断用户选择的是哪一类考试,从而获取对应分类的数据信息。关键代码如下:

```python
def games(request, s):
    """
    获取所有考试接口
    :param request: 请求对象
    :param s: 请求关键字
    :return: 返回该请求关键字对应的所有考试类别
    """
    if s == 'hot':
        # 筛选条件:完成时间大于当前时间;根据参与人数降序排序;根据创建时间降序排序;筛选10个
        kinds = CompetitionKindInfo.objects.filter(
            cop_finishat__gt=datetime.datetime.now(tz=datetime.timezone.utc),
        ).order_by('-total_partin_num').order_by('-created_at')[:10]
    elif s == 'tech':                                       # 获取所有技术类考试
        kinds = CompetitionKindInfo.objects.filter(
            kind_type=CompetitionKindInfo.IT_ISSUE,
            cop_finishat__gt=datetime.datetime.now(tz=datetime.timezone.utc)
        ).order_by('-total_partin_num').order_by('-created_at')
    elif s == 'edu':                                        # 获取所有教育类考试
        kinds = CompetitionKindInfo.objects.filter(
            kind_type=CompetitionKindInfo.EDUCATION,
            cop_finishat__gt=datetime.datetime.now(tz=datetime.timezone.utc)
        ).order_by('-total_partin_num').order_by('-created_at')
    elif s == 'culture':                                    # 获取所有文化类考试
        kinds = CompetitionKindInfo.objects.filter(
            kind_type=CompetitionKindInfo.CULTURE,
            cop_finishat__gt=datetime.datetime.now(tz=datetime.timezone.utc)
        ).order_by('-total_partin_num').order_by('-created_at')
    elif s == 'sport':                                      # 获取所有体育类考试
        kinds = CompetitionKindInfo.objects.filter(
            kind_type=CompetitionKindInfo.SPORT,
            cop_finishat__gt=datetime.datetime.now(tz=datetime.timezone.utc)
        ).order_by('-total_partin_num').order_by('-created_at')
    elif s == 'general':                                    # 获取所有常识类考试
        kinds = CompetitionKindInfo.objects.filter(
            kind_type=CompetitionKindInfo.GENERAL,
            cop_finishat__gt=datetime.datetime.now(tz=datetime.timezone.utc)
        ).order_by('-total_partin_num').order_by('-created_at')
    elif s == 'interview':                                  # 获取所有面试题考试
        kinds = CompetitionKindInfo.objects.filter(
            kind_type=CompetitionKindInfo.INTERVIEW,
            cop_finishat__gt=datetime.datetime.now(tz=datetime.timezone.utc)
        ).order_by('-total_partin_num').order_by('-created_at')
    else:
        kinds = None
    return render(request, 'competition/games.html', {
```

```
        'kinds': kinds,
    })
```

考试列表页面的运行结果如图 9.9 所示。

图 9.9 考试列表页面

9.6.2 考试详情页面

考试详情页面主要用来展示考试相关的信息，包括考试名称、出题机构、考试题目数量和题库大小等，其效果如图 9.10 所示。

图 9.10 考试详情页面

在 competition app 下添加一个 cop_render.py 文件，用来存放考试详情页面的视图渲染函数，代码如下：

```
def home(request):
    """
    考试详情页面视图
    :param request: 请求对象
    :return: 渲染视图: user_info: 用户信息; kind_info: 考试信息;
        is_show_userinfo: 是否展示用户信息表单;user_info_has_entered: 是否已经录入表单;
    userinfo_fields: 表单字段;option_fields: 表单字段中呈现为下拉框的字段;
    """
    uid = request.GET.get('uid', '')                        # 获取 uid
    kind_id = request.GET.get('kind_id', '')                # 获取 kind_id
    created = request.GET.get('created', '0')               # 获取标志位，以后会用到
```

```python
    try:
        kind_info = CompetitionKindInfo.objects.get(kind_id=kind_id)    # 获取考试数据
    except CompetitionKindInfo.DoesNotExist:
        return render(request, 'err.html', CompetitionNotFound)          # 不存在时渲染错误视图
    try:
        bank_info = BankInfo.objects.get(bank_id=kind_info.bank_id)      # 获取题库数据
    except BankInfo.DoesNotExist:
        return render(request, 'err.html', BankInfoNotFound)             # 不存在时渲染错误视图
    try:
        profile = Profile.objects.get(uid=uid)                           # 获取用户数据
    except Profile.DoesNotExist:
        return render(request, 'err.html', ProfileNotFound)              # 不存在时渲染错误视图
    if kind_info.question_num > bank_info.total_question_num:            # 考试出题数量是否小于题库大小
        return render(request, 'err.html', QuestionNotSufficient)
    show_info = get_pageconfig(kind_info.app_id).get('show_info', {})    # 从 Redis 获取页面配置信息
    # 页面配置信息, 用来控制答题前是否展示一张表单
    is_show_userinfo = show_info.get('is_show_userinfo', False)
    form_fields = collections.OrderedDict()                              # 生成一个有序的用来保存表单字段的字典
    form_regexes = []                                                    # 生成一个空的正则表达式列表
    if is_show_userinfo:
        # 从页面配置中获取 userinfo_fields
        userinfo_fields = show_info.get('userinfo_fields', '').split('#')
        for i in userinfo_fields:                                        # 将页面配置的每个正则表达式取出来放入正则表达式列表
            form_regexes.append(get_form_regex(i))
        userinfo_field_names = show_info.get('userinfo_field_names', '').split('#')
        for i in range(len(userinfo_fields)):                            # 将每个表单字段信息保存到有序的表单字段字典中
            form_fields.update({userinfo_fields[i]: userinfo_field_names[i]})
    return render(request, 'competition/index.html', {                   # 渲染页面
        'user_info': profile.data,
        'kind_info': kind_info.data,
        'bank_info': bank_info.data,
        'is_show_userinfo': 'true' if is_show_userinfo else 'false',
        'userinfo_has_enterd': 'true' if get_enter_userinfo(kind_id, uid) else 'false',
        'userinfo_fields': json.dumps(form_fields) if form_fields else '{}',
        'option_fields': json.dumps(show_info.get('option_fields', '')),
        'field_regexes': form_regexes,
        'created': created
    })
```

考试详情页面中除了返回考试的信息,还需要返回页面的配置信息。本项目中,在 business app 的数据模型中创建一个 AppConfigInfo,使其关联每个 BusinessAppInfo 的 app_id,这主要用来指定每个 AppInfo 在页面中的不同配置,以便于让整个页面多样化、可定制化。这里指定了一个配置,即:如果机构用户开启了此功能,则每个答题用户需要在参与考试之前填写一个表单,如图 9.11 所示。

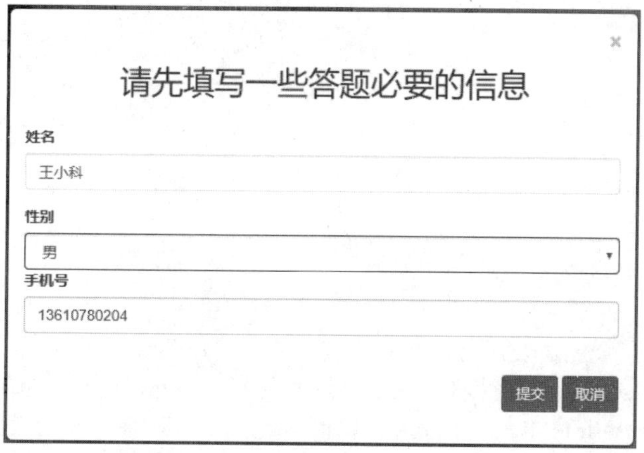

图 9.11 答题之前需要填写的表单

图 9.11 所示的表单用于收集答题用户的信息，以便日后可以联系该用户。在 business.models 模块中，添加一个名称为 AppConfigInfo 的模型类，代码如下：

```python
class AppConfigInfo(CreateUpdateMixin):
    """ 应用配置信息类 """

    app_id = models.CharField(_(u'应用id'), max_length=32, help_text=u'应用唯一标识',
                              db_index=True)
    app_name = models.CharField(_(u'应用名'), max_length=40, blank=True, null=True,
                                help_text=u'应用名')
    # 文案配置
    rule_text = models.TextField(_(u'考试规则'), max_length=255, blank=True, null=True,
                                 help_text=u'考试规则')

    # 显示信息
    is_show_userinfo = models.BooleanField(_(u'展示用户表单'), default=False,
                                           help_text=u'是否展示用户信息表单')
    userinfo_fields = models.CharField(_(u'用户表单字段'), max_length=128, blank=True, null=True,
                                       help_text=u'需要用户填写的字段#隔开')
    userinfo_field_names = models.CharField(_('用户表单label'), max_length=128, blank=True,
                                            null=True, help_text=u'用户需要填写的表单字段label名称')
    option_fields = models.CharField(_(u'下拉框字段'), max_length=128, blank=True, null=True,
                                     help_text=u'下拉框字段选项配置，#号隔开，每个字段由:h和","号'
                                               '组成。如 option1:吃饭，喝水，睡觉#option2:上班，学习，看电影')

    class Meta:
        verbose_name = _(u'应用配置信息')
        verbose_name_plural = _(u'应用配置信息')

    def __unicode__(self):
        return str(self.pk)

    # 页面配置数据
    @property
    def show_info(self):
        return {
            'is_show_userinfo': self.is_show_userinfo,
            'userinfo_fields': self.userinfo_fields,
            'userinfo_field_names': self.userinfo_field_names,
            'option_fields': self.option_fields,
        }

    @property
    def text_info(self):
        return {
            'rule_text': self.rule_text,
        }

    @property
    def data(self):
        return {
            'show_info': self.show_info,
            'text_info': self.text_info,
            'app_id': self.app_id,
            'app_name': self.app_name,
        }
```

上面的模型类中指定了页面需要进行的一些配置，本项目中将页面中的用户信息设计成了一个动态的表单，可以通过 is_show_userinfo 字段来控制它的显示和隐藏。另外，用户信息表单中要输入的字段也设计成了动态的，这可以通过 userinfo_fields 字段进行设置，userinfo_fields 字段的设置格式如下：

```
name#sex#age#phone            # 以#隔开的一个纯文本值，每一段的值代表了表单中的一个字段
```

9.6.3 答题功能的实现

当单击"开始挑战"按钮时,代表已经确认过考试信息,可以开始答题了,答题页面效果如图 9.12 所示。

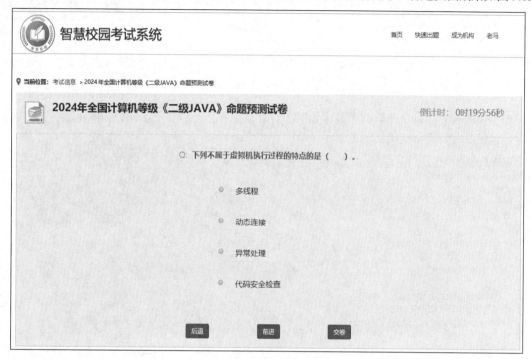

图 9.12 答题页面效果

在 competition /urls.py 中添加答题功能的 URL 路由,代码如下:

```
path('game', cop_render.game, name='game'),
```

上面路由中用到了 game()视图函数,该函数位于 competition app 的 cop_render.py 文件中,用来获取考试、题库和用户相关的信息,其详细代码如下:

```python
@check_login
@check_copstatus
def game(request):
    """
    返回考试题目信息的视图
    :param request: 请求对象
    :return: 渲染视图: user_info: 用户信息;kind_id: 考试唯一标识;
        kind_name: 考试名称;cop_finishat: 考试结束时间;rule_text: 大赛规则;
    """
    uid = request.GET.get('uid', '')                              # 获取 uid
    kind_id = request.GET.get('kind_id', '')                      # 获取 kind_id
    try:                                                          # 获取考试信息
        kind_info = CompetitionKindInfo.objects.get(kind_id=kind_id)
    except CompetitionKindInfo.DoesNotExist:                      # 未获取到,渲染错误视图
        return render(request, 'err.html', CompetitionNotFound)
    try:                                                          # 获取题库信息
        bank_info = BankInfo.objects.get(bank_id=kind_info.bank_id)
    except BankInfo.DoesNotExist:                                 # 未获取到,渲染错误视图
        return render(request, 'err.html', BankInfoNotFound)
    try:                                                          # 获取用户信息
        profile = Profile.objects.get(uid=uid)
```

```python
    except Profile.DoesNotExist:
        return render(request, 'err.html', ProfileNotFound)      # 未获取到，渲染错误视图
    if kind_info.question_num > bank_info.total_question_num:    # 检查题库大小
        return render(request, 'err.html', QuestionNotSufficient)
    pageconfig = get_pageconfig(kind_info.app_id)                # 获取页面配置信息
    return render(request, 'competition/game.html', {            # 渲染视图信息
        'user_info': profile.data,
        'kind_id': kind_info.kind_id,
        'kind_name': kind_info.kind_name,
        'cop_finishat': kind_info.cop_finishat,
        'period_time': kind_info.period_time,
        'rule_text': pageconfig.get('text_info', {}).get('rule_text', '')
    })
```

当答题页面加载时，只是获取到了基本数据。对于题目信息，需要在 game.html 页面中使用 Ajax 异步请求的方式进行获取，关键代码如下：

```javascript
var currentPage = 1;
var hasPrevious = false;
var hasNext = false;
var questionNum = 0;
var response;
var answerDict;
$(document).ready(function () {
    if({{ period_time|safe }}) {                                 # 开始计时
        startTimer1();
    }
    $('#loadingModal').modal('show');                            # 显示弹窗
    uid = '{{ user_info.uid|safe }}';                            # 获取用户 id
    kind_id = '{{ kind_id|safe }}';                              # 获取类型 id
    # Ajax 异步请求
    $.ajax({
        url: '/api/questions',                                   # 请求 URL
        type: 'get',                                             # 请求类型
        data: {                                                  # 请求数据
            'uid': uid,
            'kind_id': kind_id
        },
        dataType: 'json',                                        # 返回数据类型
        success: function (res) {                                # 回调函数
            response = res;                                      # 接收返回数据
            questionNum = res.data.kind_info.question_num;       # 获取题号
            answerDict = new Array(questionNum);                 # 获取问题数组
            # 遍历问题数组
            for(var i=0; i < questionNum; i++){
                if(response.data.questions[i].qtype === 'choice') {
                    answerDict['c_' + response.data.questions[i].pk] = '';
                }else{
                    answerDict['f_' + response.data.questions[i].pk] = '';
                }
            }
            # 选择题
            if(res.data.questions[0].qtype === 'choice') {
                $('#question').html(res.data.questions[0].question);   // currentPage - 1
                $('#item1').html(res.data.questions[0].items[0]);
                $('#item2').html(res.data.questions[0].items[1]);
                $('#item3').html(res.data.questions[0].items[2]);
                $('#item4').html(res.data.questions[0].items[3]);
                $('#itemPk').html('c_' + res.data.questions[0].pk);
                hasNext = (currentPage < questionNum);
                $('#fullinBox').hide();
            } else{
```

```javascript
                    # 填空题
                    $('#question').html(res.data.questions[0].question.replace('##', '_____'));
                    $('#answerPk').val('f_' + res.data.questions[0].pk);
                    hasNext = (currentPage < questionNum);
                    $('#choiceBox').hide();
                }
                $('#loadingModal').modal('hide');                    # 隐藏弹窗
            }
        });
```

由于需要从题库中随机抽取指定数目的题目，所以在 competition app 的 game_views.py 接口视图中添加一个 get_questions()视图函数，主要用于生成考试数据，考试数据是从题库中随机抽取的指定数目的题目。另外，需要注意答题是有限制时间的，因此需要设置开始时间戳。get_questions()视图函数代码如下：

```python
@check_login
@check_copstatus
@transaction.atomic
def get_questions(request):
    """
    获取题目信息接口
    :param request: 请求对象
    :return: 返回 json 数据: user_info: 用户信息;kind_info: 考试信息;qa_id: 考试答题记录;questions: 随机的考试题目;
    """
    kind_id = request.GET.get('kind_id', '')                         # 获取 kind_id
    uid = request.GET.get('uid', '')                                 # 获取 uid
    try:                                                             # 获取考试信息
        kind_info = CompetitionKindInfo.objects.select_for_update().get(kind_id=kind_id)
    except CompetitionKindInfo.DoesNotExist:                         # 未获取到，返回错误码 100001
        return json_response(*CompetitionError.CompetitionNotFound)
    try:                                                             # 获取题库信息
        bank_info = BankInfo.objects.get(bank_id=kind_info.bank_id)
    except BankInfo.DoesNotExist:                                    # 未获取到，返回错误码 100004
        return json_response(*CompetitionError.BankInfoNotFound)
    try:                                                             # 获取用户信息
        profile = Profile.objects.get(uid=uid)
    except Profile.DoesNotExist:                                     # 未获取到，返回错误码 200001
        return json_response(*ProfileError.ProfileNotFound)
    qc = ChoiceInfo.objects.filter(bank_id=kind_info.bank_id)        # 选择题
    qf = FillInBlankInfo.objects.filter(bank_id=kind_info.bank_id)   # 填空题
    questions = []                                                   # 将两种题型放到同一个列表中
    for i in qc.iterator():
        questions.append(i.data)
    for i in qf.iterator():
        questions.append(i.data)
    question_num = kind_info.question_num                            # 出题数
    q_count = bank_info.total_question_num                           # 总题数
    if q_count < question_num:                                       # 出题数大于总题数，返回错误码 100005
        return json_response(CompetitionError.QuestionNotSufficient)
    qs = random.sample(questions, question_num)                      # 随机分配题目
    qa_info = CompetitionQAInfo.objects.select_for_update().create(  # 创建答题 log 数据
        kind_id=kind_id,
        uid=uid,
        qsrecord=[q['question'] for q in qs],
        asrecord=[q['answer'] for q in qs],
        total_num=question_num,
        started_stamp=tc.utc_timestamp(ms=True, milli=True),         # 设置开始时间戳
        started=True
    )
    for i in qs:                                                     # 剔除答案信息
        i.pop('answer')
    return json_response(200, 'OK', {                                # 返回 JSON 数据，包括题目信息、答题 log 信息等
        'kind_info': kind_info.data,
```

```
            'user_info': profile.data,
            'qa_id': qa_info.qa_id,
            'questions': qs
        })
```

上面的 api 视图需要在 api 模块下的 urls.py 中配置路由,代码如下:

```
url(r'^questions$', game_views.get_questions, name='get_questions'),
```

9.6.4 提交答案与显示成绩排行榜

当用户完成答题后,需要判断答题剩余时间。如果剩余时间为 0,或者已经超时,则把答题的日志保存为超时,并且答题成绩不加入排行榜;如果剩余时间还很充足,用户的成绩要加入排行榜,并且将答题日志标记为已完成,用来区别未完成的答题记录。提交答案与显示成绩单页面效果如图 9.13 所示。

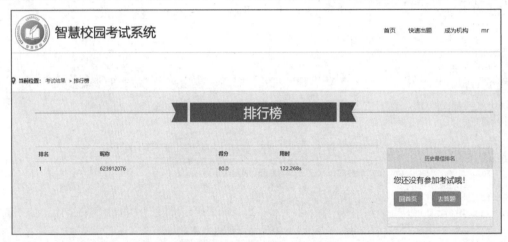

图 9.13 提交答案与显示成绩单页面

在答题过程中,前端需要记录用户的答题数据和顺序,并生成一个指定的数据形式,以便提交到后台进行答案的匹配,game.html 页面中提交答案的实现代码如下:

```
$('#answerSubmit').click(function () {                      # 单击"提交答案"按钮
    if(window.confirm("确认提交答案吗?")) {                    # 弹出确认框
        if({{ period_time|safe }}) {                        # 正常结束
            stopTimer1();                                   # 停止计时
        }
        var answer = "";
        # 组织答案
        for (var key in answerDict) {
            if (!answer) {
                answer = String(key) + "," + answerDict[key] + "#";
            }else{
                answer += String(key) + "," + answerDict[key] + "#";
            }
        }
        # Ajax 异步请求
        $.ajax({
            url: '/api/answer',                             # 请求 URL
            type: 'post',                                   # 请求类型
            data: {                                         # 请求数据
                'qa_id': response.data.qa_id,
                'uid': response.data.user_info.uid,
                'kind_id': kind_id,
```

```javascript
                        'answer': answer
                    },
                    dataType: 'json',                                       # 返回数据类型
                    success: function (res) {                               # 回调函数
                        if(res.status === 200) {                            # 请求成功，页面跳转
                            window.location.href = "/bs/result?uid=" + res.data.user_info.uid +
                                "&kind_id=" + res.data.kind_id + "&qa_id=" + res.data.qa_id;
                        }else{
                            alert('提交失败');
                        }
                    }
                })
            }else {}
        })
    });
```

上面代码中的/api/answer 接口对应的路由需要在 api 模块的 urls.py 中设置，代码如下：

```
url(r'^answer$', game_views.submit_answer, name='submit_answer'),
```

上面路由中用到了 submit_answer()视图函数，该视图函数主要用来获取提交的答题信息，并进行答案核对、保存日志数据、记录结束时间戳等操作，关键代码如下：

```python
@csrf_exempt
@check_login
@check_copstatus
@transaction.atomic
def submit_answer(request):
    """
    提交答案接口
    :param request: 请求对象
    :return: 返回 json 数据: user_info: 用户信息; qa_id: 考试答题记录标识; kind_id: 考试唯一标识
    """
    stop_stamp = tc.utc_timestamp(ms=True, milli=True)                  # 结束时间戳
    qa_id = request.POST.get('qa_id', '')                               # 获取 qa_id
    uid = request.POST.get('uid', '')                                   # 获取 uid
    kind_id = request.POST.get('kind_id', '')                           # 获取 kind_id
    answer = request.POST.get('answer', '')                             # 获取 answer
    try:                                                                # 获取考试信息
        kind_info = CompetitionKindInfo.objects.get(kind_id=kind_id)
    except CompetitionKindInfo.DoesNotExist:                            # 未获取到，返回错误码 100001
        return json_response(*CompetitionError.CompetitionNotFound)
    try:                                                                # 获取题库信息
        bank_info = BankInfo.objects.get(bank_id=kind_info.bank_id)
    except BankInfo.DoesNotExist:                                       # 未获取到，返回错误码 100004
        return json_response(*CompetitionError.BankInfoNotFound)
    try:                                                                # 获取用户信息
        profile = Profile.objects.get(uid=uid)
    except Profile.DoesNotExist:                                        # 未获取到，返回错误码 200001
        return json_response(*ProfileError.ProfileNotFound)
    try:                                                                # 获取答题 log 信息
        qa_info = CompetitionQAInfo.objects.select_for_update().get(qa_id=qa_id)
    except CompetitionQAInfo.DoesNotExist:                              # 未获取到，返回错误码 100006
        return json_response(*CompetitionError.QuestionNotFound)
    answer = answer.rstrip('#').split('#')                              # 处理答案数据
    total, correct, wrong, wrong_list, correct_list = check_correct_num(answer)  # 检查答题情况
    qa_info.aslogrecord = answer
    qa_info.finished_stamp = stop_stamp
    qa_info.expend_time = stop_stamp - qa_info.started_stamp
    qa_info.finished = True
    qa_info.correct_num = correct if total == qa_info.total_num else 0
```

```python
    qa_info.incorrect_num = wrong if total == qa_info.total_num else qa_info.total_num
    qa_info.save()                                                          # 保存答题 log
    if qa_info.correct_num == kind_info.question_num:                       # 得分处理
        score = kind_info.total_score
    elif not qa_info.correct_num:
        score = 0
    else:
        score = round((kind_info.total_score / kind_info.question_num) * correct, 3)
    qa_info.score = score                                                   # 继续保存答题 log
    qa_info.save()
    kind_info.total_partin_num += 1                                         # 保存考试数据
    kind_info.save()                                                        # 考试答题次数
    bank_info.partin_num += 1
    bank_info.save()                                                        # 题库答题次数
    if (kind_info.period_time > 0) and (qa_info.expend_time > kind_info.period_time * 60 * 1000):  # 超时，不加入排行榜
        qa_info.status = CompetitionQAInfo.OVERTIME
        qa_info.save()
    else:                                                                   # 正常完成，加入排行榜
        add_to_rank(uid, kind_id, qa_info.score, qa_info.expend_time)
        qa_info.status = CompetitionQAInfo.COMPLETED
        qa_info.save()
    return json_response(200, 'OK', {                                       # 返回 JSON 数据
        'qa_id': qa_id,
        'user_info': profile.data,
        'kind_id': kind_id,
    })
```

9.7 批量录入题库功能设计

在导航栏上单击"快速出题"超链接，在进入的页面中，单击"开始录入"按钮将进入批量录入题库页面，如图 9.14 所示。

图 9.14 批量录入题库页面

录入题库功能的实现方法是，在页面中为用户提供一个 Excel 模板，用户按照对应的模板格式来编写题

库信息。编写完成后，在页面中选择带有题库的 Excel 文件，单击“开始录入”按钮，进行题库的录入。题库 Excel 模板如图 9.15 所示。

1	问题	答案	选项一	选项二	选项三	选项四	问题图片链接	问题音频链接	问题来源
2	商女不知亡国恨下一句是？	隔江犹唱后庭花	隔江犹闻笙竹宴	铁马冰河入梦来	一弦一柱思华年	江枫渔火对愁眠			
3	补全诗句：到现在，乡愁是一湾##，我在这头，大陆在那头。	浅浅的海峡							
4	君住长江南，我住长江北，##，共饮长江水	日日思君不见君							
5	龙舟是为了纪念哪位文化名人的？	屈原	王之涣	李白	纳兰性德	屈原			
6	元宵又称为上元节，那么下元节是几月几号？	十月十五	六月十五	十二月十五	十月十五	七月十五			
7	北京是在哪个朝代建都的？	元朝	清朝	明朝	汉朝	元朝			
8	李白的《渡荆门送别》一诗中，写出诗人渡过荆门进入楚地看到江水冲出山峦向著原野奔腾而去的壮阔景色的诗句是：##，江入大荒流	山随平野尽							
9	不识##真面目，只缘身在此山中	庐山							
10	有情##含春泪，无力蔷薇卧晚枝。	芍药							
11	人闲##落，夜静春山空。	桂花							
12	采##东篱下，悠然见南山。	菊							
13	唐宋散文八大家不包括以下哪一位人物？	李白	苏轼	苏辙	李白	王安石			
14	清明上河图的作者是哪一个朝代的人？	北宋	南宋	北宋	晚唐	元朝			
15	天子呼来不上船，自称臣是酒中仙。说的是哪一位人物？	李白	杜甫	王之涣	李白	贾岛			
16									
17	说明：请按照表格对应的行和列填写题目。								
18									
19	对于题型为选择题的问题，题目填写到第一列，答案填写到第二列，四个选项分别填写到第三到第六列。如果题目中有图片需要展示，请在第七列填写图片的链接，如果题目中存在音频，请在第八列填写音频链接。								
20									
21	对于题型为填空题的问题，题目填写到第一列，答案填写到第二列。								
22	注意：问题中需要补全答案的位置，不论要补全的字数有多少，都用两个井号代替 ##，如： 慈母手中线，##身上衣。另外，填空题不存在选项，所以第三列到第六列不用填写。								
23	如果题目中有图片需要展示，请在第七列填写图片的链接，如果题目中存在音频，请在第八列填写音频链接。								
24									

图 9.15　题库 Excel 模板

在智慧校园考试系统中，录入题库主要分为以下 5 个步骤：

（1）用户下载模板文件。

（2）根据自己的题库需求修改 Excel 模板文件。

（3）输入题库名称并选择题库类型。

（4）上传文件。

（5）提交到数据库。

下面详细讲解录入题库功能的实现过程。

首先在 competition app 的 urls.py 中添加下面路由，以便为配置题库添加一个导航页，代码如下：

```
# 配置考试 url
urlpatterns += [
    path('set', set_render.index, name='set_index'),
    path('set/bank', set_render.set_bank, name='set_bank'),
    path('set/bank/tdownload', set_render.template_download, name='template_download'),
    path('set/bank/upbank', set_render.upload_bank, name='upload_bank'),
    path('set/game', set_render.set_game, name='set_game'),
]
```

在 competition app 中添加一个 render 视图模块 set_render.py，并在其中添加 index()函数，用来渲染视图和用户信息数据，关键代码如下：

```
@check_login
def index(request):
    """
    题库和考试导航页
    :param request: 请求对象
    :return: 渲染视图和 user_info 用户信息数据
    """

    uid = request.GET.get('uid', '')

    try:
        profile = Profile.objects.get(uid=uid)
    except Profile.DoesNotExist:
        return render(request, 'err.html', ProfileNotFound)

    return render(request, 'setgames/index.html', {'user_info': profile.data})
```

在 set_render.py 视图模块中添加一个 set_bank() 函数，该函数用来处理用户的请求并渲染配置题库页面，关键代码如下：

```python
@check_login
def set_bank(request):
    """
    配置题库页面
    :param request: 请求对象
    :return: 渲染页面返回 user_info 用户信息数据和 bank_types 题库类型数据
    """
    uid = request.GET.get('uid', '')
    try:
        profile = Profile.objects.get(uid=uid)                          # 检查账户信息
    except Profile.DoesNotExist:
        return render(request, 'err.html', ProfileNotFound)
    bank_types = []
    for i, j in BankInfo.BANK_TYPES:                                    # 返回所有题库类型
        bank_types.append({'id': i, 'name': j})
    return render(request, 'setgames/bank.html', {                      # 渲染模板
        'user_info': profile.data,
        'bank_types': bank_types
    })
```

上面代码中的要渲染的配置题库模板页面 bank.html 的关键代码如下：

```html
<form id="uploadFileForm" method="post" action="/bs/set/bank/upbank"
      enctype="multipart/form-data">{% csrf_token %}
<div id="uploadMainRow" class="row" style="margin-top: 120px;">
    <div class="col-md-3">
        <label>① 下载题库</label>
        <p style="color: gray;margin-top: 5px;">
         <a id="tDownload" href="/bs/set/bank/tdownload?uid={{ user_info.uid }}">下载</a>
            我们的简易模板，按照模板中的要求修改题库。
        </p>
    </div>
    <div class="col-md-3">
        <div class="form-group">
            <label for="bankName">② 题库名称</label>
            <input id="bankName" name="bank_name" type="text" class="form-control"
                   placeholder="请输入题库名称" />
        </div>
    </div>
    <div class="col-md-3">
        <label for="choicedValue">③ 题库类型</label>
        <div class="dropdown">
            <input type="button" id="choicedValue" data-toggle="dropdown" name="bank_type"
                   value="选择一个题库类型" />
            <div class="dropdown-menu">
                {% for t in bank_types %}
                    <div onclick="choiceBankType(this)">{{ t.name }}</div>
                {% endfor %}
            </div>
        </div>
    </div>
    <div class="col-md-3">
        <div class="row" style="margin-left:-1px;">
            <label for="uploadFile">④ 上传文件</label>
            <input class="form-control" name="template" type="file" id="uploadFile">
        </div>
    </div>
    <input type="hidden" name="uid" value="{{ user_info.uid }}" />
</div>
<div class="row" style="margin-top:35px;">
    <input type="submit" id="startUpload" class="btn btn-danger" value="开始录入">
```

```html
</div>
</form>
<script type="text/javascript">
    var choicedBankType;
    var responseTypes = {{ bank_types|safe }};
    var choiceBankType = function (t) {
        var cbt = $(t).html();
        for(var i in responseTypes){
            if(responseTypes[i].name === cbt){
                choicedBankType = responseTypes[i].id;
                break;
            }
        }
        $('#choicedValue').val(cbt);
    }
</script>
```

在开始录入题库前,用户需要先单击"下载"按钮,下载 Excel 题库模板文件并进行编辑后才能提交。下载题库模板功能是在 template_download()视图函数中实现的,该函数检查模板文件是否存在。如果模板文件不存在,则返回渲染后的错误页面 err.html;否则,创建一个 StreamingHttpResponse 对象,传入 iterator 生成器作为数据源,并设置内容类型为 Excel 文件(.xls)。然后设置响应头的 Content-Disposition 字段,表明该响应的内容应作为附件下载,并指定下载后的文件名为 template.xls。最后,返回这个流式响应对象,浏览器会识别响应头信息,从而触发下载操作,实现下载题库模板文件的功能。关键代码如下:

```python
@check_login
def template_download(request):
    """
    题库模板下载
    :param request: 请求对象
    :return: 返回 Excel 文件的数据流
    """
    uid = request.GET.get('uid', '')                              # 获取 uid
    try:
        Profile.objects.get(uid=uid)                              # 用户信息
    except Profile.DoesNotExist:
        return render(request, 'err.html', ProfileNotFound)
    def iterator(file_name, chunk_size=512):                      # chunk_size 大小为 512KB
        with open(file_name, 'rb') as f:                          # rb:以字节读取
            while True:
                c = f.read(chunk_size)
                if c:
                    yield c                                       # 使用 yield 返回数据,直到所有数据返回完毕才退出
                else:
                    break
    template_path = 'web/static/template/template.xls'
    file_path = os.path.join(settings.BASE_DIR, template_path)    # 希望将题库文件保存到一个单独目录
    if not os.path.exists(file_path):                             # 路径不存在
        return render(request, 'err.html', TemplateNotFound)
    # 将文件以流式响应返回到客户端
    response = StreamingHttpResponse(iterator(file_path), content_type='application/vnd.ms-excel')
    response['Content-Disposition'] = 'attachment; filename=template.xls'    # 格式为 xls
    return response
```

用户单击"开始录入"按钮时,实现上传题库功能,该功能是在 upload_bank()视图函数中实现的。在该函数中,首先将返回的 Excel 题库模板保存到指定目录,以便后期使用;然后生成一个题库 BankInfo 对象,使用一个自定义的 Python 脚本将 Excel 题库文件中的数据逐一读取出来,并保存到数据库中。upload_bank()视图函数的关键代码如下:

```python
@check_login
```

```python
@transaction.atomic
def upload_bank(request):
    """
    上传题库
    :param request:请求对象
    :return: 返回用户信息 user_info 和上传成功的个数
    """
    uid = request.POST.get('uid', '')                                         # 获取 uid
    bank_name = request.POST.get('bank_name', '')                             # 获取题库名称
    bank_type = int(request.POST.get('bank_type', BankInfo.IT_ISSUE))         # 获取题库类型
    template = request.FILES.get('template', None)                            # 获取模板文件
    if not template:
        return render(request, 'err.html', FileNotFound)                      # 模板不存在
    if template.name.split('.')[-1] not in ['xls']:
        return render(request, 'err.html', FileTypeError)                     # 模板格式为 xls
    try:                                                                      # 获取用户信息
        profile = Profile.objects.get(uid=uid)
    except Profile.DoesNotExist:
        return render(request, 'err.html', ProfileNotFound)

    bank_info = BankInfo.objects.select_for_update().create(                  # 创建题库 BankInfo
        uid=uid,
        bank_name=bank_name or '暂无',
        bank_type=bank_type
    )
    today_bank_repo = os.path.join(settings.BANK_REPO, get_today_string())    # 保存文件目录以当天时间为准
    if not os.path.exists(today_bank_repo):
        os.mkdir(today_bank_repo)                                             # 不存在该目录则创建
    final_path = os.path.join(today_bank_repo, get_now_string(bank_info.bank_id)) + '.xls'   # 生成文件名
    with open(final_path, 'wb+') as f:                                        # 保存到目录
        f.write(template.read())
    choice_num, fillinblank_num = upload_questions(final_path, bank_info)     # 使用 xlrd 读取 Excel 文件到数据库
    return render(request, 'setgames/bank.html', {                            # 渲染视图
        'user_info': profile.data,
        'created': {
            'choice_num': choice_num,
            'fillinblank_num': fillinblank_num
        }
    })
```

上面代码中用到了 upload_questions.py 脚本文件，该文件用来使用 xlrd 模块读取 Excel 题库文件中的每一行内容。在读取时，需要判断第一列中的题目信息中是否包含##，以此来区分填空题和选择题，并按照题型分别导入到两个不同的 Django 模型——ChoiceInfo（选择题）和 FillInBlankInfo（填空题）中，同时更新题库的统计信息。upload_questions.py 脚本文件的关键代码如下：

```python
import xlrd                                                                   # xlrd 库
from django.db import transaction                                             # 数据库事务
from competition.models import ChoiceInfo, FillInBlankInfo                    # 题目数据模型

def check_vals(val):                                                          # 检查值是否被转换成 float，如果是，将.0 结尾去掉
    val = str(val)
    if val.endswith('.0'):
        val = val[:-2]
    return val
@transaction.atomic
def upload_questions(file_path=None, bank_info=None):
    book = xlrd.open_workbook(file_path)                                      # 读取文件
    table = book.sheets()[0]                                                  # 获取第一张表
    nrows = table.nrows                                                       # 获取行数
    choice_num = 0                                                            # 选择题数量
    fillinblank_num = 0                                                       # 填空题数量
```

```python
        for i in range(1, nrows):
            rvalues = table.row_values(i)                              # 获取行中的值
            if (not rvalues[0]) or rvalues[0].startswith('说明'):        # 取出多余行
                break
            if '##' in rvalues[0]:                                     # 选择题
                FillInBlankInfo.objects.select_for_update().create(
                    bank_id=bank_info.bank_id,
                    question=check_vals(rvalues[0]),
                    answer=check_vals(rvalues[1]),
                    image_url=rvalues[6],
                    source=rvalues[7]
                )
                fillinblank_num += 1                                    # 填空题数加 1
            else:                                                       # 填空题
                ChoiceInfo.objects.select_for_update().create(
                    bank_id=bank_info.bank_id,
                    question=check_vals(rvalues[0]),
                    answer=check_vals(rvalues[1]),
                    item1=check_vals(rvalues[2]),
                    item2=check_vals(rvalues[3]),
                    item3=check_vals(rvalues[4]),
                    item4=check_vals(rvalues[5]),
                    image_url=rvalues[6],
                    source=rvalues[7]
                )
                choice_num += 1                                         # 选择题数加 1
        bank_info.choice_num = choice_num
        bank_info.fillinblank_num = fillinblank_num
        bank_info.save()
        return choice_num, fillinblank_num
```

说明

（1）如果题目中包含##，表示该题目是填空题，在答题时，页面会将##解读为 4 条下画线（＿＿＿），方便用户答题。

（2）本项目的后台（http://127.0.0.1:8000/admin）主要利用 Django 框架根据相应数据模型和自定义配置自动生成，其每个管理模块的后台配置代码在相应 app 的 admin.py 代码中，读者可以在资源包中查看其详细实现代码。

9.8 项目运行

通过前述步骤，设计并完成了"智慧校园考试系统"项目的开发。下面运行该项目，检验一下我们的开发成果。运行"智慧校园考试系统"项目的步骤如下。

（1）打开 Exam\config\local_settings.py 文件，根据自己的 MySQL 数据库、Redis 数据库及邮箱信息对下面配置代码进行修改：

```
# MySQL 配置
DATABASES = {
    'default': {
        'ENGINE': 'django.db.backends.mysql',
        'NAME': 'xx',
        'USER': 'xxx',
        'PASSWORD': 'xxx'
    }
```

```
}
# Redis 配置
REDIS = {
    'default': {
        'HOST': '127.0.0.1',
        'PORT': 6379,
        'USER': '',
        'PASSWORD': '',
        'db': 0,
    }
}
BANK_REPO = ' F:/PythonProject/exam/backup '              # 修改为存放 Excel 题库的位置，用来保存题库
BASE_NUM_ID = 100000
INIT_PASSWORD = 'p@ssw0rd'
DOMAIN = "http://xxx.xx.xx.xxx"                            # 需要修改此处域名
WEB_INDEX_URI = "{}/web/index".format(DOMAIN)              # 首页

# 发送邮件
EMAIL_BACKEND = 'django.core.mail.backends.smtp.EmailBackend'   # 邮箱验证后台
EMAIL_USE_TLS = True                                       # 使用 TLS
EMAIL_USE_SSL = False                                      # 使用 SSL
EMAIL_SSL_CERTFILE = None                                  # SSL 证书
EMAIL_SSL_KEYFILE = None                                   # SSL 文件
EMAIL_TIMEOUT = None                                       # 延时
EMAIL_HOST = 'xxx.xxx@xx.xxx'                              # SMTP 地址
EMAIL_PORT = 465                                           # 端口
EMAIL_HOST_USER = 'xxx@xxx.xx'                             # 发件邮箱
EMAIL_HOST_PASSWORD = 'password'                           # 密码
SERVER_EMAIL = EMAIL_HOST_USER                             # 服务器邮箱
DEFAULT_FROM_EMAIL = EMAIL_HOST_USER                       # 默认发件人
ADMINS = [('Admin', 'xxx@xxx.xx')]                         # 管理员邮箱

MANAGERS = ADMINS
```

（2）打开命令提示符对话框，进入 Exam 项目文件夹所在目录，在命令提示符对话框中输入如下命令来创建 venv 虚拟环境：

```
virtualenv venv
```

（3）在命令提示符对话框中输入如下命令来启动 venv 虚拟环境：

```
venv\Scripts\activate
```

（4）在虚拟环境下使用如下命令来安装 Django 等依赖包：

```
pip install -r requirements.txt
```

（5）使用 MySQL 命令行方式或 MySQL 可视化管理工具（如 Navicat）创建数据库。在命令提示符对话框中使用命令行方式时输入如下命令：

```
create database exam default character set utf8;
```

（6）使用 MySQL 命令行方式或 MySQL 可视化管理工具（如 Navicat）将 Exam\exam.sql 文件导入数据库中。在命令提示符对话框中使用命令行方式时输入如下命令：

```
USE exam;
SOURCE D:\Code\Exam\Exam.sql;
```

在上面的命令中，"D:\Code\Exam\Exam.sql" 为数据库脚本文件的绝对路径，需要根据实际情况填写。

（7）在 PyCharm 的左侧项目结构中展开"智慧校园考试系统"的项目文件夹 Exam，在其中选中 manage.py 文件，单击鼠标右键，在弹出的快捷菜单中选择 Modify Run Configuration…命令，如图 9.16 所示。

（8）在打开的对话框中的 Parameters 文本框中输入 runserver，并且单击 Apply 按钮，如图 9.17 所示。

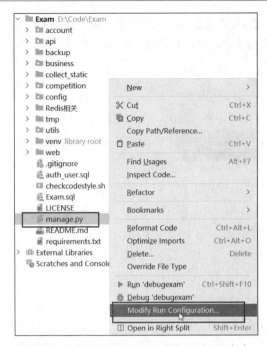 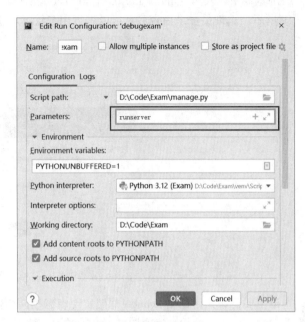

图 9.16　选择 Modify Run Configuration…命令　　　　图 9.17　输入 runserver

（9）在浏览器中输入网址 http://127.0.0.1:8000，即可进入智慧校园考试系统的前台首页，效果如图 9.18 所示。在前台中，普通用户可以选择考试类别并进行考试，以及查看排行榜，而注册为机构，则可以录入题库并配置考试。

图 9.18　智慧校园考试系统前台首页

在浏览器中输入网址 http://127.0.0.1:8000/admin，即可进入智慧校园考试系统的后台登录页面，效果如图 9.19 所示。管理员登录后台后，可以进行用户管理、机构管理、题库管理及考试类别管理。

图 9.19　智慧校园考试系统后台登录页面

本章主要讲解如何使用 Django 框架实现智慧校园考试系统项目，包括网站的系统设计、数据库设计以及主要的功能模块设计等。希望通过本章的学习，读者能够熟悉 Python Web 项目开发流程，并掌握 Django Web 开发技术的应用，为今后的项目开发积累经验。

9.9　源码下载

本章虽然详细地讲解了如何编码实现"智慧校园考试系统"的各个功能，但给出的代码都是代码片段，而非完整的源代码。为了方便读者学习，本书提供了该项目的完整源代码，读者可以通过扫描右侧的二维码进行下载。

源码下载

第 10 章 吃了么外卖网

——Django + MySQL + Redis

近年来，随着电子设备、网络的普及，外卖行业因其方便、快捷等特点，得以迅速发展，外卖网站成了大家日常生活中密不可分的一部分，只要有网络和相应的设备，足不出户即可解决一日三餐。本章将使用 Python Web 框架 Django，结合 MySQL 和 Redis 数据库等技术开发一个外卖网。

项目微视频

本项目的核心功能及实现技术如下：

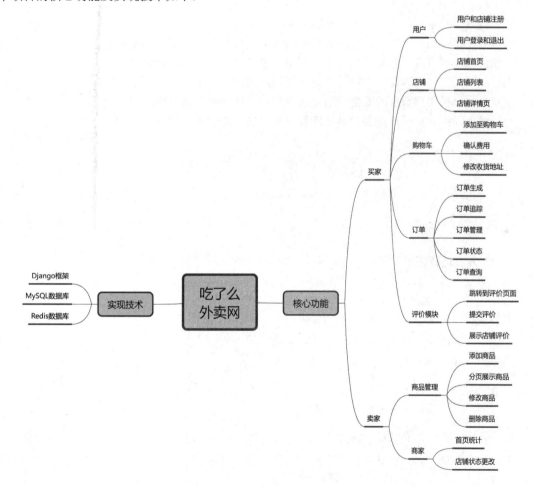

10.1 开发背景

在这个快节奏的时代，人们对于时间和效率的要求越来越高，外卖行业应运而生，并迅速成为现代生活中不可或缺的一部分。为了满足现代人的饮食需求，提高餐饮业的服务质量和效率，开发一个功能完善、操作简便、用户体验良好的外卖网站显得尤为重要。本项目选用 Django 框架，结合 MySQL 和 Redis 数据库等技术进行开发，主要基于以下考虑：

- ☑ Django 框架：Django 遵循"快速开发"的理念，提供丰富的开箱即用的组件和工具，可以加速外卖网站从原型到部署的整个过程，使开发团队能更专注于业务逻辑的实现。
- ☑ MySQL 数据库：MySQL 提供了强大的数据存储和查询功能。外卖网站需要处理大量的餐厅信息、用户账号、订单数据等，MySQL 能够提供稳定的数据存储服务，确保数据的一致性和完整性。
- ☑ Redis 缓存系统：Redis 是一个高性能的键值存储系统，常用于缓存和实时数据处理。在外卖网站中，可以利用 Redis 来缓存热点数据，以减少数据库的访问压力，提高网站响应速度。

本项目的实现目标如下：

- ☑ 注册与登录：提供买家/卖家注册功能。
- ☑ 菜单浏览：买家可以浏览各种餐厅的菜单。
- ☑ 下单功能：买家可以选择商品，填写配送地址（支持默认和新增）和联系方式，提交订单。
- ☑ 订单跟踪：买家可以实时查看订单状态。
- ☑ 评价与反馈：用户可以对已完成的订单进行评价和反馈，以改进服务质量。
- ☑ 卖家管理：可自定义店铺信息、上传菜单、设置营业时间和起送价等。
- ☑ 性能：需要能够快速响应用户请求，保证良好的用户体验。
- ☑ 安全性：保护用户数据的安全，防止未经授权的访问和数据泄露。
- ☑ 易用性：界面设计简洁直观，操作流程简单易懂，方便用户快速上手。

10.2 系统设计

10.2.1 开发环境

本项目的开发及运行环境如下：

- ☑ 操作系统：推荐 Windows 10、Windows 11 或更高版本。
- ☑ 开发工具：PyCharm 2024（向下兼容）。
- ☑ 开发语言：Python 3.12。
- ☑ 数据库：MySQL 8.0+PyMySQL 驱动、Redis。
- ☑ Python Web 框架：Django 5.0。

10.2.2 业务流程

在启动项目后，首先需要让用户进行登录。这里的用户有两种身份：一种是买家，可以选择店铺，并进行点外卖和评价操作；另一种是卖家，可以创建店铺并管理自己店铺的商品，还可以对自己店铺的订单进行处理。本项目的业务流程如图 10.1 所示。

> **说明**
>
> 本章主要讲解吃了么外卖网中主要功能模块的实现逻辑，其完整功能实现可以参考资源包中的源代码。

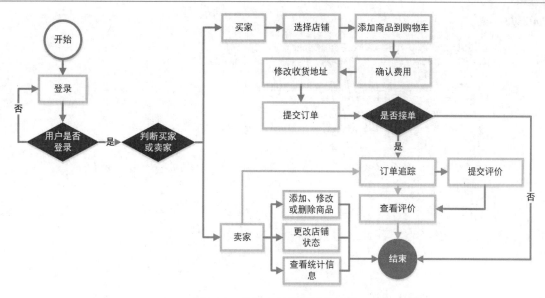

图 10.1 吃了么外卖网业务流程

10.2.3 功能结构

本项目的功能结构已经在章首页中给出。主要分为两个部分，分别是买家部分和卖家部分。本项目实现的具体功能如下：

☑ 买家：
  ➤ 用户模块：用于买家和卖家登录网站，包括用户和店铺注册、用户登录和退出等功能。
  ➤ 店铺模块：用于选择查看店铺信息，包括店铺首页、店铺列表和店铺详情页。
  ➤ 购物车模块：用于选择要下单的商品，包括添加至购物车、确认费用和修改收货地址。
  ➤ 订单模块：用于完成订单，包括订单生成、订单追踪、订单管理、订单状态和订单查询。
  ➤ 评价模块：用于对订单进行评价，包括跳转到评价页面、提交评价和展示店铺评价。

☑ 卖家：
  ➤ 商品管理：用于卖家对商品进行管理，包括添加商品、分页展示商品、修改商品和删除商品。
  ➤ 卖家模块：用于对店铺数据进行统计和更改店铺状态，包括首页统计和店铺状态更改。

10.3 技术准备

本项目的主体框架采用 Django 框架实现，关于 Django 框架的基本用法请参考本书的 6.3.2 节和 7.3.2 节。另外，本项目中的数据存储主要使用了 MySQL 数据库及 Redis 数据库，其中操作 MySQL 数据库时使用了 PyMySQL 模块，关于该知识在《Python 从入门到精通（第 3 版）》中有详细的讲解，对该知识不太熟悉的读者可以参考该书对应的内容。下面将对使用 django-redis 模块操作 Redis 数据库进行详细介绍。

使用 django-redis 模块操作 Redis 数据库时，需要先使用 pip install 命令安装它，具体的命令如下：

```
pip install django-redis
```

安装完成后，还需要在 Django 项目的配置文件 settings.py 中配置 Redis，让其作为后端缓存。主要指定 Redis 服务器的 IP 地址、端口号以及缓存键的前缀，避免与其他应用冲突。具体配置代码如下：

```
CACHES = {
    "default": {
        "BACKEND": "django_redis.cache.RedisCache",
        "LOCATION": "redis://127.0.0.1:6379/0",   # 根据实际情况修改地址和端口
        "OPTIONS": {
            "CLIENT_CLASS": "django_redis.client.DefaultClient",
            # 如果 Redis 设置了密码，需要在这里添加
            # "PASSWORD": "your_redis_password",
            # 可以根据需要添加其他选项，如序列化方式等
        }
    },
    # 如果有特殊需求，可以配置多个缓存实例，比如用于存储验证码
    "verify_codes": {
        "BACKEND": "django_redis.cache.RedisCache",
        "LOCATION": "redis://127.0.0.1:6379/1",
        "OPTIONS": {
            "CLIENT_CLASS": "django_redis.client.DefaultClient",
        }
    },
}
```

例如，本项目中对应的配置代码如下：

```
CACHES = {
    "default": {
        "BACKEND": "django_redis.cache.RedisCache",
        "LOCATION": "redis://127.0.0.1:6379/1",
        "OPTIONS": {
            "CLIENT_CLASS": "django_redis.client.DefaultClient",
            # 提升 Redis 解析性能
            "PARSER_CLASS": "redis.connection._HiredisParser",
        }
    }
}
```

接下来就可以在 Django 视图函数中使用 django-redis 提供的缓存功能，例如，使用 cache.set()存储数据（键值对），或者使用 cache.get()检索缓存中的值。示例代码如下：

```
from django_redis import get_redis_connection

redis_conn = get_redis_connection()            # 获取默认的 Redis 连接
redis_conn.set('key', 'value')                 # 存储数据
value = redis_conn.get('key')                  # 获取数据
redis_conn.delete('key')                       # 删除数据
```

10.4 数据库设计

10.4.1 数据库设计概要

吃了么外卖网使用 MySQL 数据库来存储数据，数据库名为 clmwm，共包含 24 张数据表（包括 Django 默认的 7 张数据表），其对应的中文表名及主要作用如表 10.1 所示。

表 10.1 clmwm 数据库中的数据表及作用

英 文 表 名	中 文 表 名	作　　用
auth_group	授权组表	Django 默认的授权组

续表

英文表名	中文表名	作用
auth_group_permissions	授权组权限表	Django 默认的授权组权限信息
auth_permission	授权权限表	Django 默认的权限信息
df_user	用户表	用于存储用户的信息
df_goods	商品表	用于存储商品信息
df_shop	店铺表	用于存储店铺信息
df_order_info	订单表	用于存储订单信息
df_order_goods	订单明细表	用于存储订单明细信息
df_goods_type	商品分类表	用于存储商品分类信息
df_goods_sku	具体商品表	用于存储商品详细信息
df_address	地址表	用于存储店铺、用户地址信息
df_user_image	用户图片地址表	用于存储用户图片在服务器上的地址信息
df_shop_image	店铺图片地址表	用于存储店铺图片在服务器上的地址信息
df_order_image	订单评论图片地址表	用于存储订单评论图片在服务器上的地址信息
df_goods_image	商品图片地址表	用于存储商品图片在服务器上的地址信息
df_index_type_goods	首页分类展示商品表	用于存储首页分类展示商品信息
df_index_promotion	首页促销活动表	用于存储首页促销活动信息
df_index_banner	首页轮播商品表	用于存储首页轮播商品信息
df_order_track	订单轨迹表	用于存储订单轨迹信息
df_shop_type	店铺类型表	用于存储店铺类型信息
django_admin_log	Django 日志表	保存 Django 管理员登录日志
django_content_type	Django contenttype 表	保存 Django 默认的 content type
django_migrations	Django 迁移表	保存 Django 的数据库迁移记录
django_session	Django session 表	保存 Django 默认的授权等 session 记录

10.4.2 数据表结构

df_user 用户表的表结构如表 10.2 所示。

表 10.2 df_user 用户表的表结构

字段	类型	长度	是否允许为空	含义
id	INT	默认	否	主键，编号
password	VARCHAR	128	否	密码（加密）
last_login	DATETIME	6	是	最后登录时间
is_superuser	TINYINT	1	否	是否管理员（值为 0 表示非管理员；值为 1 表示管理员）
username	VARCHAR	150	否	用户名
first_name	VARCHAR	30	否	名字
last_name	VARCHAR	150	否	姓氏
email	VARCHAR	254	否	E-mail
is_staff	TINYINT	1	否	身份（值为 1 表示卖家；值为 0 表示买家）
is_active	TINYINT	1	否	是否激活（值为 0 表示未激活；值为 1 表示激活）
date_joined	DATETIME	6	否	加入时间

续表

字　段	类　型	长　度	是否允许为空	含　义
create_time	DATETIME	6	否	创建时间
update_time	DATETIME	6	否	更新时间
is_delete	TINYINT	1	否	是否删除（值为1表示删除；值0为未删除）
sex	VARCHAR	2	否	性别
phone	VARCHAR	100	否	联系电话
real_name	VARCHAR	20	否	真实姓名

df_goods 商品表的表结构如表 10.3 所示。

表 10.3　df_goods 商品表的表结构

字　段	类　型	长　度	是否允许为空	含　义
id	INT	默认	否	主键，编号
create_time	DATETIME	6	否	创建时间
update_time	DATETIME	6	否	更新时间
is_delete	TINYINT	1	否	是否删除（值为1表示删除；值0为未删除）
name	VARCHAR	20	否	商品名称
detail	VARCHAR	200	否	详细信息
shop_id	INT	1	否	店铺ID

df_shop 店铺表的表结构如表 10.4 所示。

表 10.4　df_shop 店铺表的表结构

字　段	类　型	长　度	是否允许为空	含　义
id	INT	默认	否	主键，编号
create_time	DATETIME	6	否	创建时间
update_time	DATETIME	6	否	更新时间
is_delete	TINYINT	1	否	是否删除（值为1表示删除；值0为未删除）
shop_name	VARCHAR	20	否	店铺名字
shop_addr	VARCHAR	256	否	店铺地址
shop_type	VARCHAR	256	否	店铺类型
type_detail	VARCHAR	256	否	类型详细信息
shop_score	DECIMAL	(10,1)	否	店铺评分
shop_price	DECIMAL	(10,2)	否	起送价格
shop_sale	INT	默认	否	店铺销量
shop_image	VARCHAR	100	否	店铺头像
receive_start	TIME	默认	否	接单开始时间
receive_end	TIME	默认	否	接单结束时间
business_do	TINYINT	1	否	是否营业（值为1表示营业；值0为未营业）
high_opinion	VARCHAR	20	否	好评度
user_id	INT	1	否	卖家ID

df_order_info 订单表的表结构如表 10.5 所示。

表 10.5 df_order_info 订单表的表结构

字段	类型	长度	是否允许为空	含义
order_id	VARCHAR	128	否	主键，订单编号
create_time	DATETIME	6	否	创建时间
update_time	DATETIME	6	否	更新时间
is_delete	TINYINT	1	否	是否删除（值为 1 表示删除；值 0 为未删除）
pay_method	SMALLINT	默认	否	支付方式
total_count	INT	默认	否	商品数量
total_price	DECIMAL	(10,2)	否	商品总价
transit_price	DECIMAL	(10,2)	否	订单运费
order_status	SMALLINT	默认	否	订单状态
trade_no	VARCHAR	128	否	支付编号
comment	VARCHAR	256	否	评论
invoice_head	VARCHAR	256	否	发票抬头
taxpayer_number	VARCHAR	256	否	纳税人识别号
remarks	VARCHAR	256	否	订单备注
score	SMALLINT	默认	否	订单综合评分
addr_id	INT	1	否	地址 ID
shop_id	INT	1	否	店铺 ID
user_id	INT	1	否	用户 ID
transit_time	INT	1	否	配送时间

df_order_goods 订单明细表的表结构如表 10.6 所示。

表 10.6 df_order_goods 订单明细表的表结构

字段	类型	长度	是否允许为空	含义
id	INT	默认	否	主键，编号
create_time	DATETIME	6	否	创建时间
update_time	DATETIME	6	否	更新时间
is_delete	TINYINT	1	否	是否删除（值为 1 表示删除；值 0 为未删除）
count	INT	默认	否	商品数目
price	DECIMAL	(10,2)	否	商品价格
order_id	VARCHAR	128	否	订单 ID
sku_id	INT	默认	否	商品 SKUID
comment	VARCHAR	256	否	评论

> **说明**
> 限于篇幅，其他数据表的表结构这里就不再介绍了，对应字段的类型及含义可以参考对应的数据模型。

10.4.3 数据表关系

为了更好地理解数据表之间的关系，这里给出梳理后的数据表关系模型图，如图 10.2 所示。

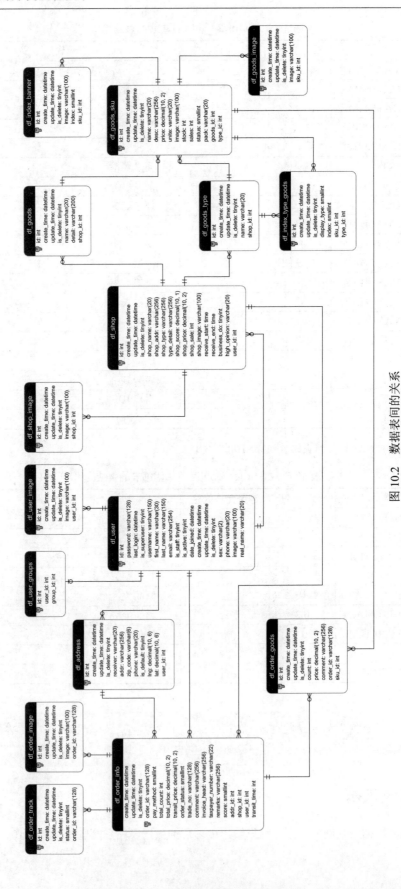

图 10.2　数据表间的关系

10.5　商品管理模块设计

当用户访问首页时，看到的所有店铺、商品信息都是由卖家添加至数据库的，所以应该先添加店铺、商品信息，首页才会查询到数据。商品管理模块主要包括添加商品、分页展示商品、修改商品和删除商品4部分功能。下面分别介绍每个功能的实现。

10.5.1　添加商品

为了降低代码的耦合度，使逻辑更加清晰，我们将新建一个商品管理模块 goodsmanage，具体命令如下：

```
(clmwmvenv) D:\Code\clmwm>python manage.py startapp goodsmanage
```

将新生成的模块拖动到 apps 文件夹下，并新建 urls.py 文件。在 base.py 文件中，对新建的模块完成注册，由于会应用到商品模块的相关数据表，所以此处也要将商品模块进行注册，具体代码如下：

```python
INSTALLED_APPS = [
'django.contrib.admin',
    'django.contrib.auth',
    'django.contrib.contenttypes',
    'django.contrib.sessions',
    'django.contrib.messages',
    'django.contrib.staticfiles',
    'apps.user',                          # 用户模块
    'apps.goods',                         # 商品模块
    'apps.goodsmanage',                   # 商品管理模块
]
```

注册完成后，在 clmwm 文件夹下的 urls.py 文件中配置请求路由，具体代码如下：

```python
from django.contrib import admin
from django.urls import path, include

urlpatterns = [
    path('admin/', admin.site.urls),
    path('user/', include(('apps.user.urls', 'apps.user'), namespace='user')),  # 用户模块
    # 商品管理模块
    path('goodsmanage/', include(('apps.goodsmanage.urls', 'apps.goodsmanage'), namespace='goodsmanage')),
]
```

上述配置是新建模块的通用配置，以后新建模块时将不再赘述。接下来还需要配置 goodsmanage 下的 urls.py 文件的请求路由信息，关键代码如下：

```python
from django.conf.urls import re_path

from apps.goodsmanage.views import  SjcpglView

urlpatterns = [
    re_path(r'^sj_cpgl/$', SjcpglView.as_view(), name='sj_cpgl'),        # 商品管理
]
```

配置完请求的 URL 信息后，还需在 views.py 文件中渲染出添加商品页面，具体代码如下：

```python
class SjcpglView(View):
    """卖家商品管理"""
```

```
def get(self, request):

    return render(request, 'sj_cpgl.html')
```

在 views.py 中可以发现,其通过 render() 渲染的是一个 sj_cpgl.html 页面,所以还需将 static 下的 sj_cpgl.html 文件复制到 templates 文件夹下,并且继承已经抽象出来的父模板 sj.html,具体代码如下:

```
{% extends 'sj.html' %}
{% load staticfiles %}
{% block title %}吃了么-商品管理{% endblock title %}
{% block topfiles %}
<script src="{% static 'js/jquery.min.js' %}"></script>
<script type="text/javascript">
$(function () {
    $('.sj_order_man ul li').hover(function(){
        // 获得当前被点击的元素索引值
          var Index = $(this).index();
        var line=300*Index-300;
        // 给菜单添加选择样式
        $(this).addClass('active').siblings().removeClass('active');
        $(".line").stop(true,true).animate({left:line},200);
        $('.sj_order_man').children('div').eq(Index).show().siblings('div').hide();
    });
});
</script>
{% endblock topfiles %}
{% block body %}

    <div class="add_cp">
        <h3>添加商品</h3>
        <form method="post" enctype="multipart/form-data" id="sj_cogl" name="sj_cpgl">
            {% csrf_token %}
            <ul>
                <li><span>商品名称<input type="text" id="name" name="name"></span><span>商品价格<input type="text" id="price" name="price"></span></li>
                <li><span>餐盒价格<input type="text" id="pack" name="pack"></span><span>商品库存<input type="text" id="stock" name="stock"></span></li>
                <li><span>商品类型<input type="text" id="type" name="type"></span></li>
                <li><span>商品图片</span></li>
                <li>
                    <div class="img_yulan">
                        <img id="preview" />
                    </div>
                    <script type="text/javascript">
                        function imgPreview(fileDom){
                            // 判断是否支持 FileReader
                            if (window.FileReader) {
                                var reader = new FileReader();// 创建 filereader 对象
                            } else {
                                alert("您的设备不支持图片预览功能,如需该功能请升级您的设备!");
                            }
                            // 获取文件
                            var file = fileDom.files[0];
                            // 读取完成
                            reader.onload = function(e) {
                                // 获取图片 dom
                                var img = document.getElementById("preview");
                                // 图片路径设置为读取的图片
                                img.src = e.target.result;
                            };
                            reader.readAsDataURL(file);
                        }
```

```
            </script>
                <input type="file" name="file" onchange="imgPreview(this)">
            </li>
        </ul>
        {#        <input type="submit" class="agree_btn" value="点击上架">#}
    </form>
    <button class="agree_btn">点击上架</button>
    <script language="javascript" type="text/javascript">
        $(".agree_btn").click(function(){
            $("#sj_cogl").submit();
        });
    </script>
    {{ errmsg }}
    {{ success }}
</div>
{% endblock body %}
```

模板修改完成后，启动项目，在浏览器中访问http://127.0.0.1:8000/goodsmanage/sj_cpgl/就会出现添加商品页面，如图10.3所示。

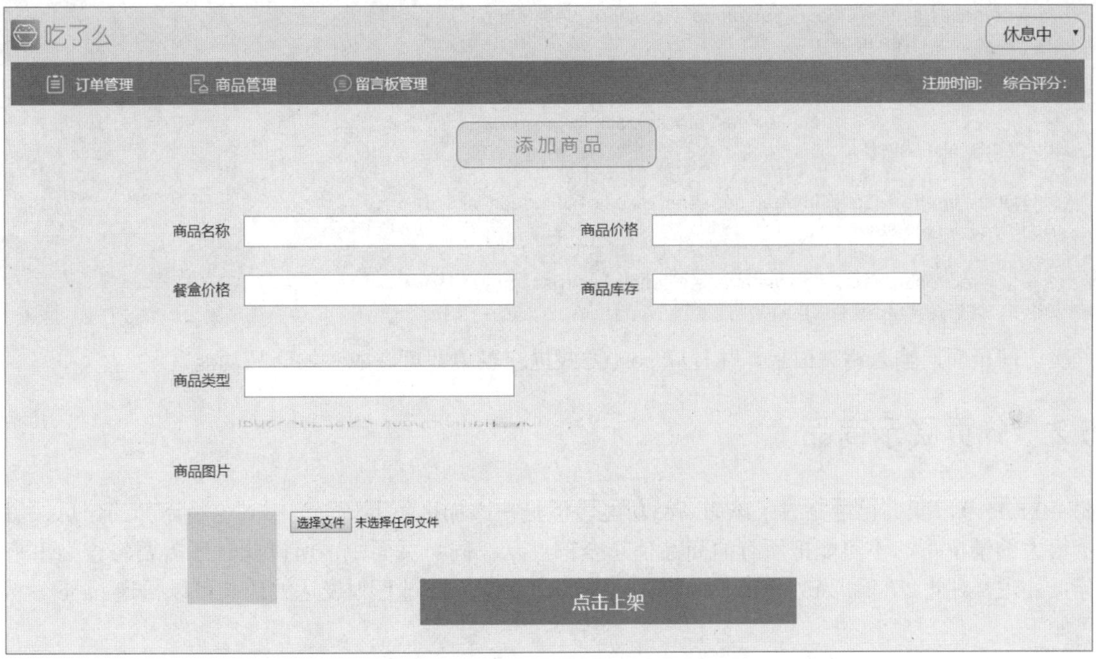

图 10.3　添加商品

在该页面输入完信息后，提交一个form表单给后台即可完成添加商品功能。在views.py文件中，刚刚添加的类SjcpglView中新增一个post()方法处理这个请求，具体代码如下：

```
def post(self, request):
    """处理商品上架"""

    # 接收数据
    name = request.POST.get('name')
    price = request.POST.get('price')
    stock = request.POST.get('stock')
    pack = request.POST.get('pack')
    typ = request.POST.get('type')
    img = request.FILES.get('file')

    # 校验基本数据
```

```python
        if not all([img, name, price, stock, pack, typ]):
            # 数据不完整
            return render(request, 'sj_cpgl.html', {'errmsg': '缺少相关数据'})

        # FDFS 上传图片
        rec = FastDFSStorage().save(img.name, img)

        # 业务表添加数据
        try:
            df_shop = Shop.objects.get(user_id=request.user.id)
        except Shop.DoesNotExist:
            return render(request, 'login.html', {'errmsg': '用户登录信息已失效,请重新登录!'})

        df_goods = Goods(name=name, shop_id=df_shop.id)
        df_goods.save()

        # 确定是否为新增类型
        try:
            df_goods_type = GoodsType.objects.get(name=typ, shop_id=df_shop.id)
        except GoodsType.DoesNotExist:
            df_goods_type = GoodsType(name=typ, shop_id=df_shop.id)
            df_goods_type.save()

        df_goods_sku = GoodsSKU(goods=df_goods, type=df_goods_type, name=name, price=price,
                                unite='per', stock=stock, pack=pack)
        df_goods_sku.save()

        df_goods_image = GoodsImage(image=rec, sku_id=df_goods_sku.id)
        df_goods_image.save()

        # return redirect(reverse('goodsmanage:sj_cpgl', kwargs={'page': 1}))
        return render(request, 'sj_cpgl.html')
```

再次访问页面,输入相关信息,并且以 post 方式提交表单即可完成添加商品功能。

10.5.2 分页展示商品

在 10.5.1 节,我们已经完成了添加商品功能,但是在添加页面上却查询不到已经添加的信息,假设在数据量很大的情况下,不可能把所有的商品信息全部展示在同一页面中,所以此时还需要将查询出来的数据做分页处理,对此我们需要改写 10.5.1 节添加的 URL 信息,通过正则表达式让其可以传递 page 参数,具体代码如下:

```python
from django.conf.urls import re_path

from apps.goodsmanage.views import SjcpglView

urlpatterns = [
    re_path(r'^sj_cpgl(?P<page>\d+)/$', SjcpglView.as_view(), name='sj_cpgl'),  # 商品管理
]
```

由于匹配的 URL 增加了 page 变量,所以 views.py 也要做出相应的修改,否则将不能匹配该方法,并且变量名称要相同,具体代码如下:

```python
from django.shortcuts import render, redirect, reverse
from django.views.generic import View

from apps.goods.models import GoodsType, GoodsSKU, Goods, GoodsImage
from apps.user.models import Shop
# 调用了 3 个公共方法
```

```python
from utils.common import goods_item, page_item, shop_is_new

from utils.fdfs.storage import FastDFSStorage


class SjcpglView(View):
    """卖家商品管理"""

    def get(self, request, page):

        # 获取店铺信息
        try:
            df_shop = Shop.objects.get(user_id=request.user.id)
        except Shop.DoesNotExist:
            return render(request, 'login.html', {'errmsg': '用户登录信息已失效，请重新登录！'})

        # 获取店铺下的所有商品
        goods_sku_info = GoodsSKU.objects.filter(goods__shop_id=df_shop.id).order_by('-update_time')

        # 为商品添加图片
        goods_item(goods_sku_info)

        # 对数据进行分页
        context = page_item(goods_sku_info, page, 10)

        shop_is_new(df_shop)

        context['shop'] = df_shop

        return render(request, 'sj_cpgl.html', context)

    def post(self, request, page):
        """处理商品上架"""

        # 接收数据
        name = request.POST.get('name')
        price = request.POST.get('price')
        stock = request.POST.get('stock')
        pack = request.POST.get('pack')
        typ = request.POST.get('type')
        img = request.FILES.get('file')

        # 校验基本数据
        if not all([img, name, price, stock, pack, typ]):
            # 数据不完整
            return render(request, 'sj_cpgl.html', {'errmsg': '缺少相关数据'})

        # FDFS 上传图片
        rec = FastDFSStorage().save(img.name, img)

        # 业务表添加数据
        try:
            df_shop = Shop.objects.get(user_id=request.user.id)
        except Shop.DoesNotExist:
            return render(request, 'login.html', {'errmsg': '用户登录信息已失效，请重新登录！'})

        df_goods = Goods(name=name, shop_id=df_shop.id)
        df_goods.save()

        # 确定是否为新增类型
        try:
            df_goods_type = GoodsType.objects.get(name=typ, shop_id=df_shop.id)
```

```python
    except GoodsType.DoesNotExist:
        df_goods_type = GoodsType(name=typ, shop_id=df_shop.id)
        df_goods_type.save()

    df_goods_sku = GoodsSKU(goods=df_goods, type=df_goods_type, name=name, price=price,
                            unite='per', stock=stock, pack=pack)
    df_goods_sku.save()

    df_goods_image = GoodsImage(image=rec, sku_id=df_goods_sku.id)
    df_goods_image.save()

    # return render(request, 'sj_cpgl.html')
    return redirect(reverse('goodsmanage:sj_cpgl', kwargs={'page': 1}))
```

在改写后的 views.py 中,我们从 utils 下的 common.py 文件引入了 3 个公共的方法: goods_item()方法,用于获取商品图片; page_item()方法,用于控制分页; shop_is_new()方法,用于判断店铺是否为新店。具体代码如下:

```python
def goods_item(item):
    """获取商品图片"""
    from collections import Iterable
    from apps.goods.models import GoodsImage

    # 判断是否为可迭代对象
    if isinstance(item, Iterable):
        for info in item:
            image = GoodsImage.objects.get(sku_id=info.id).image
            info.image = image
    else:
        item.image = GoodsImage.objects.get(sku_id=item.id).image
    return item

def shop_is_new(df_shop):
    """判断店铺是否为新店"""
    import datetime
    from collections import Iterable

    # 获取 30 天前的时间
    one_month_ago = (datetime.datetime.now() - datetime.timedelta(30)).replace(tzinfo=None)

    # 判断传入数据是否可以遍历
    if isinstance(df_shop, Iterable):
        for info in df_shop:
            if info.create_time.replace(tzinfo=None) > one_month_ago:
                info.shop_score = '新店'
            info.create_time = info.create_time.strftime("%Y-%m-%d")
    else:
        if df_shop.create_time.replace(tzinfo=None) > one_month_ago:
            df_shop.shop_score = '新店'
        df_shop.create_time = df_shop.create_time.strftime("%Y-%m-%d")

    return df_shop

def page_item(info, page, page_number):
    """控制分页"""
    from django.core.paginator import Paginator

    # 对数据进行分页
    paginator = Paginator(info, page_number)

    # 获取第 page 页的内容
    try:
```

```python
        page = int(page)
    except Exception as e:
        page = 1
    if page > paginator.num_pages:
        page = 1

    # 获取第 page 页的 Page 实例对象
    info = paginator.page(page)

    # 进行页码的控制,页面上最多显示5个页码
    # 1.总页数小于5页,页面上显示所有页码
    # 2.如果当前页是前3页,显示1-5页
    # 3.如果当前页是后3页,显示后5页
    # 4.其他情况,显示当前页的前2页,当前页,当前页的后2页
    num_pages = paginator.num_pages
    if num_pages < 5:
        pages = range(1, num_pages + 1)
    elif page <= 3:
        pages = range(1, 6)
    elif num_pages - page <= 2:
        pages = range(num_pages - 4, num_pages + 1)
    else:
        pages = range(page - 2, page + 3)

    # 整合数据
    context = {'info': info, 'pages': pages}
    return context
```

至此,我们已经对数据进行了分页,接下来便是将分页后的数据渲染在模板中,并且添加页码进行控制,具体代码如下:

```
{% extends 'sj.html' %}
{% load staticfiles %}
{% block title %}吃了么-商品管理{% endblock title %}
{% block topfiles %}
<script src="{% static 'js/jquery.min.js' %}"></script>
<script type="text/javascript">
$(function () {
    $('.sj_order_man ul li').hover(function(){
        // 获得当前被点击的元素索引值
        var Index = $(this).index();
      var line=300*Index-300;
       // 给菜单添加选择样式
        $(this).addClass('active').siblings().removeClass('active');
        $(".line").stop(true,true).animate({left:line},200);
        $('.sj_order_man').children('div').eq(Index).show().siblings('div').hide();
    });
});
</script>
{% endblock topfiles %}
{% block body %}

    <div class="add_cp">
        <h3>添加商品</h3>
        <form method="post" enctype="multipart/form-data" id="sj_cogl" name="sj_cpgl">
            {% csrf_token %}
            <ul>
                <li><span>商品名称<input type="text" id="name" name="name"></span><span>商品价格<input type="text" id="price" name="price"></span></li>
                <li><span>餐盒价格<input type="text" id="pack" name="pack"></span><span>商品库存<input type="text" id="stock" name="stock"></span></li>
                <li><span>商品类型<input type="text" id="type" name="type"></span></li>
```

```html
            <li><span>商品图片</span></li>
            <li>
                <div class="img_yulan">
                    <img id="preview" />
                </div>
                <script type="text/javascript">
                    function imgPreview(fileDom){
                        // 判断是否支持 FileReader
                        if (window.FileReader) {
                            var reader = new FileReader();//创建 filereader 对象
                        } else {
                            alert("您的设备不支持图片预览功能，如需该功能请升级您的设备！");
                        }
                        // 获取文件
                        var file = fileDom.files[0];
                        // 读取完成
                        reader.onload = function(e) {
                            // 获取图片 dom
                            var img = document.getElementById("preview");
                            // 图片路径设置为读取的图片
                            img.src = e.target.result;
                        };
                        reader.readAsDataURL(file);
                    }
                </script>
                <input type="file" name="file" onchange="imgPreview(this)">
            </li>
    </ul>
    {#        <input type="submit" class="agree_btn" value="点击上架">#}
</form>
<button class="agree_btn">点击上架</button>
<script language="javascript" type="text/javascript">
    $(".agree_btn").click(function(){
        $("#sj_cogl").submit();
    });
</script>
{{ errmsg }}
{{ success }}
<div class="cp_list">
    <table>
        <tr>
            <td width="17%">商品名称</td>
            <td width="7%">商品价格</td>
            <td width="7%">餐盒价格</td>
            <td width="7%">库存</td>
            <td width="15%">商品分类</td>
            <td width="15%">图片展示</td>
            <td width="32%">商品管理</td>
        </tr>
        {% for item in info %}
        <tr>
            <td>{{ item.name }}</td>
            <td>{{ item.price }}</td>
            <td>{{ item.pack }}</td>
            <td>{{ item.stock }}</td>
            <td>{{ item.type }}
            </td>
            <td>
                <img src="{{ item.image}}">
            </td>
            <td>
                {# <button class="cpup_btn"><a href="{% url 'goodsmanage:sj_cpgl_update' item.id %}">点击修改</a></button>#}
```

```
            {# <button><a href="{% url 'goodsmanage:update_del' item.id %}" onclick="if(confirm('是否确认删除?')){return
      true;}else{return false;}">删   除</a></button>#}
                </td>
            </tr>
            {% endfor %}

    </table>

    <div class="pagenation">
        {% if info.has_previous %}
        <a href="{% url 'goodsmanage:sj_cpgl' info.previous_page_number %}"> < 上一页</a>
        {% endif %}

        {% for pindex in pages %}
            {% if pindex == info.number %}
            <a href="{% url 'goodsmanage:sj_cpgl' pindex %}" class="active">{{ pindex }}</a>
            {% else %}
            <a href="{% url 'goodsmanage:sj_cpgl' pindex %}">{{ pindex }}</a>
            {% endif %}
        {% endfor %}

        {% if info.has_next %}
          <a href="{% url 'goodsmanage:sj_cpgl' info.next_page_number %}"> 下一页 ></a>
            {% endif %}
    </div>
      </div>
   </div>
{% endblock body %}
```

模板渲染完成后，启动项目，再次访问页面可查看到分页后的数据，如图 10.4 所示。

图 10.4　分页后的商品数据

> **说明**
>
> 由于修改商品和删除商品与添加商品和分页展示商品类似，限于篇幅，这里不再赘述，具体功能代码请参见本书资源包中提供的源码。

10.6 店铺模块设计

在 10.5 节我们已经添加了商品信息,这样在用户访问店铺首页时就会出现卖家添加的数据。这时,我们需要实现店铺模块。店铺模块主要包括店铺首页、店铺列表、和店铺详情页 3 部分功能。下面分别介绍每个功能的实现。

10.6.1 店铺首页

当用户在浏览器中输入绑定的域名后,我们应该默认指向首页,所以在新建 goods 模块后,在 clmwm\urls.py 文件中包含店铺模块首页,具体代码如下:

```python
from django.contrib import admin
from django.urls import path, include

urlpatterns = [
    path('admin/', admin.site.urls),
    path('', include(('apps.goods.urls', 'apps.goods'), namespace='goods')),  # 店铺模块首页(买家)
]
```

如果此时启动项目,会抛出一个 django.core.exceptions.ImproperlyConfigured 错误,因为我们还没在 goods\urls.py 文件中添加任何信息,在该文件中设置一个空的 urlpatterns,例如,urlpatterns = [ ],就可以解决此类报错。为了直接匹配域名,在 goods\urls.py 文件中所有匹配信息最下方添加如下代码:

```python
from django.urls import path, re_path
from apps.goods import views

urlpatterns = [
    path('', views.index, name='index'),                   # 店铺首页(买家浏览)
]
```

接下来,我们还需要在 views.py 文件中将所有的店铺信息查询出来,并将查询出的数据做分页处理,具体代码如下:

```python
from django.shortcuts import render, reverse, redirect

from apps.user.models import Shop, User

from utils.common import shop_is_new, for_item, calculate_distance_duration
# Create your views here.

def index(request):
    """店铺首页(买家浏览)"""

    if request.user.is_staff:
        return redirect(reverse('goods:sj_index'))      # 判断是否为卖家

    # 由于需要获取用户地址信息,所以必须为登录状态
    try:
        user = User.objects.get(id=request.user.id)
    except User.DoesNotExist:
        return redirect(reverse('user:login'))

    # 获得所有店铺,数字为店铺大类(中餐、西餐、水果、饮品)
    shop_info0 = index_again(Shop.objects.filter(shop_type='0')[:5], user)
```

```python
shop_info1 = index_again(Shop.objects.filter(shop_type='1')[:5], user)
shop_info2 = index_again(Shop.objects.filter(shop_type='2')[:4], user)
shop_info3 = index_again(Shop.objects.filter(shop_type='3')[:4], user)

context = {'shop_info0': shop_info0, 'shop_info1': shop_info1,
           'shop_info2': shop_info2,  'shop_info3': shop_info3}
return render(request, 'index.html', context)
def index_again(shop_info, user):
    """处理买家首页，抽出公共方法"""
    for_item(shop_info)                                    # 为店铺插入图片
    shop_is_new(shop_info)                                 # 判断店铺是否为新店
    calculate_distance_duration(shop_info, user)           # 调用百度地图API计算配送费和时间
    return shop_info
```

在view.py文件中调用了两个common.py中的公共方法，所以在common.py中，还需定义相应的方法，具体代码如下：

```python
def for_item(item):
    """获取店铺图片"""
    from collections import Iterable
    from apps.user.models import ShopImage

    # 判断是否为可迭代对象
    if isinstance(item, Iterable):
        for info in item:
            image = ShopImage.objects.get(shop_id=info.id).image
            info.shop_image = image
    else:
        item.shop_image = ShopImage.objects.get(shop_id=item.id).image
    return item

def calculate_distance_duration(df_shop, user, address_id=None):
    """调用百度地图计算配送费、时间"""
    from collections import Iterable
    from apps.user.models import Address

    # 判断传入数据是否可以遍历
    if isinstance(df_shop, Iterable):
        for info in df_shop:
            if address_id is None:
                address_shop = Address.objects.get(is_default=True, user=info.user)
                address_user = Address.objects.get(is_default=True, user=user)
            else:
                address_shop = Address.objects.get(is_default=True, user=info.user)
                address_user = Address.objects.get(id=address_id)
            distance_duration = get_distance_duration({'lat': address_shop.lat, 'lng': address_shop.lng},
                                                     {'lat': address_user.lat, 'lng': address_user.lng})
            info.send_price = int(distance_duration['distance']/1000)
            info.duration = int(distance_duration['duration']/60)
    else:
        if address_id is None:
            address_shop = Address.objects.get(is_default=True, user=df_shop.user)
            address_user = Address.objects.get(is_default=True, user=user)
        else:
            address_shop = Address.objects.get(is_default=True, user=df_shop.user)
            address_user = Address.objects.get(id=address_id)
        distance_duration = get_distance_duration({'lat': address_shop.lat, 'lng': address_shop.lng},
                                                 {'lat': address_user.lat, 'lng': address_user.lng})
        df_shop.send_price = int(distance_duration['distance']/1000)
        df_shop.duration = int(distance_duration['duration']/60)

    return df_shop
```

将店铺首页的模板文件 index.html 复制到 templates 文件夹内,并且让其继承抽象出来的 mj.html 买家模板文件,继承模板后代码如下:

```
{% extends 'mj.html' %}
{% load staticfiles %}
{% block title %}吃了么{% endblock title %}
{% block topfiles %}
<script src="{% static 'js/jquery.min.js' %}"></script>
<script src="{% static 'js/Popt.js' %}"></script>
<script src="{% static 'js/cityJson.js' %}"></script>
<script src="{% static 'js/citySet.js' %}"></script>
{% endblock topfiles %}
{% block body %}
<div id="banner">
    <nav>
        <ul>
            <li><a href=""><img src="{% static 'image/zc.png' %}">中餐</a></li>
            <li><a href=""><img src="{% static 'image/xc.png' %}">西餐</a></li>
            <li><a href=""><img src="{% static 'image/sg.png' %}">水果</a></li>
            <li><a href=""><img src="{% static 'image/yp.png' %}">饮品</a></li>
        </ul>
    </nav>
    <div class="banner">
    </div>
</div>
<input id="input-id" type="number" class="rating" min=0 max=5 step=0.5 data-size="lg" >
<div id="dinner">
    <div class="dinner_title">
        <h2>中餐</h2><a href="">查看全部></a>
    </div>
    <div class="dinner_con_bg">
        <div class="dinner_con">

            <ul>
              {% for item in shop_info0 %}
                <li>
                <a href="">
                    <img src="{{ item.shop_image }}">
                    <div class="cg_inf">
                        <h3>{{ item.shop_name }}</h3>
                        <div class="cg_eva">
                            <ul>
                                <li><img src="{% static 'image/eva.png' %}"></li>
                                <li><img src="{% static 'image/eva.png' %}"></li>
                                <li><img src="{% static 'image/eva.png' %}"></li>
                                <li><img src="{% static 'image/eva.png' %}"></li>
                                <li><img src="{% static 'image/eva.png' %}"></li>
                            </ul>
                            <span>{{ item.shop_score }}</span>
                        </div>
                        <div class="food_sc">
                            <span>起送:{{ item.shop_price }}</span><span>配送费:{{ item.send_price }}</span>
                            <span>时间:{{ item.duration }}分钟</span>
                        </div>
                    </div>
                </a>
                </li>
              {% endfor %}
            </ul>
        </div>
    </div>
</div>
```

```html
<div id="dinner">
    <div class="dinner_title">
        <h2>西餐</h2><a href="">查看全部</a>
    </div>
    <div class="dinner_con_bg">
        <div class="dinner_con">
            <ul>
                {% for item in shop_info1 %}
                <li>
                    <a href="">
                        <img src="{{ item.shop_image }}">
                        <div class="cg_inf">
                            <h3>{{ item.shop_name }}</h3>
                            <div class="cg_eva">
                                <ul>
                                    <li><img src="{% static 'image/eva.png' %}"></li>
                                    <li><img src="{% static 'image/eva.png' %}"></li>
                                    <li><img src="{% static 'image/eva.png' %}"></li>
                                    <li><img src="{% static 'image/eva.png' %}"></li>
                                    <li><img src="{% static 'image/eva.png' %}"></li>
                                </ul>
                                <span>{{ item.shop_score }}</span>
                            </div>
                            <div class="food_sc">
                                <span>起送:{{ item.shop_price }}</span><span>配送费:{{ item.send_price }}</span>
                                <span>时间:{{ item.duration }}分钟</span>
                            </div>
                        </div>
                    </a>
                </li>
                {% endfor %}
            </ul>
        </div>
    </div>
</div>
<div id="fruit">
    <div class="fruit_title">
        <h2>水果</h2>
    </div>
    <div class="fruit_con">
        <a href="" class="a_href">查看全部</a>
        <div class="fruit_con_ul">
            <ul>
                {% for item in shop_info2 %}
                <li>
                    <a href=""><img src="{{ item.shop_image }}">
                        <div class="sgd_inf">
                            <h3>{{ item.shop_name }}</h3>
                            <div class="cg_eva">
                                <ul>
                                    <li><img src="{% static 'image/eva.png' %}"></li>
                                    <li><img src="{% static 'image/eva.png' %}"></li>
                                    <li><img src="{% static 'image/eva.png' %}"></li>
                                    <li><img src="{% static 'image/eva.png' %}"></li>
                                    <li><img src="{% static 'image/eva.png' %}"></li>
                                </ul>
                                <span>{{ item.shop_score }}</span>
                            </div>
                            <div class="food_sc">
                                <span>起送:{{ item.shop_price }}</span><span>配送费:{{ item.send_price }}</span>
                                <span>时间:{{ item.duration }}分钟</span>
                            </div>
```

```html
                </div>
            </a>
        </li>
        {% endfor %}
    </ul>
</div>
</div>
</div>
<div id="drink">
    <div class="drink_title">
        <h2>饮品</h2>
    </div>
    <div class="drink_con">
        <a href="" class="a_href">查看全部></a>
        <div class="drink_con_ul">
            <ul>
                {% for item in shop_info3 %}
                <li>
                    <a href=""><img src="{{ item.shop_image }}">
                        <div class="ypd_inf">
                            <h3>{{ item.shop_name }}</h3>
                            <div class="cg_eva">
                                <ul>
                                    <li><img src="{% static 'image/eva.png' %}"></li>
                                    <li><img src="{% static 'image/eva.png' %}"></li>
                                    <li><img src="{% static 'image/eva.png' %}"></li>
                                    <li><img src="{% static 'image/eva.png' %}"></li>
                                    <li><img src="{% static 'image/eva.png' %}"></li>
                                </ul>
                                <span>{{ item.shop_score }}</span>
                            </div>
                            <div class="food_sc">
                                <span>起送:{{ item.shop_price }}</span><span>配送费:{{ item.send_price }}</span><span>时间:{{ item.duration }}分钟</span>
                            </div>
                        </div>
                    </a>
                </li>
                {% endfor %}
            </ul>
        </div>
    </div>
</div>
<div id="service_int">
    <div class="ser_title">
        <h2>服务介绍</h2>
    </div>
    <div class="ser_con">
        <ul>
            <li><a href=""><img src="{% static 'image/ser_1.jpg' %}"><span>中餐</span></a></li>
            <li><a href=""><img src="{% static 'image/ser_2.jpg' %}"><span>西餐</span></a></li>
            <li><a href=""><img src="{% static 'image/ser_3.jpg' %}"><span>水果</span></a></li>
            <li><a href=""><img src="{% static 'image/ser_4.jpg' %}"><span>饮品</span></a></li>
        </ul>
    </div>
</div>
{% endblock body %}
```

由于必须登录成功之后才能访问首页信息，而现在登录成功之后虽然跳转到了首页，但是浏览器的地址栏并没有改变，所以还需要对登录页面代码做一些调整，让其可以匹配到首页。对 user\views.py 文件中 LoginView 类下的 post() 方法做出如下修改：

```
# 判断卖家或买家，跳转不同首页，1 为卖家
if user.is_staff:
    response = redirect(reverse('goods:sj_index'))
    # response = render(request, 'sj_index.html')
else:
    response = redirect(reverse('goods:index'))    # HttpResponseRedirect
    # response = render(request, 'index.html')
return response
```

最后，启动项目，在浏览器中访问 http://127.0.0.1:8000/就会自动跳转至登录页面，登录成功后即可访问到店铺首页信息，如图 10.5 所示。

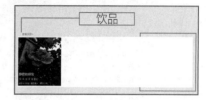

图 10.5 临时首页

10.6.2 店铺列表

店铺首页仅展示了一些热门店铺，并未展示出所有的店铺信息，所以还应增加相应的列表页将所有的店铺根据具体的类型展示出来，并添加分页功能。

在 urls.py 文件中，在首页匹配的 URL 信息上，新增匹配信息，具体代码如下：

```python
from django.urls import path, re_path

from apps.goods import views

urlpatterns = [
    re_path(r'^wm_index/(?P<code>[0-4])/(?P<page>\d+)/$', views.WmIndexView.as_view(), name='wm_index'), # 店铺列表
    path('', views.index, name='index') # 店铺首页（买家浏览）
]
```

由于可以通过 index.html 首页和 mj.html 模板访问到列表页，所以还需在首页和模板之间增加跳转到列表页的链接，关键代码如下：

```html
<ul>
    <li><a href="{% url 'goods:wm_index' '0' '1'%}"><img src="{% static 'image/zc.png' %}">中餐</a></li>
    <li><a href="{% url 'goods:wm_index' '1' '1'%}"><img src="{% static 'image/xc.png' %}">西餐</a></li>
    <li><a href="{% url 'goods:wm_index' '2' '1'%}"><img src="{% static 'image/sg.png' %}">水果</a></li>
    <li><a href="{% url 'goods:wm_index' '3' '1'%}"><img src="{% static 'image/yp.png' %}">饮品</a></li>
</ul>
```

根据以上匹配的 URL 信息对首页和模板做出修改后，接下来处理对应的业务逻辑即可。在 goods\views.py 文件中新增一个类 WmIndexView，具体代码如下：

```python
from django.shortcuts import render, reverse, redirect
from django.views.generic import View
from django.core.paginator import Paginator

from apps.user.models import Shop, User, ShopTypeDetail

from utils.common import shop_is_new, for_item, calculate_distance_duration


class WmIndexView(View):
    """店铺分类展示"""

    def get(self, request, code, page):

        try:
            user = User.objects.get(id=request.user.id)
        except User.DoesNotExist:
            return render(request, 'login.html', {'errmsg': '用户登录信息已失效，请重新登录！'})
```

```python
    type_info = None
    if code == '4':
        shop_info = Shop.objects.all()
    else:
        shop_info = Shop.objects.filter(shop_type=code)
        if code == '0':
            type_info = ShopTypeDetail.objects.filter(type_code__contains='C')
        if code == '1':
            type_info = ShopTypeDetail.objects.filter(type_code__contains='W')

    for_item(shop_info)                                      # 获取店铺图片
    calculate_distance_duration(shop_info, user)             # 计算配送费和时间
    shop_is_new(shop_info)                                   # 判断店铺是否为新店
    paginator = Paginator(shop_info, 16)                     # 对数据进行分页

    # 获取第 page 页的内容
    try:
        page = int(page)
    except Exception as e:
        page = 1

    if page > paginator.num_pages:
        page = 1
    skus_page = paginator.page(page)                         # 获取第 page 页的 Page 实例对象

    # 进行页码的控制，页面上最多显示 5 个页码
    # 1.总页数小于 5 页，页面上显示所有页码
    # 2.如果当前页是前 3 页，显示 1-5 页
    # 3.如果当前页是后 3 页，显示后 5 页
    # 4.其他情况，显示当前页的前 2 页，当前页，当前页的后 2 页
    num_pages = paginator.num_pages
    if num_pages < 5:
        pages = range(1, num_pages + 1)
    elif page <= 3:
        pages = range(1, 6)
    elif num_pages - page <= 2:
        pages = range(num_pages - 4, num_pages + 1)
    else:
        pages = range(page - 2, page + 3)

    context = {'code': code, 'type_info': type_info, 'skus_page': skus_page, 'pages': pages}
    return render(request, 'wm_index.html', context)
```

接下来只需要将后台传递过来的信息通过 render() 渲染的模板显示出来即可，将 static 下的 wm_index.html 文件复制到 templates 文件夹，具体代码如下：

```html
{% extends 'mj.html'%}
{% load staticfiles %}
{% block title %}吃了么-外卖{% endblock title %}
{% block topfiles %}
<script src="{% static 'js/jquery.min.js' %}"></script>
<script type="text/javascript">
$(function () {
    $('.wm_list li').click(function(){
        // 获得当前被点击的元素索引值
        var Index = $(this).index();
        var line=300*Index-300;
        // 给菜单添加选择样式
        $(this).addClass('active').siblings().removeClass('active');
        $(".line").stop(true,true).animate({left:line},200);
        $('.wm_list_con').children('div').eq(Index).show().siblings('div').hide();
```

```
    });
});
</script>
<script src="{% static 'js/Popt.js' %}"></script>
<script src="{% static 'js/cityJson.js' %}"></script>
<script src="{% static 'js/citySet.js' %}"></script>
{% endblock topfiles %}

{% block body %}
<div id="main">
    <ul class="wm_list">
        <li class="active">全部</li>
        {% for type in type_info %}
            <li>{{ type.type_name }}</li>
        {% endfor %}
    </ul>
    <div class="wm_list_con">
        <div class="list_1">
            <ul>
                {% for item in skus_page %}
                    <li>
                        <a href="">
                            <div class="wm_img"><img src="{{ item.shop_image }}"></div>
                            <div class="cg_inf">
                                <h3>{{ item.shop_name }}</h3>
                                <div class="cg_eva">
                                    <ul>
                                        <li><img src="{% static 'image/eva.png' %}"></li>
                                        <li><img src="{% static 'image/eva.png' %}"></li>
                                        <li><img src="{% static 'image/eva.png' %}"></li>
                                        <li><img src="{% static 'image/eva.png' %}"></li>
                                        <li><img src="{% static 'image/eva.png' %}"></li>
                                    </ul>
                                    <span>{{ item.shop_score }}</span>
                                </div>
                                <div class="food_sc">
                                    <span>起送：{{ item.shop_price }}</span><span>配送费：{{ item.send_price }}</span><span>时间：{{ item.duration }}分钟</span>
                                </div>
                            </div>
                        </a>
                    </li>
                {% endfor %}
            </ul>

            <div class="pagenation">
                {% if skus_page.has_previous %}
                    <a href="{% url 'goods:wm_index' code skus_page.previous_page_number %}">< 上一页</a>
                {% endif %}

                {% for pindex in pages %}
                    {% if pindex == skus_page.number %}
                        <a href="{% url 'goods:wm_index' code pindex %}" class="active">{{ pindex }}</a>
                    {% else %}
                        <a href="{% url 'goods:wm_index' code pindex %}">{{ pindex }}</a>
                    {% endif %}
                {% endfor %}

                {% if skus_page.has_next %}
                    <a href="{% url 'goods:wm_index' code skus_page.next_page_number %}">下一页 ></a>
                {% endif %}
            </div>
```

```
            </div>
        </div>
    </div>
{% endblock body %}
```

修改完模板代码后，启动项目，单击首页或者模板页面相关的链接，就可以查看店铺的列表页。

10.6.3 店铺详情页

用户在首页或店铺的列表页，如果单击某一店铺，就会跳转到店铺的详情页面，此时只需要在前端将店铺的 id 信息传递给后台即可。通常发送的是 GET 请求，所以 urls.py 匹配的具体信息如下：

```
from django.urls import path, re_path

from apps.goods import views

urlpatterns = [
    re_path(r'^goods/(?P<goods_id>\d+)/(?P<a_page>\d+)/(?P<b_page>\d+)/(?P<c_page>\d+)$',
        views.ShopDetailView.as_view(), name='shop_detail'),       # 店铺详情页
    path('', views.index, name='index')                             # 买家首页
]
```

根据匹配的 URL 信息，在可以访问店铺详情的页面，为其添加一个 a 标签，关键代码如下：

```
<a href="{% url 'goods:shop_detail' item.id '1'  '1' '1' %}">
```

接下来在 views.py 文件中添加具体的方法，查询出店铺所有的商品信息、评价信息（在评论模块处理），具体代码如下：

```python
from django.shortcuts import render, reverse, redirect
from django.views.generic import View
from django.core.paginator import Paginator

from apps.user.models import Shop, User, ShopTypeDetail
from apps.goods.models import GoodsType, GoodsSKU

from utils.common import shop_is_new, for_item, calculate_distance_duration, goods_item, get_comment_confusion, 
    good_rate
class ShopDetailView(View):
    """店铺详情页"""

    def get(self, request, goods_id, a_page, b_page, c_page):
        """显示详情页"""

        try:
            shop = Shop.objects.get(id=goods_id)
            for_item(shop)
        except Shop.DoesNotExist:
            return render(request, 'wm_index.html', {'errmsg': '店铺不存在'})

        try:
            user = User.objects.get(id=request.user.id)
        except User.DoesNotExist:
            return render(request, 'login.html', {'errmsg': '用户登录信息已失效，请重新登录！'})

        type_info = GoodsType.objects.filter(shop_id=goods_id)      # 获取店铺下所售商品种类
        sku_info = GoodsSKU.objects.filter(goods__shop_id=goods_id) # 获取该店铺所有商品
        goods_item(sku_info)                                         # 添加图片路径
        shop_is_new(shop)                                            # 是否显示为新店
```

```python
        calculate_distance_duration(shop, user)           # 调用百度地图API计算配送费和时间
        order_info_a = get_comment_confusion(6, 11, shop, a_page)   # 获取该店铺的好评
        order_info_b = get_comment_confusion(3, 6, shop, b_page)    # 获取该店铺的中评
        order_info_c = get_comment_confusion(0, 3, shop, c_page)    # 或取该店铺的差评

        rate = good_rate(shop)

        # page_item(order_info_a, 1, 10)

        # 整合数据
        context = {'shop': shop, 'sku_info': sku_info, 'type_info': type_info, 'rate': rate,
                   'order_info_a': order_info_a,
                   'order_info_b': order_info_b,
                   'order_info_c': order_info_c}
        return render(request, 'wm_shop.html', context)
```

在 view.py 的 ShopDetailView 类中，将公共的方法抽象出来，添加到 common.py 文件中，具体代码如下：

```python
def get_comment_confusion(score_begin, score_end, shop, page):
    """获取各种评价的公共方法"""
    from collections import Iterable
    from apps.order.models import OrderInfo, CommentImage, OrderGoods
    from apps.user.models import UserImage

    order_info = OrderInfo.objects.filter(score__gte=score_begin, score__lt=score_end, shop=shop,
       order_status=7).order_by('-create_time')

    get_image(order_info, UserImage, 'user_id')

    get_image(order_info, CommentImage, 'order_id', foreign_key=True)

    if isinstance(order_info, Iterable):
        for order in order_info:
            goods = OrderGoods.objects.filter(order=order)
            order.goods = goods
    else:
        goods = OrderGoods.objects.filter(order=order_info)
        order_info.goods = goods

    # 将数据进行分页
    context = page_item(order_info, page, 4)

    return context

def get_image(item, object_name, field, foreign_key=False):
    """获取图片"""
    from collections import Iterable

    # 判断是否为可迭代对象
    if isinstance(item, Iterable):
        for info in item:
            # 订单表无id特殊处理
            if not foreign_key and not hasattr(info, 'id'):
                info = info.user
            elif not hasattr(info, 'id'):
                info.id = info.order_id

            try:
                image = object_name.objects.get(**{field: info.id}).image
                info.image = image
            except Exception as e:
                image_list = object_name.objects.filter(**{field: info.id})
                info.image_list = image_list
```

```python
        else:
            # 订单表特殊处理
            if not foreign_key and not hasattr(item, 'id'):
                item = item.user
            elif not hasattr(item, 'id'):
                item.id = item.order_id
            try:
                image = object_name.objects.get(**{field: item.id}).image
                item.image = image
            except Exception as e:
                image_list = object_name.objects.filter(**{field: item.id})
                item.image_list = image_list
    return item

def good_rate(shop):
    """计算好评率，return % 形式字符串"""
    from apps.order.models import OrderInfo

    a_count = OrderInfo.objects.filter(score__gte=6, score__lte=10, shop=shop, order_status=7).count()
    total_count = OrderInfo.objects.filter(shop=shop, order_status=7).count()
    if total_count != 0:
        rate = '{:.2%}'.format(a_count / total_count)
    else:
        rate = '0.00%'
    return rate
```

业务逻辑处理完成之后，由于 views.py 文件渲染的是 wm_shop.html，所以将 static 文件夹下的该文件复制到 templates 文件夹，并修改成如下代码：

```
{% extends 'mj.html' %}
{% load staticfiles %}
{% block title %}吃了么-卖家店铺{% endblock title %}
{% block topfiles %}
<script src="{% static 'js/jquery.min.js' %}"></script>
<script src="{% static 'js/Popt.js' %}"></script>
<script src="{% static 'js/cityJson.js' %}"></script>
<script src="{% static 'js/citySet.js' %}"></script>
<script src="{% static 'js/comment.js' %}"></script>
<script type="text/javascript" src="{% static 'js/floor.js' %}"></script>
<script type="text/javascript">
$(function () {
    $('.shop_tab ul li').hover(function(){
        // 获得当前被点击的元素索引值
        var Index = $(this).index();
        var line=300*Index-300;
        // 给菜单添加选择样式
        $(this).addClass('shop_active').siblings().removeClass('shop_active');
        $(".line").stop(true,true).animate({left:line},200);
        $('.shop_tab_con').children('div').eq(Index).show().siblings('div').hide();
    });
});
</script>
<script type="text/javascript">
$(function () {
    $('.eval_nav ul li').click(function(){
        // 获得当前被点击的元素索引值
        var Index = $(this).index();
        var line=300*Index-300;
        // 给菜单添加选择样式
        $(this).addClass('eval_active').siblings().removeClass('eval_active');
        $(".line").stop(true,true).animate({left:line},200);
        $('.eval_list_con').children('div').eq(Index).show().siblings('div').hide();
```

```
    });
});
</script>
<script>
    $(function () {
        var obj = new commentMove('.ph1', '.phview1');
        obj.init()
    })

    $(function () {
        var obj = new commentMove('.ph2', '.phview2');
        obj.init()
    })
</script>
{% endblock topfiles %}

{% block body %}
<div id="main">
    <div class="loc_nav"><a href="{% url 'goods:index' %}">首页</a>><a href="">{{ shop.shop_name }}</a></div>
    <div class="shop_inf">
        <img src="{{ shop.shop_image }}">
        <div class="shop_inf_wz">
            <h2>{{ shop.shop_name }}<span>证</span></h2>
            <b>综合评分：{{ shop.shop_score }}</b>
            <span>接单时间：{{ shop.receive_start }}-{{ shop.receive_end }}
                <font>
                    {% if shop.business_do %}
                        营业中
                    {% else %}
                        休息中
                    {% endif %}
                </font>
            </span>
            <span><font>商户地址：</font>    {{ shop.shop_addr }}</span>
            <ul>
                <li><span>{{ shop.duration }}分钟</span><span>最快送达时间</span></li>
                <li><span>￥{{ shop.shop_price }}</span><span>起送价</span></li>
                <li><span>￥{{ shop.send_price }}</span><span>配送费</span></li>
            </ul>
        </div>
    </div>
    <div class="shop_tab">
        <ul>
            <li class="shop_active">菜单</li>
            <li>评价</li>
        </ul>
    </div>
    <div class="shop_tab_con">
        <div class="shop_bill">
            <div class="shop_bill_nav">
                <ul class="fixedmeau">
                    {% for type in type_info %}
                        <li class="bill_active">{{ type.name }}</li>
                    {% endfor %}
                </ul>
            </div>

            <div class="louceng_box">
                {% for type in type_info %}
                    <div class="louceng">
                        <p>{{ type.name }}</p>
                        <ul class="menu_li">
                            {% for sku in sku_info %}
```

```html
                {% if type.id == sku.type_id %}
                <li>
                    <div>
                        <img src="{{ sku.image }}">
                    </div>
                    <span id="sku_id" name="sku_id" style="display: none">{{ sku.id }}</span>
                    <span>{{ sku.name }}</span>
                    <b>￥{{sku.price}}</b>
                <div class="bill_btn">
                        <button class="minus">
                            <strong></strong>
                        </button>
                        <i class="num">0</i>{# 没有 num 这个样式 #}
                        <button class="add">
                            <strong></strong>
                        </button>
                    </div>
                </li>
                {% endif %}
            {% endfor %}
            </ul>
        </div>
        {% endfor %}

        <div class="shop_car">
            <div class="shop_car_title">
                <span>购物车</span><a href="javascript:void(0)" class="clear_shopcar">清空购物车</a>
                <div class="clr"></div>
            </div>
            <ul class="shop_car_con">
                <p>总计：￥<span id="totalpriceshow">0</span>元</p>
                <button class="shop_btn" style="display:none">立即下单</button>
                <button class="shop_btn1">￥{{ shop.shop_price }}元起送</button>
            </ul>
        </div>

    </div>
    <script language="javascript" type="text/javascript">
        $(function() {
// 加的效果
$(".add").click(function () {
    $(this).prevAll().css("display", "inline-block");
    var n = $(this).prev().text();
    var num = parseInt(n) + 1;
    if (num == 0) { return; }
    $(this).prev().text(num);
    var danjia = ($(this).parent().prev().text()).substring(1);    // 获取单价
    var a = $("#totalpriceshow").html();                           // 获取当前所选总价
    $("#totalpriceshow").html(a * 1 + danjia * 1);                 // 计算当前所选总价
    jss();
});
// 减的效果
$(".minus").click(function () {
    var n = $(this).next().text();
    var num = parseInt(n) - 1;
    $(this).next().text(num);//减 1
    var danjia = ($(this).parent().prev().text()).substring(1);    // 获取单价
    var a = $("#totalpriceshow").html();                           // 获取当前所选总价
    $("#totalpriceshow").html(a * 1 - danjia * 1);                 // 计算当前所选总价
    if (num <= 0) {
        $(this).next().css("display", "none");
        $(this).css("display", "none");
        jss();                                                     // 改变按钮样式
```

```
            return
        }
        if ($("#totalpriceshow").html()<= 0) {
            $(this).next().css("display", "none");
            $(this).css("display", "none");
                $("#totalpriceshow").html(0)
            jss();                                           // 改变按钮样式
            return
        }
        jss();
    });
        })

$(".shop_btn").click(function(){
        // window.location.href="wm_plaorder.html";
        var menu = $(".menu_li").children();                 // 获取所有商品
            var array_id = [];
            var array_count = [];
                if(menu.length > 0){
                    $.each(menu, function(i,ele){
                        $(ele).find(".num").html();
                        // alert($(ele).find(".num").html())
                            if ($(ele).find(".num").html()>0){
                                array_id.push($(ele).find("#sku_id").html());
                                array_count.push($(ele).find(".num").html());
                            }
                    });
                    // var csrf = $('input[name="csrfmiddlewaretoken"]').val();

                    post('/cart/wm_start',{cm1:array_id ,cm2:array_count});
                }

    })

function post(URL, PARAMS) {
    var temp = document.createElement("form");
    temp.action = URL;
    temp.method = "get";
    temp.style.display = "none";
    for (var x in PARAMS) {
        var opt = document.createElement("textarea");
        opt.name = x;
        opt.value = PARAMS[x];
        // alert(opt.name)
        temp.appendChild(opt);
    }
    document.body.appendChild(temp);
    temp.submit();
    return temp;
}

function jss() {
    var m = $("#totalpriceshow").html();
    if (m >= {{ shop.shop_price }}) {
        $(".shop_btn").css("display","inline-block");
        $(".shop_btn1").css("display","none");
    } else {
        $(".shop_btn").css("display","none");
        $(".shop_btn1").css("display","inline-block");
    }
};
    $(".clear_shopcar").click(function () {
```

```
                $(".bill_btn .minus").css("display","none");
                $(".bill_btn i").text(0);
                $(".bill_btn i").css("display","none");
                $(".shop_btn").css("display","none");
                $(".shop_btn1").css("display","inline-block");
                 $("#totalpriceshow").html(0);
              });
           </script>
           <script type="text/javascript" language="javascript">
               /*
                  totop                                 // 返回顶部按钮
                  fixedevery                            // 左侧固定导航的每一项
                  louceng                               // 模块的每一项
                  header                                // 头部
               */
               $(function(){
                    var obj = new floor('.totop','.fixedmeau>li','.louceng_box>.louceng','.header_box');
                    obj.init()
               })
           </script>
        </div>
        <div class="shop_eval">
            <div class="Praise_degree">
                <p><span>好评度</span><span>{{ rate }}</span></p>
                <div class="clr"></div>
            </div>
        </div>
    </div>
</div>
{% endblock body %}
```

在 wm_shop.html 文件中，包含了大量购物车的 JS 代码，在 10.7 节会详细介绍。

启动项目，买家在浏览器中访问 http://127.0.0.1:8000/goods/1/1/1/1 就可以查看到店铺的详细信息，如图 10.6 所示。

图 10.6　店铺详情

10.7 购物车模块设计

买家登录吃了么外卖网后，可以选择想要下单的店铺并将想要下单的商品添加到购物车，这可以通过购物车模块实现。在购物车模块中，主要包括添加至购物车、确认费用和修改收货地址 3 部分功能，下面分别进行介绍。

10.7.1 添加至购物车

在店铺的详情页功能中，我们可以将商品添加至购物车中，或者修改购物车中的商品，简单地计算总价等。此类需求由于没有数据库的相关操作，因此通过前端代码即可实现。但是我们也要再添加一个新模块，用来处理其他的功能需求。

通过命令"python manage.py startapp cart"新建一个购物车模块，并将其拖动到 apps 文件夹下，新增 urls.py 文件，并且在其中设置 urlpatterns = [ ]，在 base.py 文件中的 INSTALLED_APPS 下注册新增的模块，关键代码如下：

```
INSTALLED_APPS = [
    'django.contrib.admin',
    'django.contrib.auth',
    'django.contrib.contenttypes',
    'django.contrib.sessions',
    'django.contrib.messages',
    'django.contrib.staticfiles',
    'haystack',                          # 全文检索
    'apps.user',                         # 用户模块
    'apps.goods',                        # 商品模块
    'apps.cart',                         # 购物车模块
]
```

在 clmwm\urls.py 文件中设置路由信息，具体代码如下：

```
from django.contrib import admin
from django.urls import path, include

urlpatterns = [
    path('admin/', admin.site.urls),
    path('user/', include(('apps.user.urls', 'apps.user'), namespace='user')),    # 用户模块
    path('cart/', include(('apps.cart.urls', 'apps.cart'), namespace='cart')),    # 购物车模块
]
```

在店铺详情页面，已经把所有的商品信息查询出来。当单击商品的"+"按钮时，就会将 1 件商品添加至购物车中，多次单击将会添加多个商品，并且显示"—"功能，以及显示商品数量，购物车中显示商品总价并判断是否符合配送价格。关键代码如下：

```
{% extends 'mj.html' %}
{% load staticfiles %}
{% block title %}吃了么-卖家店铺{% endblock title %}
{% block topfiles %}
<script src="{% static 'js/jquery.min.js' %}"></script>
<script src="{% static 'js/Popt.js' %}"></script>
<script src="{% static 'js/cityJson.js' %}"></script>
<script src="{% static 'js/citySet.js' %}"></script>
<script src="{% static 'js/comment.js' %}"></script>
```

```html
<script type="text/javascript" src="{% static 'js/floor.js' %}"></script>
<script type="text/javascript">
$(function () {
    $('.shop_tab ul li').hover(function(){
          // 获得当前被点击的元素索引值
          var Index = $(this).index();
        var line=300*Index-300;
        // 给菜单添加选择样式
            $(this).addClass('shop_active').siblings().removeClass('shop_active');
            $(".line").stop(true,true).animate({left:line},200);
        $('.shop_tab_con').children('div').eq(Index).show().siblings('div').hide();
    });
});
</script>
<script type="text/javascript">
$(function () {
    $('.eval_nav ul li').click(function(){
          // 获得当前被点击的元素索引值
          var Index = $(this).index();
        var line=300*Index-300;
        // 给菜单添加选择样式
            $(this).addClass('eval_active').siblings().removeClass('eval_active');
            $(".line").stop(true,true).animate({left:line},200);
        $('.eval_list_con').children('div').eq(Index).show().siblings('div').hide();
    });
});
</script>
<script>
    $(function () {
        var obj = new commentMove('.ph1', '.phview1');
        obj.init()
    })

    $(function () {
        var obj = new commentMove('.ph2', '.phview2');
        obj.init()
    })
</script>
{% endblock topfiles %}

{% block body %}
 {% if is_none %}
      {{ message }}
 {% endif   %}
<div id="main">

    <div class="loc_nav"><a href="{% url 'goods:index' %}">首页</a>><a href="">{{ shop.shop_name }}</a></div>
    <div class="shop_inf">
        <img src="{{ shop.shop_image }}">
        <div class="shop_inf_wz">
            <h2>{{ shop.shop_name }}<span>证</span></h2>
            <b>综合评分：{{ shop.shop_score }}</b>
            <span>接单时间: {{ shop.receive_start }}-{{ shop.receive_end }}
                <font>
                    {% if shop.business_do %}
                        营业中
                    {% else %}
                        休息中
                    {% endif %}
                </font>
            </span>
            <span><font>商户地址: </font>　{{ shop.shop_addr }}</span>
```

```html
            <ul>
                <li><span>{{ shop.duration }}分钟</span><span>最快送达时间</span></li>
                <li><span>¥ {{ shop.shop_price }}</span><span>起送价</span></li>
                <li><span>¥ {{ shop.send_price }}</span><span>配送费</span></li>
            </ul>
        </div>
    </div>
    <div class="shop_tab">
        <ul>
            <li class="shop_active">菜单</li>
            <li>评价</li>
        </ul>
    </div>
    <div class="shop_tab_con">
        <div class="shop_bill">
            <div class="shop_bill_nav">
                <ul class="fixedmeau">
                    {% for type in type_info %}
                        <li class="bill_active">{{ type.name }}</li>
                    {% endfor %}
                </ul>
            </div>

            <div class="louceng_box">
                {% for type in type_info %}
                    <div class="louceng">
                        <p>{{ type.name }}</p>
                        <ul class="menu_li">

                            {% for sku in sku_info %}
                                {% if type.id == sku.type_id %}
                            <li>
                                <div>
                                    <img src="{{ sku.image }}">
                                </div>
                                <span id="sku_id" name="sku_id" style="display: none">{{ sku.id }}</span>
                                <span>{{ sku.name }}</span>
                                <b> ¥ {{sku.price}}</b>
                                <div class="bill_btn">
                                    <button class="minus">
                                        <strong></strong>
                                    </button>
                                    <i class="num">0</i>{# 没有 num 这个样式 #}
                                    <button class="add">
                                        <strong></strong>
                                    </button>
                                </div>
                            </li>
                                {% endif %}
                            {% endfor %}
                        </ul>
                    </div>
                {% endfor %}

                <div class="shop_car">
                    <div class="shop_car_title">
                        <span>购物车</span><a href="javascript:void(0)" class="clear_shopcar">清空购物车</a>
                        <div class="clr"></div>
                    </div>
                    <ul class="shop_car_con">
                        <p>总计：¥ <span id="totalpriceshow">0</span>元</p>
                        <button class="shop_btn" style="display:none">立即下单</button>
```

```html
                <button class="shop_btn1"> ¥ {{ shop.shop_price }}元起送</button>
            </ul>
        </div>

    </div>
    <script language="javascript" type="text/javascript">
        $(function() {
// 加的效果
$(".add").click(function () {
    $(this).prevAll().css("display", "inline-block");
    var n = $(this).prev().text();
    var num = parseInt(n) + 1;
    if (num == 0) { return; }
    $(this).prev().text(num);
    var danjia = ($(this).parent().prev().text()).substring(1);   // 获取单价
    var a = $("#totalpriceshow").html();                          // 获取当前所选总价
    $("#totalpriceshow").html(a * 1 + danjia * 1);                // 计算当前所选总价
    jss();
});
// 减的效果
$(".minus").click(function () {
    var n = $(this).next().text();
    var num = parseInt(n) - 1;
    $(this).next().text(num);                                     // 减1
    var danjia = ($(this).parent().prev().text()).substring(1);   // 获取单价
    var a = $("#totalpriceshow").html();                          // 获取当前所选总价
    $("#totalpriceshow").html(a * 1 - danjia * 1);                // 计算当前所选总价
    if (num <= 0) {
        $(this).next().css("display", "none");
        $(this).css("display", "none");
        jss();                                                    // 改变按钮样式
        return
    }
    if ($("#totalpriceshow").html()<= 0) {
        $(this).next().css("display", "none");
        $(this).css("display", "none");
        $("#totalpriceshow").html(0)
        jss();                                                    // 改变按钮样式
        return
    }
    jss();
});
        })

$(".shop_btn").click(function(){
    // window.location.href="wm_plaorder.html";
    var menu = $(".menu_li").children();                          // 获取所有商品
        var array_id = [];
        var array_count = [];
            if(menu.length > 0){
                $.each(menu, function(i,ele){
                    $(ele).find(".num").html();
                    // alert($(ele).find(".num").html())
                        if ($(ele).find(".num").html()>0){
                            array_id.push($(ele).find("#sku_id").html());
                            array_count.push($(ele).find(".num").html());
                        }
                });
                    // var csrf = $('input[name="csrfmiddlewaretoken"]').val();

            post('/cart/wm_start',{cm1:array_id ,cm2:array_count});
        }
```

```
    })
        function post(URL, PARAMS) {
            var temp = document.createElement("form");
            temp.action = URL;
            temp.method = "get";
            temp.style.display = "none";
            for (var x in PARAMS) {
                var opt = document.createElement("textarea");
                opt.name = x;
                opt.value = PARAMS[x];
                // alert(opt.name)
                temp.appendChild(opt);
            }
            document.body.appendChild(temp);
            temp.submit();
            return temp;
        }
        function jss() {
            var m = $("#totalpriceshow").html();
            if (m >= {{ shop.shop_price }}) {
                $(".shop_btn").css("display","inline-block");
                $(".shop_btn1").css("display","none");
            } else {
                $(".shop_btn").css("display","none");
                $(".shop_btn1").css("display","inline-block");
            }
        };
        $(".clear_shopcar").click(function () {
            $(".bill_btn .minus").css("display","none");
            $(".bill_btn i").text(0);
            $(".bill_btn i").css("display","none");
            $(".shop_btn").css("display","none");
            $(".shop_btn1").css("display","inline-block");
             $("#totalpriceshow").html(0);
        });
        </script>
        <script type="text/javascript" language="javascript">
            /*
                totop                                        // 返回顶部按钮
                fixedevery                                   // 左侧固定导航的每一项
                louceng                                      // 模块的每一项
                header                                       // 头部
            */
            $(function(){
                var obj = new floor('.totop','.fixedmeau>li','.louceng_box>.louceng','.header_box');
                obj.init()
            })
        </script>
      </div>
        <div class="shop_eval">
            <div class="Praise_degree">
                <p><span>好评度</span><span>{{ rate }}</span></p>
                <div class="clr"></div>
            </div>
        </div>
    </div>
  </div>
</div>
{% endblock body %}
```

启动项目，重新访问店铺的详情页就可以查看商品的添加和修改功能。

10.7.2 确认费用

在店铺的详细信息页面选购完商品后，单击"立即下单"按钮，会将商品数据以 POST 请求的方式传递到后台，关键代码如下：

```javascript
$(".shop_btn").click(function(){
    // window.location.href="wm_plaorder.html";
    var menu = $(".menu_li").children();                // 获取所有商品
        var array_id = [];
        var array_count = [];
            if(menu.length > 0){
                $.each(menu, function(i,ele){
                    $(ele).find(".num").html();
                        if ($(ele).find(".num").html()>0){
                            array_id.push($(ele).find("#sku_id").html());
                            array_count.push($(ele).find(".num").html());
                        }
                });
                get('/cart/wm_start',{cm1:array_id ,cm2:array_count});
            }

})
function get(URL, PARAMS) {
    var temp = document.createElement("form");
    temp.action = URL;
    temp.method = "get";
    temp.style.display = "none";
    for (var x in PARAMS) {
        var opt = document.createElement("textarea");
        opt.name = x;
        opt.value = PARAMS[x];
        temp.appendChild(opt);
    }
    document.body.appendChild(temp);
    temp.submit();
    return temp;
}
```

为了匹配请求的路由信息并指定处理方法，还需要在 cart\urls.py 文件中添加如下代码：

```python
from django.urls import path

from apps.cart.views import WmStartView

urlpatterns = [
    path('wm_start/', WmStartView.as_view(), name='wm_start'),
]
```

接下来，在 views.py 文件中新增 WmStartView 类，在其中接收前台传递过来的数据，具体代码如下：

```python
from django.shortcuts import render, redirect, reverse
from django.views.generic import View

from apps.goods.models import GoodsSKU, Goods
from apps.user.models import Shop, User, Address
from utils.common import calculate_distance_duration
```

```python
class WmStartView(View):
    """订单确认"""
    def get(self, request):
        array_id = request.GET.get('cm1').split(',')
        array_count = request.GET.get('cm2').split(',')
        if len(array_id) != len(array_count):
            return redirect(reverse('goods:wm_index'), {'errmsg': '数据错误'})

        sku_info = GoodsSKU.objects.filter(id__in=array_id)

        # 遍历出所有商品信息
        flag, total, total_goods = 0, 0, 0
        for sku in sku_info:
            sku.unite = array_count[flag]
            total_goods = total_goods + (sku.price + int(sku.pack)) * int(sku.unite)
            goods = Goods.objects.get(id=sku.goods_id)
            shop = Shop.objects.get(id=goods.shop_id)
            flag += 1

        if request.user.id:
            print('该用户的 id 为： %s ' % request.user.id)
        else:
            return redirect(reverse('user:login'))

        user = User.objects.get(id=request.user.id)
        distance = calculate_distance_duration(shop, user)                    # 计算运费
        total = total_goods + int(distance.send_price)                        # 一共支付
        address_info = Address.objects.order_by('-is_default').filter(user_id=request.user.id)  # 设置顺序

        # 整合数据
        context = {'sku_info': sku_info, 'total_goods': total_goods, 'total': total, 'shop': shop, 'user': user,
                   'address_info': address_info}

        return render(request, 'wm_plaorder.html', context)

    def post(self, request):
        pass
```

将 static 文件夹下的 wm_plaorder.html 文件复制到 templates 文件夹，并修改成如下代码：

```
{% extends 'mj.html' %}
{% load staticfiles %}
{% block title %}立即下单{% endblock title %}
{% block topfiles %}
<style type="text/css">.window{
    width:920px;
    background-color:#d0def0;
    position:absolute;
    padding:2px;
    margin:5px;
    display:none;
    }
.content{
    height:420px;
    background-color:#FFF;
    font-size:14px;
    overflow:auto;
    }
    .title{
        padding:2px;
```

```
            color:#0CF;
            font-size:14px;
        }
    .title img{
            float:right;
        }
</style>
<link href="{% static 'css/add_address.css' %}" rel="stylesheet" type="text/css">
<script src="{% static 'js/jquery.min.js' %}"></script>
<script src="{% static 'js/Popt.js' %}"></script>
<script src="{% static 'js/cityJson.js' %}"></script>
<script src="{% static 'js/citySet.js' %}"></script>
<script type="text/javascript">
 var windowHeight;                                         // 获取窗口的高度
 var windowWidth;                                          // 获取窗口的宽度
 var popWidth;                                             // 获取弹窗的宽度
 var popHeight;                                            // 获取弹窗的高度
 function init(){
    windowHeight=$(window).height();
    windowWidth=$(window).width();
    popHeight=$(".window").height();
    popWidth=$(".window").width();
 }
 // 关闭窗口的方法
 function closeWindow(){
    $(".title img").click(function(){
        $(this).parent().parent().hide("slow");
    });
 }
 // 定义弹出居中窗口的方法
 function popCenterWindow(){
    init();
    var popY=(windowHeight-popHeight)/2;                    // 计算弹出窗口的左上角 Y 的偏移量
    var popX=(windowWidth-popWidth)/2;
    //alert('jihua.cnblogs.com');
    $("#center").css("top",popY).css("left",popX).slideToggle("slow");  // 设定窗口的位置
    closeWindow();
 }
</script>
<script type="text/javascript" language="javascript">
    $(document).ready(function () {
        $("#btn_center").click(function () {
            popCenterWindow();
        });

    });

    function address_change(address_id, shop_id){
        $.get('/cart/change/', {'address_id': address_id, 'shop_id': shop_id}, function (data) {
            if (data.res == 3){
                $('#send_price').html('￥'+ data.mesg);
                var total = {{ total_goods }} + data.mesg;
                $('#con_count').html('￥' + parseFloat(total).toFixed(2));
                $('#pay_count').html('￥' + parseFloat(total).toFixed(2))
            }else{
                alert(data.errmsg);
            }
        });

    }
```

```
</script>
{% endblock topfiles %}

{% block body %}
<div id="main">
    <div class="loc_nav"><a href="{% url 'goods:index' %}">首页</a>><a href="/goods/ {{ shop.id }}/1/1/1">{{ shop.shop_name }}</a>><a href="">确认购买</a>    </div>
    <div class="sure_or">

        <div class="sure_list">
            <div></div>
            <ul>
                <li><span>商品</span><span>（价格 + 包装费）* 份数</span></li>
                {% for sku in sku_info %}
                    <li><span>{{ sku.name }}</span><span>¥（{{ sku.price }} + {{ sku.pack }}）* {{ sku.unite }}</span></li>
                {% endfor %}
                <li><span>配送费</span><span id="send_price" name="send_price">¥ {{ shop.send_price }}</span></li>
                <li><span>合计</span><span id="con_count" name="con_count">¥ {{ total }}</span></li>
            </ul>
        </div>

        <div class="window" id="center">
            <div id="title" class="title"><img src=" {% static 'image/close.jpg' %} " alt="关闭" />新增收货地址</div>
            <div class="content">
                <form method="post" action="/cart/address/" id="save_address" name="save_address">
                  <ul>
                    <li><span>收货人</span><input type="text" name="receiver" id="receiver"></li>
                    <li><span>所在地区</span><input type="text" name="region" id="region"></li>
                    <li><span>详细地址</span><input type="text" name="addr" id="addr"></li>
                    <li><span>手机号码</span><input type="text" name="phone" id="phone"></li>
                    <li><span>是否设为默认地址</span><input id="default" name="default" type="radio" value="1" checked/><span>是</span><input id="default" name="default" type="radio" value="0"/><span>否</span></li>
                  </ul>
                </form>
                <button class="agree_btn">保存</button>
                <script language="javascript" type="text/javascript">
                    $(".agree_btn").click(function(){
                        csrf = $('input[name="csrfmiddlewaretoken"]').val();
                        params = {
                            'receiver':$('#receiver').val(),
                            'region':$('#region').val(),
                            'addr':$('#addr').val(),
                            'phone':$('#phone').val(),
                            'default':$('#default').val(),
                            'csrfmiddlewaretoken':csrf
                        }
                        // 发起ajax post请求，访问/order/pay,传递参数:order_id
                        $.post('/cart/address/', params, function (data) {
                            if (data.res == 1){
                                alert(data.errmsg);
                            }else{
                                window.location.reload();
                            }

                        });
                    });
                </script>
            </div>
```

```html
            </div>
            <div class="sure_xx">
                <p>送餐详情</p>
                <div class="row" id="myVue"   v-cloak>
                    <ul>
                        <li><span class="addAddress" type="button" id="btn_center" >新增收货地址</span></li>
                    </ul>
                </div>
                <script type="text/javascript" src="{% static 'js/vue/vue.js' %}"></script>
                <script type="text/javascript" src="{% static 'js/eleme-ui/index.js' %}"></script>
                <script type="text/javascript" src="{% static 'js/ShoppingCart.js' %}"></script>
                <select id="addresses"  name="addresses"  onchange="address_change(this.options[this.options.selectedIndex].value, {{ shop.id }});">
                    {% for address in address_info %}
                    <option id="select_addr" name="select_addr" value="{{ address.id }}" >{{ address.addr }} {{ address.receiver }} {{ address.phone }}</option>
                    {% endfor %}
                </select>
                 <form id="order_generate" name="order_generate" method="post" action="/order/generate/">
                    {% csrf_token %}
                    <input style="display: none" id="address" name="address" value="">
                    <input style="display: none" id="sku_ids" name="sku_ids" value="">
                    <input style="display: none" id="shop_id" name="shop_id" value="{{ shop.id }}">
                    <ul>
                        <li>
                            <span>我要留言：</span><input id="remarks" name="remarks" type="text"
                                        placeholder="少辣 加米饭">
                        </li>
                        <li>
                            <span>发票信息：</span><input id="invoice_head" name="invoice_head" type="text"
                                        placeholder="输入发票抬头">
                        </li>
                        <li>
                            <span>发票信息：</span><input id="taxpayer_number" name="taxpayer_number" type="text"
                                        placeholder="输入纳税人识别号">
                        </li>
                    </ul>
                </form>
                <div class="go_pay">
                    <span>您需要支付： <b id="pay_count" name="pay_count">￥{{ total }}</b></span><button class="go_pay">
                                    去付款</button>
                </div>
            </div>
            <div class="clr"></div>

    </div>
    <script language="javascript" type="text/javascript">
        $(".go_pay").click(function(){
            document.getElementById("address").value = $("#select_addr").val()
            document.getElementById("sku_ids").value = window.location.search
            $("#order_generate").submit();
            // window.location.href="wm_pay.html";
        });
    </script>
</div>
{% endblock %}
```

启动项目，单击购物车中的"立即购买"按钮，就会跳转到订单确认页面，如图10.7所示。

图 10.7　订单确认

10.7.3　修改收货地址

在弹出订单确认页面后，默认收货地址为用户注册时填写的收货地址，但部分用户填写的地址并不精准，所以此处应该增加修改收货地址功能，并且确认是否设为默认地址，由于地址信息改变，所以还需要重新计算配送信息。

当单击"新增收货地址"时，会触发以下方法，弹出页面：

```
$(document).ready(function () {
    $("#btn_center").click(function () {
        popCenterWindow();
    });
});
<div class="window" id="center">
    <div id="title" class="title"><img src=" {% static 'image/close.jpg' %} " alt="关闭" />新增收货地址</div>
    <div class="content">
        <form method="post" action="/cart/address/" id="save_address" name="save_address">
            <ul>
                <li><span>收货人</span><input type="text" name="receiver" id="receiver"></li>
                <li><span>所在地区</span><input type="text" name="region" id="region"></li>
                <li><span>详细地址</span><input type="text" name="addr" id="addr"></li>
                <li><span>手机号码</span><input type="text" name="phone" id="phone"></li>
                <li><span>是否设为默认地址</span><input id="default" name="default" type="radio" value="1"
                    checked/><span>是</span><input id="default" name="default" type="radio" value="0"/><span>否
                </span></li>
            </ul>
        </form>
        <button class="agree_btn">保存</button>
        <script language="javascript" type="text/javascript">
```

```javascript
            $(".agree_btn").click(function(){
                csrf = $('input[name="csrfmiddlewaretoken"]').val();
                params = {
                    'receiver':$('#receiver').val(),
                    'region':$('#region').val(),
                    'addr':$('#addr').val(),
                    'phone':$('#phone').val(),
                    'default':$('#default').val(),
                    'csrfmiddlewaretoken':csrf
                }
                $.post('/cart/address/', params, function (data) {
                    if (data.res == 1){
                        alert(data.errmsg);
                    }else{
                        window.location.reload();
                    }
                });
            });
        </script>
    </div>
</div>
```

在 urls.py 文件中，添加保存收货地址的路由信息，具体代码如下：

```python
from django.urls import path
from apps.cart import views

urlpatterns = [
    path('address/', views.save_address, name='address')
]
```

在 views.py 文件中查询出地址信息，确认是否设为默认地址，将新增地址数据保存在地址表中，具体代码如下：

```python
from django.shortcuts import render, redirect, reverse
from django.views.generic import View
from django.http import JsonResponse

from apps.goods.models import GoodsSKU, Goods
from apps.user.models import Shop, User, Address
from utils.common import calculate_distance_duration, get_lng_lat

# Create your views here.

def save_address(request):
    """处理新增地址"""

    try:
        user = User.objects.get(id=request.user.id)
    except User.DoesNotExist:
        return render(request, 'login.html', {'errmsg': '用户登录信息已失效，请重新登录！'})

    receiver = request.POST.get('receiver')
    addr_region = request.POST.get('region')
    addr_factor = request.POST.get('addr')
    phone = request.POST.get('phone')
    default = request.POST.get('default')

    # 校验数据
    if not all([addr_region, default, receiver, phone, addr_factor]):
        # 缺少相关数据
        return JsonResponse({'res': 1, 'errmsg': '缺少相关数据'})
```

```python
# 整合地址数据
addr = addr_region + addr_factor

if int(default) > 0:
    default = True
    # 查询出默认地址,设为非默认状态
    address_user = Address.objects.get(is_default=True, user=user)
    address_user.is_default = False
    address_user.save()
else:
    default = False

# 调用方法,获取经纬度
lat_lng = get_lng_lat(addr)
lat = lat_lng['lat']
lng = lat_lng['lng']

# 地址表新增数据
address = Address(user=user, receiver=receiver, addr=addr, phone=phone, is_default=default, lat=lat, lng=lng)
address.save()

return JsonResponse({'res': 2, 'errmsg': '保存成功'})
```

启动项目后,单击"新增收货地址"按钮,就可以正常添加地址信息。

10.8 订单模块设计

在购物车模块中,添加想要下单的商品后,单击"去付款"按钮,就可以进入订单模块进行订单的处理。订单模块主要包括订单生成、订单追踪、订单管理、订单状态和订单查询 5 部分,下面分别进行介绍。

10.8.1 订单生成

当用户决定购买并单击"去付款"按钮时,前端需要把商品数据、地址数据传递到后台,出于安全考虑,后台需要重新处理商品信息、地址信息,而不是直接使用传递过来的数据。根据这些数据,将会生成一张未支付状态的订单。

此处,需要通过命令"python manage.py startapp order"新建一个订单模块,然后完成模块注册和新建 urls.py 文件,并且设置 urlpatterns,即在 clmwm\urls.py 文件中,新增一条路由信息,具体代码如下:

```python
from django.contrib import admin
from django.urls import path, include

urlpatterns = [
    path('admin/', admin.site.urls),
    path('order/', include(('apps.order.urls', 'apps.order'), namespace='order')),  # 订单模块
]
```

在 order\urls.py 文件中,新增一条 URL 信息,用来匹配订单生成,具体代码如下:

```python
from django.urls import path
from apps.order.views import OrderGenerateView
urlpatterns = [
    path('generate/', OrderGenerateView.as_view(), name='generate')
]
```

在 views.py 文件中新增一个类 OrderGenerateView,在其中控制订单的并发生成,并处理业务逻辑。根

据并发量，本项目中采用的是乐观锁。具体代码如下：

```python
from django.shortcuts import render, redirect, HttpResponse, reverse
from django.views.generic import View
from datetime import datetime
from django.db import transaction

from apps.user.models import Shop, User
from apps.goods.models import GoodsSKU, Goods
from apps.order.models import OrderInfo, OrderGoods, OrderTrack
from utils.common import calculate_distance_duration

class OrderGenerateView(View):
    """处理订单生成"""

    def get(self, request):
        pass

    @transaction.atomic
    def post(self, request):

        try:
            user = User.objects.get(id=request.user.id)
        except User.DoesNotExist:
            return render(request, 'login.html', {'errmsg': '用户登录信息已失效，请重新登录！'})

        sku_str = request.POST.get('sku_ids')[1:]                           # 接收数据
        addr = request.POST.get('address')
        remarks = request.POST.get('remarks')
        invoice_head = request.POST.get('invoice_head')
        taxpayer_number = request.POST.get('taxpayer_number')

        try:
            sku_ids = sku_str[4:sku_str.find('cm2=') - 1].split('%2C')
            sku_counts = sku_str[sku_str.find('cm2=') + 4:].split('%2C')
            sku_info = GoodsSKU.objects.filter(id__in=sku_ids)
            for index, sku in enumerate(sku_info):
                goods = Goods.objects.get(id=sku.goods_id)
                shop = Shop.objects.get(id=goods.shop_id)
        except Exception:
            return redirect(reverse('goods:wm_index'), {'errmsg': '数据错误'})
        if not all([sku_str, shop, addr, sku_ids, sku_counts, sku_info]):   # 校验基本数据
            return render(request, 'sj_cpgl.html', {'errmsg': '缺少相关数据'})  # 数据不完整

        order_id = datetime.now().strftime('%Y%m%d%H%M%S') + str(request.user.id)  # 订单 id: 20200802181630+用户 id

        total_price, total_count, range_flag = 0, 0, 0

        save_id = transaction.savepoint()                                   # 设置事务保存点
        try:

            distance = calculate_distance_duration(shop, user, address_id=addr).send_price  # 重新计算运费

            # 订单表添加数据
            order = OrderInfo.objects.create(order_id=order_id, user_id=request.user.id, addr_id=addr, shop=shop,
                                             remarks=remarks, invoice_head=invoice_head, total_price=0,
                                             total_count=0,
                                             taxpayer_number=taxpayer_number, transit_price=distance)

            order_track = OrderTrack.objects.create(order=order, status=1)  # 订单轨迹表添加一条数据

            # 乐观锁尝试 3 次
            for i in range(3):
```

```python
                    if range_flag:
                        break
                # 生成订单明细
                for index, item in enumerate(sku_info):
                    # 判断商品库存
                    if int(item.stock) < int(sku_counts[index]):
                        return HttpResponse('商品库存不足')
                    # 插入数据
                    order_goods = OrderGoods.objects.create(order=order, sku=item, price=item.price,
                                                            count=sku_counts[index])
                    # 更新库存，返回受影响的行数
                    stock = item.stock - int(sku_counts[index])
                    res = GoodsSKU.objects.filter(id=item.id, stock=item.stock).update(
                        stock=stock, sales=item.sales+int(sku_counts[index]))
                    if res == 0:
                        if i == 2:
                            # 尝试的第3次
                            transaction.savepoint_rollback(save_id)
                            return HttpResponse('下单失败')
                        continue

                    # 累加计算订单商品的总数量和总价格
                    total_price += (item.price + int(item.pack)) * int(sku_counts[index])
                    total_count += int(sku_counts[index])

                    a = len(sku_info)
                    if index == len(sku_info)-1:
                        range_flag = 1

                # 更新订单信息表中的商品的总数量和总价格
                order.total_count = total_count
                order.total_price = total_price
                order.save()

        except Exception as e:
            transaction.savepoint_rollback(save_id)
            # return JsonResponse({'res': 7, 'errmsg': '下单失败'})

        # 提交事务
        transaction.savepoint_commit(save_id)

        total_all = int(order.transit_price) + total_price

        return render(request, 'wm_pay.html', {'order': order, 'shop': shop, 'total_all': total_all})
```

render()渲染的文件为wm_pay.html，所以将static文件夹下的wm_pay.html文件复制到templates文件夹，并修改成如下代码：

```html
<!doctype html>
{% extends 'mj.html' %}
{% load staticfiles %}
{% block title %}吃了么{% endblock title %}
{% block topfiles %}
<link href="{% static 'css/add_address.css' %}" rel="stylesheet" type="text/css">
<script src="{% static 'js/jquery.min.js' %}"></script>
<script src="{% static 'js/Popt.js' %}"></script>
<script src="{% static 'js/cityJson.js' %}"></script>
<script src="{% static 'js/citySet.js' %}"></script>
```

```html
<script>
window.onload = function(){
    var endTime = new Date().getTime() + 900*1000;        // 最终毫秒
    setInterval(clock,1000);                               // 开启定时器
    function clock(){
      var nowTime = new Date();
      var second = parseInt((endTime - nowTime.getTime()) / 1000);
      var m = parseInt(second / 60 );
      var s = parseInt(second % 60); // 当前的秒
      console.log(s);
      m<10 ? m="0"+m : m;
      s<10 ? s="0"+s : s;
      document.getElementById("time_down").innerHTML = "<img src=\"{% static 'image/warn.png' %}\">
      请在<b>"+m+":"+s+"</b>内完成支付，超时订单会自动取消";
    }
}

function order_check() {
    order_id = $(this).attr('order_id');
    csrf = $('input[name="csrfmiddlewaretoken"]').val();
    params = {'order_id':order_id, 'csrfmiddlewaretoken':csrf};
    $.post('/order/check/', params, function (data){
        if (data.res == 3){
            // 重定向页面
            window.location.href="http://127.0.0.1:8000/order/success/" + order_id;
        }
        else{
            alert(data.errmsg);
        }
    })
}
</script>
{% endblock topfiles %}

{% block body %}
<div id="main">
    <div class="pay_warn">
        <span id="time_down"></span>
    </div>
    <div class="pay_title">
        <span>店铺:{{ shop.shop_name }}    订单号:{{ order.order_id }}</span>
        <span>应付金额:<b>￥{{ total_all }}</b></span>
    </div>
    <div class="pay_con">
        <form action="/order/pay/" method="post" id="wm_pay" name="wm_pay">
            {% csrf_token %}
            <input type="text" name="order_id" id="order_id" style="display: none" value="{{ order.order_id }}">
            <input type="radio" name="pay" value="微信支付" id="wx_pay">
                <label for="wx_pay"><img src="{% static 'image/wx_icon.png' %}">微信支付</label>
            <input type="radio" name="pay" value="支付宝支付" checked id="zfb_pay">
                <label for="zfb_pay"><img src="{% static 'image/zfb.png' %}">支付宝支付</label>
        </form>
        <div class="pay_state">
            <span>支付<b>￥{{ total_all }}</b></span>
            <div class="pay_a">
                <a href="{% url 'goods:index' %}">回到首页</a>
                <a href="/goods/{{ shop.id }}/1/1/1">重新下单</a>
                <button id="go_pay" name="go_pay" order_id="{{ order.order_id }}" status="{{ order.order_status }}">
                    去付款</button>
                {# <button id="order_check" name="order_check" onclick="order_check();" order_id="{{ order.order_id }}">
                    已支付</button>#}
                <script language="javascript" type="text/javascript">
                    $("#go_pay").click(function () {
```

```
                // 获取 status
                status = $(this).attr('status');
                // 获取订单 id
                order_id = $(this).attr('order_id');
                if (status == 1){
                    // 此处省略了支付功能的处理代码
                    csrf = $('input[name="csrfmiddlewaretoken"]').val();
                    // 组织参数
                    params = {'order_id':order_id, 'csrfmiddlewaretoken':csrf};
                    // 发起 ajax post 请求，访问/order/pay, 传递参数:order_id
                    $.post('/order/pay/', params, function (data) {
                        // 浏览器访问/order/check, 获取支付交易的结果
                        $.post('/order/check/', params, function (data){
                            if (data.res == 3){
                                alert('支付成功')
                                // 重定向页面
                                window.location.href="http://127.0.0.1:8000/order/success/" + order_id;
                            }
                            else{
                                alert(data.errmsg)
                            }
                        })

                    })
                }
                else if (status == 4){
                    // 其他情况
                    alert('跳转到评价页面')
                    location.href = '/order/comment/'+order_id
                }
            })
        </script>            </div>
    </div>
</div>
{% endblock %}
```

启动项目，添加商品，选择地址，单击"去付款"按钮，即可生成一张订单信息，如图 10.8 所示。

图 10.8　待支付订单生成

10.8.2 订单追踪

在 10.8.1 节的运行界面中，单击"去付款"按钮，会将当前页面重定向到订单追踪页面。这时需要在 views.py 文件中添加 OrderBuySuccessView 类，用来查询订单信息，并跳转到订单追踪页面，具体代码如下：

```python
class OrderBuySuccessView(View):
    """支付成功，跳转详情页面"""
    def get(self, request, order_id, is_comment=False):

        # 用户是否登录
        user = request.user
        if not user.is_authenticated:
            return JsonResponse({'res': 0, 'errmsg': '用户未登录'})

        try:
            order_info = OrderInfo.objects.get(order_id=order_id, user=user)
        except OrderInfo.DoesNotExist:
            return JsonResponse({'res': 2, 'errmsg': '订单错误 2'})

        order_goods = OrderGoods.objects.filter(order=order_info)
        pay_price = int(order_info.total_price) + int(order_info.transit_price)
        order_track = OrderTrack.objects.filter(order=order_info, status__gt=1)
        # 预计送达时间
        arrive_time = order_info.create_time + datetime.timedelta(minutes=order_info.transit_time)

        context = {'order': order_info, 'order_goods': order_goods, 'pay_price': pay_price, 'order_track': order_track, 'arrive_time': arrive_time}

        # 控制页面跳转
        if is_comment:
            return context
        else:
            return render(request, 'wm_ordertrack.html', context)
    def post(self):
        pass
```

将 static 文件夹下的 wm_ordertrack.html 文件复制到 templates 文件夹中，并修改成如下格式：

```html
<!doctype html>
{% extends 'mj.html' %}
{% load staticfiles %}
{% block title %}吃了么-订单跟踪{% endblock title %}
{% block topfiles %}
<script src="{% static 'js/jquery.min.js' %}"></script>
<script src="{% static 'js/Popt.js' %}"></script>
<script src="{% static 'js/cityJson.js' %}"></script>
<script src="{% static 'js/citySet.js' %}"></script>
{% endblock topfiles %}
{% block body %}
<body>
<div id="main">
    <div class="order_tra">
        <div class="order_t_title">
            <span>预计送达时间:{{ arrive_time }} </span>
        </div>
        <div class="order_t_con">
            <ul>

                {% if order.order_status > 1 %}
```

```
            <li>
              <div class="count_turn">
                <b>1</b><span>卖家待接单</span><font></font><span>{{ order_track.0.create_time }}</span>
                <div>卖家待接单</div>
              </div>
            </li>
        {% endif %}

        {% if order.order_status > 2 %}
            <li>
              <div class="count_turn">
                <b>2</b><span>卖家已接单</span><font></font><span>{{ order_track.0.create_time }}</span>
                <div>卖家已接单</div>
              </div>
            </li>
        {% elif order.order_status == 0   %}
            <li>
              <div class="count_turn">
                <b>1</b><span>卖家拒绝接单</span><font></font><span>{{ order_track.0.create_time }}</span>
                <div>支付金额两小时内退回原账户</div>
              </div>
            </li>
        {% endif %}

        {% if order.order_status > 3 %}
            <li>
              <div class="count_turn">
                <b>3</b><span>骑手取货中</span><font></font><span>{{ order_track.0.create_time }}</span>
                <div>骑手取货中</div>
              </div>
            </li>
        {% endif %}

        {% if order.order_status > 4 %}
            <li>
              <div class="count_turn">
                <b>4</b><span>订单配送中</span><font></font><span>{{ order_track.0.create_time }}</span>
                <div>订单配送中</div>
              </div>
            </li>
        {% endif %}

        {% if order.order_status > 5 %}
            <li>
              <div class="count_turn">
                <b>5</b><span>订单已送达</span><font></font><span>{{ order_track.0.create_time }}</span>
                <div>订单已送达</div>
              </div>
            </li>
              <button style="float: left;color: #fff9e5ff;background: #00d36e;width: 15%;height: 50px;font-size: large">去评价</button>
        {% endif %}

        </ul>
        <div class="wm_order_btn">
            <button onclick="history.go(0)">刷新订单</button>
            <button style="display: none">催单</button>
            <button style="display: none">取消订单</button>
        </div>
    </div>
  </div>
</div>
{% endblock body %}
```

修改完成后，启动项目，刷新页面，即可追踪进行中的订单，如图 10.9 所示。

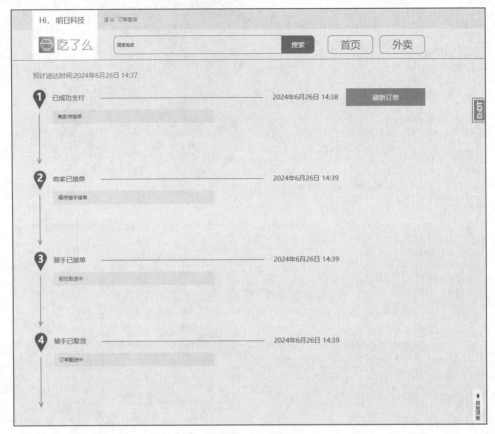

图 10.9 订单追踪

10.8.3 订单管理

卖家在订单管理页面中可查询到用户已完成支付的订单信息，并选择是否接受该订单。为区分卖家与买家的业务处理，此处应该新建一个 ordermanage 模块，用来实现卖家管理订单。执行命令"python manage.py startapp ordermanage"在 base.py 文件中创建该模块，并且在 clmwm\urls.py 文件中添加路由信息，具体代码如下：

```
from django.contrib import admin
from django.urls import path, include
urlpatterns = [
    path('admin/', admin.site.urls),
    path('ordermanage/',include(('apps.ordermanage.urls', 'apps.ordermanage'), namespace='ordermanage')),# 订单管理模块
]
```

在 ordermanage\urls.py 文件中配置具体的 URL 信息，代码如下：

```
from django.urls import path, re_path
from apps.ordermanage import views

urlpatterns = [
    re_path(r'^sj_order/(?P<page>\d+)/$', views.SjOrderView.as_view(), name='sj_order'),   # 卖家订单管理
]
```

在 views.py 文件中，查询出属于该店铺的所有订单，并分类展示，具体代码如下：

```python
from django.shortcuts import render
from django.views.generic import View

from utils.common import order_detail, shop_is_new

from apps.user.models import Shop

class SjOrderView(View):
    """卖家订单管理"""

    def get(self, request, page):

        try:
            shop = Shop.objects.get(user_id=request.user.id)
        except Shop.DoesNotExist:
            return render(request, 'login.html', {'errmsg': '用户登录信息已失效,请重新登录!'})

        order_exam = order_detail(2, page, shop)
        order_pass = order_detail(3, page, shop)
        order_conduct = order_detail(4, page, shop)
        order_delivery = order_detail(5, page, shop)
        order_finish = order_detail(7, page, shop)
        order_cancel = order_detail(0, page, shop)

        shop_is_new(shop)

        context = {'shop': shop, 'order_exam': order_exam, 'order_pass': order_pass, 'order_conduct': order_conduct,
                   'order_delivery': order_delivery, 'order_finish': order_finish, 'order_cancel': order_cancel}

        return render(request, 'sj_order.html', context)
```

在 common.py 文件中新增一个 order_detail()方法,用来查询订单明细,具体代码如下:

```python
def order_detail(order_status, page, shop):
    """
    根据状态查询明细
    以字典的形式返回
    order_info 与 order_goods
    """
    from collections import Iterable
    from django.db.models import F, Q
    from apps.order.models import OrderInfo, OrderGoods

    if isinstance(order_status, Iterable):
        order_info = page_item(OrderInfo.objects.filter(
            Q(order_status=order_status[0], shop=shop) | Q(order_status=order_status[1], shop=shop)).order_by('-update_time'), page, 10)
        goods = OrderGoods.objects.filter(
            order__in=OrderInfo.objects.values_list('order_id').filter(
                Q(order_status=order_status[0], shop=shop) | Q(order_status=order_status[1], shop=shop)).order_by('-update_time'))\
            .order_by('-update_time')
        order_info.update({'goods': goods})
    else:
        order_info = page_item(OrderInfo.objects.filter(order_status=order_status, shop=shop).order_by('-update_time'),
                               page, 10)
        goods = OrderGoods.objects.filter(
            order__in=OrderInfo.objects.values_list('order_id').filter(
                order_status=order_status, shop=shop).order_by('-update_time')).order_by('-update_time')
        order_info.update({'goods': goods})
    return order_info
```

将 static 文件夹下的 sj_order.html 文件复制到 templates 文件夹中，并修改成如下所示代码：

```
{% extends 'sj.html' %}
{% load staticfiles %}
{% block title %}吃了么-订单管理{% endblock title %}
{% block topfiles %}
<script src="{% static 'js/jquery.min.js' %}"></script>
<script type="text/javascript">
$(function () {
    $('.sj_order_man ul li').hover(function(){
        // 获得当前被点击的元素索引值
        var Index = $(this).index();
        var line=300*Index-300;
        // 给菜单添加选择样式
        $(this).addClass('active').siblings().removeClass('active');
        $(".line").stop(true,true).animate({left:line},200);
        $('.sj_order_man').children('div').eq(Index).show().siblings('div').hide();
    });
});
</script>
{% endblock topfiles %}
{% block body %}

    <div class="sj_name">
        <h2><span>店铺名称：</span>{{ shop.shop_name }}</h2>
    </div>
    <div class="sj_order_man">
        <ul class="sj_menu">
            <li  class="active">待审核订单</li><li>已审核订单</li><li>进行中订单</li><li>配送中订单</li><li>已完成订单</li><li>已取消订单</li>
        </ul>

        <div class="order_list1">
          <table>
            <tr>
                <td width="11%">下单时间</td>
                <td width="5%">姓名</td>
                <td width="9%">电话</td>
                <td width="15%">地址</td>
                <td width="10%">餐品</td>
                <td width="6%">金额+运费</td>
                <td width="22%">备注</td>
                <td width="22%">是否接单</td>
            </tr>
            {% for exam in order_exam.info %}
            <tr>
                <td>{{ exam.create_time }}</td>
                <td>{{ exam.addr.receiver }}</td>
                <td>{{ exam.addr.phone }}</td>
                <td>{{ exam.addr.addr }}</td>
                <td>
                    {% for good in order_exam.goods %}
                        {% if good.order == exam %}
                            <li>{{ good.sku.name }}X{{ good.count }}</li>
                        {% endif %}
                    {% endfor %}
                </td>
                <td>{{ exam.total_price }}+{{ exam.transit_price }}</td>
                <td>{{ exam.remarks }}</td>
                <td>
                    <button><a href="">接受</a></button>
                    <button><a href="">拒绝</a></button>
```

```
            </td>
        </tr>
        {% endfor %}
    </table>
    <div class="pagenation">
        {% if order_exam.info.has_previous %}
            <a href="{% url 'ordermanage:sj_order' order_exam.info.previous_page_number %}">< 上一页 </a>
        {% endif %}

        {% for pindex in order_exam.pages %}
            {% if pindex == order_exam.info.number %}
                <a href="{% url 'ordermanage:sj_order' pindex %}" class="active">{{ pindex }}</a>
            {% else %}
                <a href="{% url 'ordermanage:sj_order' pindex %}">{{ pindex }}</a>
            {% endif %}
        {% endfor %}

        {% if order_exam.info.has_next %}
            <a href="{% url 'ordermanage:sj_order' order_exam.info.next_page_number %}"> 下一页 ></a>
        {% endif %}
    </div>
</div>

<div class="order_list6">
    <table>
        <tr>
            <td width="11%">下单时间</td>
            <td width="5%">姓名</td>
            <td width="9%">电话</td>
            <td width="15%">地址</td>
            <td width="10%">餐品</td>
            <td width="10%">金额+运费</td>
            <td width="23%">备注</td>
            <td width="15%">状态</td>
        </tr>
        {% for pass in order_pass.info %}
        <tr>
            <td>{{ pass.create_time }}</td>
            <td>{{ pass.addr.receiver }}</td>
            <td>{{ pass.addr.phone }}</td>
            <td>{{ pass.addr.addr }}</td>
            <td>
                {% for good in order_pass.goods %}
                    {% if good.order == pass %}
                        <li>{{ good.sku.name }}X{{ good.count }}</li>
                    {% endif %}
                {% endfor %}
            </td>
            <td>{{ pass.total_price }}+{{ pass.transit_price }}</td>
            <td>{{ pass.remarks }}</td>
            <td>已审核，待骑手接单</td>
        </tr>
        {% endfor %}
    </table>
    <div class="pagenation">
        {% if order_pass.info.has_previous %}
            <a href="{% url 'ordermanage:sj_order' order_pass.info.previous_page_number %}">< 上一页 </a>
        {% endif %}

        {% for pindex in order_pass.pages %}
            {% if pindex == order_pass.info.number %}
                <a href="{% url 'ordermanage:sj_order' pindex %}" class="active">{{ pindex }}</a>
```

```
                {% else %}
                    <a href="{% url 'ordermanage:sj_order' pindex %}">{{ pindex }}</a>
                {% endif %}
            {% endfor %}

            {% if order_pass.info.has_next %}
                <a href="{% url 'ordermanage:sj_order' order_pass.info.next_page_number %}"> 下一页 ></a>
            {% endif %}
        </div>
    </div>

    <div class="order_list2">
        <table>
            <tr>
                <td width="11%">下单时间</td>
                <td width="5%">姓名</td>
                <td width="9%">电话</td>
                <td width="15%">地址</td>
                <td width="10%">餐品</td>
                <td width="6%">金额+运费</td>
                <td width="22%">备注</td>
                <td width="22%">状态</td>
            </tr>
            {% for conduct in order_conduct.info %}
            <tr>
                <td>{{ conduct.create_time }}</td>
                <td>{{ conduct.addr.receiver }}</td>
                <td>{{ conduct.addr.phone }}</td>
                <td>{{ conduct.addr.addr }}</td>
                <td>
                    {% for good in order_conduct.goods %}
                        {% if good.order == conduct %}
                            <li>{{ good.sku.name }}X{{ good.count }}</li>
                        {% endif %}
                    {% endfor %}
                </td>
                <td>{{ conduct.total_price }}+{{ conduct.transit_price }}</td>
                <td>{{ conduct.remarks }}</td>
                <td>骑手接单，取货中</td>
            </tr>
            {% endfor %}
        </table>
        <div class="pagenation">
            {% if order_conduct.info.has_previous %}
            <a href="{% url 'ordermanage:sj_order' order_conduct.info.previous_page_number %}">< 上一页 </a>
            {% endif %}

            {% for pindex in order_conduct.pages %}
                {% if pindex == order_conduct.info.number %}
                    <a href="{% url 'ordermanage:sj_order' pindex %}" class="active">{{ pindex }}</a>
                {% else %}
                    <a href="{% url 'ordermanage:sj_order' pindex %}">{{ pindex }}</a>
                {% endif %}
            {% endfor %}

            {% if order_conduct.info.has_next %}
                <a href="{% url 'ordermanage:sj_order' order_conduct.info.next_page_number %}"> 下一页 ></a>
            {% endif %}
        </div>
    </div>

    <div class="order_list3">
```

```django
<table>
    <tr>
        <td width="11%">下单时间</td>
        <td width="5%">姓名</td>
        <td width="9%">电话</td>
        <td width="15%">地址</td>
        <td width="10%">餐品</td>
        <td width="6">金额+运费</td>
        <td width="22">备注</td>
        <td width="5%">催单</td>
        <td width="17%">订单状态</td>
    </tr>
    {% for delivery in order_delivery.info %}
    <tr>
        <td>{{ delivery.create_time }}</td>
        <td>{{ delivery.addr.receiver }}</td>
        <td>{{ delivery.addr.phone }}</td>
        <td>{{ delivery.addr.addr }}</td>
        <td>
            {% for good in order_delivery.goods %}
                {% if good.order == delivery %}
                    <li>{{ good.sku.name }}X{{ good.count }}</li>
                {% endif %}
            {% endfor %}
        </td>
        <td>{{ delivery.total_price }}+{{ delivery.transit_price }}</td>
        <td>{{ delivery.remarks }}</td>
        <td>催单 0 次</td>
        <td>配送中，等待客户签收</td>
    </tr>
    {% endfor %}
</table>
<div class="pagenation">
    {% if order_delivery.info.has_previous %}
    <a href="{% url 'ordermanage:sj_order' order_delivery.info.previous_page_number %}">< 上一页 </a>
    {% endif %}

    {% for pindex in order_delivery.pages %}
        {% if pindex == order_delivery.info.number %}
        <a href="{% url 'ordermanage:sj_order' pindex %}" class="active">{{ pindex }}</a>
        {% else %}
        <a href="{% url 'ordermanage:sj_order' pindex %}">{{ pindex }}</a>
        {% endif %}
    {% endfor %}

    {% if order_delivery.info.has_next %}
    <a href="{% url 'ordermanage:sj_order' order_delivery.info.next_page_number %}"> 下一页></a>
    {% endif %}
</div>
</div>

<div class="order_list4">
    <table>
        <tr>
            <td width="11%">完成时间</td>
            <td width="5%">姓名</td>
            <td width="9%">电话</td>
            <td width="15%">地址</td>
            <td width="10%">餐品</td>
            <td width="6%">金额+运费</td>
            <td width="5%">状态</td>
            <td width="14%">备注</td>
```

```django
                <td width="15%">评价</td>
            </tr>
            {% for finish in order_finish.info %}
                <tr>
                    <td>{{ finish.create_time }}</td>
                    <td>{{ finish.addr.receiver }}</td>
                    <td>{{ finish.addr.phone }}</td>
                    <td>{{ finish.addr.addr }}</td>
                    <td>
                        {% for good in order_finish.goods %}
                            {% if good.order == finish %}
                                <li>{{ good.sku.name }}X{{ good.count }}</li>
                            {% endif %}
                        {% endfor %}
                    </td>
                    <td>{{ finish.total_price }}+{{ finish.transit_price }}</td>
                    <td>已完成</td>
                    <td>{{ finish.remarks }}</td>
                    <td>{{ finish.comment }}</td>
                </tr>
            {% endfor %}
        </table>
        <div class="pagenation">
            {% if order_finish.info.has_previous %}
                <a href="{% url 'ordermanage:sj_order' order_finish.info.previous_page_number %}">< 上一页 </a>
            {% endif %}

            {% for pindex in order_finish.pages %}
                {% if pindex == order_finish.info.number %}
                    <a href="{% url 'ordermanage:sj_order' pindex %}" class="active">{{ pindex }}</a>
                {% else %}
                    <a href="{% url 'ordermanage:sj_order' pindex %}">{{ pindex }}</a>
                {% endif %}
            {% endfor %}

            {% if order_finish.info.has_next %}
                <a href="{% url 'ordermanage:sj_order' order_finish.info.next_page_number %}"> 下一页 ></a>
            {% endif %}
        </div>
    </div>

    <div class="order_list5">
        <table>
            <tr>
                <td width="11%">下单时间</td>
                <td width="5%">姓名</td>
                <td width="9%">电话</td>
                <td width="15%">地址</td>
                <td width="10%">餐品</td>
                <td width="6%">金额+运费</td>
                <td width="14">备注</td>
                <td width="5%">状态</td>
                <td width="10%">支付状态</td>
            </tr>
            {% for cancel in order_cancel.info %}
                <tr>
                    <td>{{ cancel.create_time }}</td>
                    <td>{{ cancel.addr.receiver }}</td>
                    <td>{{ cancel.addr.phone }}</td>
                    <td>{{ cancel.addr.addr }}</td>
                    <td>
                        {% for good in order_cancel.goods %}
```

```
                    {% if good.order == cancel %}
                        <li>{{ good.sku.name }}X{{ good.count }}</li>
                    {% endif %}
                {% endfor %}
                </td>
                <td>{{ cancel.total_price }}+{{ cancel.transit_price }}</td>
                <td>{{ cancel.remarks }}</td>
                <td>已拒接</td>
                <td>已退款</td>
            </tr>
            {% endfor %}
        </table>
        <div class="pagenation">
            {% if order_cancel.info.has_previous %}
                <a href="{% url 'ordermanage:sj_order' order_cancel.info.previous_page_number %}">< 上一页 </a>
            {% endif %}

            {% for pindex in order_cancel.pages %}
                {% if pindex == order_cancel.info.number %}
                    <a href="{% url 'ordermanage:sj_order' pindex %}" class="active">{{ pindex }}</a>
                {% else %}
                    <a href="{% url 'ordermanage:sj_order' pindex %}">{{ pindex }}</a>
                {% endif %}
            {% endfor %}

            {% if order_cancel.info.has_next %}
                <a href="{% url 'ordermanage:sj_order' order_cancel.info.next_page_number %}"> 下一页 ></a>
            {% endif %}
        </div>
    </div>
</div>

{% endblock body %}
```

启动项目，在浏览器中访问 http://127.0.0.1:8000/ordermanage/sj_order/1/即可查看到已经支付的订单，如图 10.10 所示。

图 10.10　订单管理

10.8.4 订单状态

在订单管理页面中,可以接受或拒绝待审核的订单。接受该订单时,把订单 ID 传送至后台,并将订单状态改为"3",拒绝时把订单状态改为"0",并退款。在 sj_order.html 文件中为"接受"和"拒绝"按钮设置 URL 信息,关键代码如下:

```html
<td>
    <button><a href="{% url 'ordermanage:receive' exam.order_id %}">接受</a></button>
    <button><a href="{% url 'ordermanage:refuse' exam.order_id %}">拒绝</a></button>
</td>
```

在 urls.py 文件中,设置匹配的路由信息,具体代码如下:

```python
from django.urls import path, re_path
from apps.ordermanage import views

urlpatterns = [
    re_path(r'^sj_order/(?P<page>\d+)/$', views.SjOrderView.as_view(), name='sj_order'),   # 卖家订单管理
    path('receive/<order_id>', views.OrderReceive.as_view(), name='receive'),              # 卖家接受订单
    path('refuse/<order_id>', views.OrderRefuse.as_view(), name='refuse'),                 # 卖家拒绝订单
]
```

在 views.py 文件中,分别对接受订单和拒绝订单的业务逻辑进行处理,具体代码如下:

```python
import os
from django.shortcuts import render, redirect, reverse
from django.views.generic import View
from django.conf import settings
from django.http import JsonResponse

from utils.common import order_detail, shop_is_new

from apps.user.models import Shop
from apps.order.models import OrderInfo, OrderTrack

from alipay import AliPay

class OrderReceive(View):
    """卖家接受订单"""

    def get(self, request, order_id):

        try:
            shop = Shop.objects.get(user_id=request.user.id)
        except Shop.DoesNotExist:
            return render(request, 'login.html', {'errmsg': '用户登录信息已失效,请重新登录!'})

        # 校验参数
        if not order_id:
            return redirect(reverse('ordermanage:sj_order', kwargs={'page': 1}))

        try:
            order = OrderInfo.objects.get(order_id=order_id, order_status=2)
        except OrderInfo.DoesNotExist:
            return redirect(reverse('ordermanage:sj_order', kwargs={'page': 1}))

        # 改变订单状态,轨迹表生成一条数据
        order.order_status = 3
```

```python
            order_track = OrderTrack.objects.create(order=order, status=3)
            order.save()

        return redirect(reverse('ordermanage:sj_order', kwargs={'page': 1}))


class OrderRefuse(View):
    """卖家拒绝订单"""

    def get(self, request, order_id):

        try:
            shop = Shop.objects.get(user_id=request.user.id)
        except Shop.DoesNotExist:
            return render(request, 'login.html', {'errmsg': '用户登录信息已失效，请重新登录！'})

        # 校验参数
        if not order_id:
            return redirect(reverse('ordermanage:sj_order', kwargs={'page': 1}))

        try:
            order = OrderInfo.objects.get(order_id=order_id, order_status=2)
        except OrderInfo.DoesNotExist:
            return redirect(reverse('ordermanage:sj_order', kwargs={'page': 1}))

        # 初始化
        alipay = AliPay(
            appid=settings.ALIPAY_APPID,              # 应用 id
            app_notify_url=None,                      # 默认回调 url
            app_private_key_path=os.path.join(settings.BASE_DIR, 'apps/order/app_private_key.pem'),
            # 支付宝的公钥，验证支付宝回传消息
            alipay_public_key_path=os.path.join(settings.BASE_DIR, 'apps/order/alipay_public_key.pem'),
            sign_type="RSA2",                         # RSA 或者 RSA2
            debug=True                                # 默认 False
        )

        order_string = alipay.api_alipay_trade_refund(
            trade_no=order.trade_no,
            refund_amount=str(order.total_price + order.transit_price),
            notify_url=None
        )

        code = order_string.get('code')

        if code == '10000' and order_string.get('msg') == 'Success':
            # 改变订单状态，轨迹表生成一条数据
            order.order_status = 0
            order_track = OrderTrack.objects.create(order=order, status=0)
            order.save()
        else:
            sub_msg = order_string.get('sub_msg')
            return render(request, 'sj_order.html', {'errmsg': sub_msg})

        return redirect(reverse('ordermanage:sj_order', kwargs={'page': 1}))
```

此时，我们已经可以接受或拒绝订单，如果订单状态想要继续改变就应该在移动端开发配送 app。在一般公司中，这往往是独立的一个子程序，所以本项目暂未开发，取而代之的是在页面添加类似"接受"的按钮来直接改变订单状态，在 sj_order.html 文件中，部分修改代码如下：

```html
<td>
```

```html
            <button><a href="{% url 'ordermanage:receive' '4' pass.order_id %}">有骑手接单</a></button>
        </td>

        <td>
            <button><a href="{% url 'ordermanage:receive' '5' conduct.order_id %}">骑手已取货</a></button>
        </td>

        <td>
            <button><a href="{% url 'ordermanage:receive' '6' delivery.order_id %}">已送达</a></button>
        </td>
```

由于新增的按钮与"接受"方法类似，都是改变订单状态，所以我们可以继续使用接受的方法，只需对其稍加修改即可。urls.py 文件中具体代码如下：

```python
from django.urls import path, re_path

from apps.ordermanage import views

urlpatterns = [
    re_path(r'^sj_order/(?P<page>\d+)/$', views.SjOrderView.as_view(), name='sj_order'),        # 卖家订单管理
    path('receive/<order_status>/<order_id>', views.OrderReceive.as_view(), name='receive'),    # 卖家接受或改变订单
]
```

在 views.py 文件中，获取传递过来的待改变状态的信息，具体代码如下：

```python
class OrderReceive(View):
    """卖家修改订单状态"""

    def get(self, request, order_status, order_id):

        try:
            shop = Shop.objects.get(user_id=request.user.id)
        except Shop.DoesNotExist:
            return render(request, 'login.html', {'errmsg': '用户登录信息已失效，请重新登录！'})

        # 校验参数
        if not order_id:
            return redirect(reverse('ordermanage:sj_order', kwargs={'page': 1}))

        try:
            order = OrderInfo.objects.get(order_id=order_id, order_status=int(order_status)-1)
        except OrderInfo.DoesNotExist:
            return redirect(reverse('ordermanage:sj_order', kwargs={'page': 1}))

        # 改变订单状态，轨迹表生成一条数据
        order.order_status = order_status
        order_track = OrderTrack.objects.create(order=order, status=order_status)
        order.save()

        return redirect(reverse('ordermanage:sj_order', kwargs={'page': 1}))
```

修改完成后，启动项目，浏览器访问 http://127.0.0.1:8000/ordermanage/sj_order/1/，单击相应的按钮，订单状态就会改变，当骑手已送达后，只有用户做出评价，订单才算已完成。

10.8.5 订单查询

当用户下完订单后，如果关闭了浏览器或者还想继续下单，就会丢失订单追踪页面，所以应该在导航栏中添加订单的查询功能，在 mj.html 文件中增加如下代码：

```html
<div class="top_lore">
{% if user.is_authenticated %}
            <a href="{% url 'user:logout' %}">退出</a>
            <a href="{% url 'order:query' '1'%}" style="color:#0da2f8">查询订单</a>
        {% else %}
<a href="{% url 'user:login' %}">请登录</a> <a href="{% url 'user:register' %}">注册</a>
        {% endif %}
</div>
```

在 order\urls.py 文件中新增匹配的路由信息，具体代码如下：

```python
from django.urls import path
from apps.order.views import QueryOrderView

urlpatterns = [
    path('query/<page>/', QueryOrderView.as_view(), name='query')
]
```

配置完路由信息后，在 order\views.py 文件中查询出进行中和已完成的订单信息，并做分页处理，具体代码如下：

```python
class QueryOrderView(View):
    """查询订单"""

    def get(self, request, page):
        """查询订单"""

        # 用户是否登录
        user = request.user
        if not user.is_authenticated:
            return JsonResponse({'res': 0, 'errmsg': '用户未登录'})
        # 进行中的订单
        order_going = OrderInfo.objects.filter(user=user, order_status__gte=2, order_status__lte=6)
        # 进行中的订单商品明细
        order_going_goods = OrderGoods.objects.filter(order__in=order_going)
        # 已完成的订单
        order_finish = OrderInfo.objects.filter(user=user, order_status__gte=7)
        # 已完成的订单商品明细
        order_finish_goods = OrderGoods.objects.filter(order__in=order_going)
        # 分页处理
        order_going = page_item(order_going, page, 10)
        order_finish = page_item(order_finish, page, 10)
        order_going.update({'order_going_goods': order_going_goods})
        order_finish.update({'order_finish_goods': order_finish_goods})
        return render(self.request, 'wm_query_order.html', {'order_going': order_going, 'order_finish': order_finish})
```

在 templates 文件夹中手动创建 wm_query_order.html 文件，在该文件中显示订单列表，具体代码如下：

```html
<!doctype html>
{% extends 'mj.html' %}
{% load staticfiles %}
{% block title %}订单查询{% endblock title %}
{% block topfiles %}
<script src="{% static 'js/jquery.min.js' %}"></script>
<script src="{% static 'js/Popt.js' %}"></script>
<script src="{% static 'js/cityJson.js' %}"></script>
<script src="{% static 'js/citySet.js' %}"></script>
<script type="text/javascript">
$(function () {
    $('.sj_order_man ul li').hover(function(){
        // 获得当前被点击的元素索引值
```

```
            var Index = $(this).index();
        var line=300*Index-300;
            // 给菜单添加选择样式
            $(this).addClass('active').siblings().removeClass('active');
            $(".line").stop(true,true).animate({left:line},200);
            $('.sj_order_man').children('div').eq(Index).show().siblings('div').hide();
    });
});
</script>
{% endblock topfiles %}
{% block body %}
<div id="main">
  <div class="sj_order_man">

        <ul class="sj_menu">
            <li  class="active" style="width: 50%">进行中的订单</li><li style="width: 50%">已完成的订单</li>
        </ul>
        <div class="order_list1">
          <table>
            <tr>
                <td width="11%">下单时间</td>
                <td width="5%">姓名</td>
                <td width="9%">电话</td>
                <td width="15%">地址</td>
                <td width="10%">餐品</td>
                <td width="6%">金额+运费</td>
                <td width="22%">备注</td>
                <td width="22%">订单状态</td>
            </tr>
            {% for going in order_going.info %}
            <tr>
                <td>{{ going.create_time }}</td>
                <td>{{ going.addr.receiver }}</td>
                <td>{{ going.addr.phone }}</td>
                <td>{{ going.addr.addr }}</td>
                <td>
                    {% for good in order_going.order_going_goods %}
                        {% if good.order == going %}
                            <li>{{ good.sku.name }}X{{ good.count }}</li>
                        {% endif %}
                    {% endfor %}
                </td>
                <td>{{ going.total_price }}+{{ going.transit_price }}</td>
                <td>{{ going.remarks }}</td>
                <td>
                    <a href="{% url 'order:success' going.order_id %}">
                        {% if going.order_status == 2 %}
                            订单已支付，待卖家接单
                        {% elif going.order_status == 3 %}
                            卖家已接单，待骑手接单
                        {% elif going.order_status == 0 %}
                            卖家拒绝接单，支付金额两小时内退回原账户
                        {% elif going.order_status == 4 %}
                            骑手已接单，前往取货中
                        {% elif going.order_status == 5 %}
                            骑手已取货，订单配送中
                        {% elif going.order_status == 6 %}
                            订单已送达，请前往评价
                        {% endif %}
                    </a>
```

```
                </td>
            </tr>
        {% endfor %}
    </table>
    <div class="pagenation">
        {% if order_going.has_previous %}
        <a href="{% url 'order:query' order_going.previous_page_number %}">< 上一页 </a>
        {% endif %}

        {% for pindex in order_going.pages %}
            {% if pindex == order_going.number %}
            <a href="{% url 'order:query' pindex %}" class="active">{{ pindex }}</a>
            {% else %}
            <a href="{% url 'order:query' pindex %}">{{ pindex }}</a>
            {% endif %}
        {% endfor %}

        {% if order_going.has_next %}
        <a href="{% url 'order:query' order_going.next_page_number %}"> 下一页 ></a>
        {% endif %}
    </div>
</div>

<div class="order_list6">
    <table>
        <tr>
            <td width="11%">下单时间</td>
            <td width="5%">姓名</td>
            <td width="9%">电话</td>
            <td width="15%">地址</td>
            <td width="10%">餐品</td>
            <td width="6%">金额+运费</td>
            <td width="22%">备注</td>
            <td width="22%">订单状态</td>
        </tr>
        {% for finish in order_finish.info %}
        <tr>
            <td>{{ finish.create_time }}</td>
            <td>{{ finish.addr.receiver }}</td>
            <td>{{ finish.addr.phone }}</td>
            <td>{{ finish.addr.addr }}</td>
            <td>
                {% for good in order_finish.order_finish_goods %}
                    {% if good.order == finish %}
                    <li>{{ good.sku.name }}X{{ good.count }}</li>
                    {% endif %}
                {% endfor %}
            </td>
            <td>{{ finish.total_price }}+{{ finish.transit_price }}</td>
            <td>{{ finish.remarks }}</td>
            <td>
                <a>订单已完成</a>
            </td>
        </tr>
        {% endfor %}
    </table>
    <div class="pagenation">
        {% if order_finish.has_previous %}
        <a href="{% url 'order:query' order_finish.previous_page_number %}">< 上一页 </a>
        {% endif %}
```

```
            {% for pindex in order_finish.pages %}
                {% if pindex == order_finish.number %}
                <a href="{% url 'order:query' pindex %}" class="active">{{ pindex }}</a>
                {% else %}
                <a href="{% url 'order:query' pindex %}">{{ pindex }}</a>
                {% endif %}
            {% endfor %}

                {% if order_finish.has_next %}
                 <a href="{% url 'order:query' order_finish.next_page_number %}"> 下一页 ></a>
                {% endif %}
            </div>
         </div>
 </div>
</div>

{% endblock body %}
```

由于在模板文件中添加了查询订单的方法，所以继承该模板的任何页面都可以访问该方法。

启动项目，在浏览器中访问 http://127.0.0.1:8000/order/query/1/即可查询出订单信息，如图 10.11 所示。

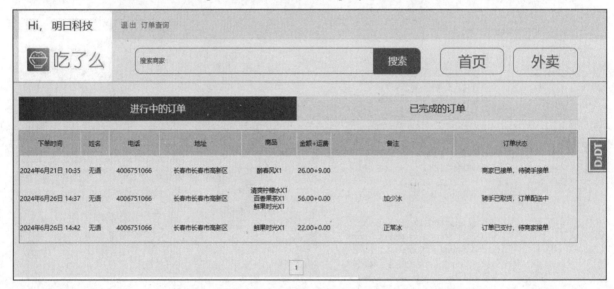

图 10.11 订单查询

10.9 项目运行

通过前述步骤，设计并完成了"吃了么外卖网"项目的开发。下面运行该项目，检验一下我们的开发成果。运行"吃了么外卖网"项目的步骤如下。

（1）打开 clmwm\clmwm\settings\develop.py 文件，根据自己的 MySQL 数据库、Redis 数据库、百度地图 AK 及 FDFS 文件存储服务器对下面配置代码进行修改：

```
DATABASES = {
    'default': {
        'ENGINE': 'django.db.backends.mysql',
        'NAME': 'clmwm',
```

```python
            'USER': 'root',
            'PASSWORD': 'root',
            'HOST': '127.0.0.1',
            'PORT': 3306,
        }
}

# Django 的缓存配置
CACHES = {
    "default": {
        "BACKEND": "django_redis.cache.RedisCache",
        "LOCATION": "redis://127.0.0.1:6379/1",
        "OPTIONS": {
            "CLIENT_CLASS": "django_redis.client.DefaultClient",
            # 提升 Redis 解析性能
            "PARSER_CLASS": "redis.connection._HiredisParser",
        }
    }
}

# 百度地图 AK，申请服务端
# 申请网址：http://lbsyun.baidu.com/apiconsole/key/create
BAIDU_AK = '61deBXb0SRMdfjBi0SJeBkmPlxzCy5hT'

# Django 文件存储
# DEFAULT_FILE_STORAGE = 'clmwm.utils.fastdfs.fdfs_storage.FastDFSStorage'
# FastDFS
# FDFS_URL = 'http://域名:端口'
# FDFS_CLIENT_CONF = os.path.join(BASE_DIR, 'utils/fastdfs/client.conf')

# 设置 Django 的文件存储类
DEFAULT_FILE_STORAGE = 'utils.fdfs.storage.FastDFSStorage' #修改的
# 设置 fdfs 使用的 client.conf 文件路径
FDFS_CLIENT_CONF = './utils/fdfs/client.conf'
# 设置 FDFS 存储服务器上 nginx 的 IP 和端口号
FDFS_URL = 'http://192.168.94.129:80/'
```

（2）打开命令提示符对话框，进入 clmwm 项目文件夹所在目录，在命令提示符对话框中输入如下命令来创建 venv 虚拟环境：

```
virtualenv venv
```

（3）在命令提示符对话框中输入如下命令来启动 venv 虚拟环境：

```
venv\Scripts\activate
```

（4）在虚拟环境下使用如下命令来安装项目所依赖的包：

```
pip install -r requirements.txt
```

（5）在虚拟环境下，从本地安装 py-fdfs-client 模块。在安装过程中，可能会报如图 10.12 所示的错误。这时需要先安装 vc_redist.x64.exe 工具。安装成功后，再从本地计算机中安装 py-fdfs-client 模块（该模块的安装文件可以在资源包中找到）。对应的命令如下：

```
pip install D:\Code\clmwm\fdfs_client-py-master.zip
```

在上面的命令中，D:\Code\clmwm\fdfs_client-py-master.zip 为安装包的绝对路径，需要根据实际情况填写。

（6）使用 MySQL 命令行方式或 MySQL 可视化管理工具（如 Navicat）创建数据库。在命令提示符对话框中使用命令行方式时输入如下命令：

```
create database clmwm default character set utf8;
```

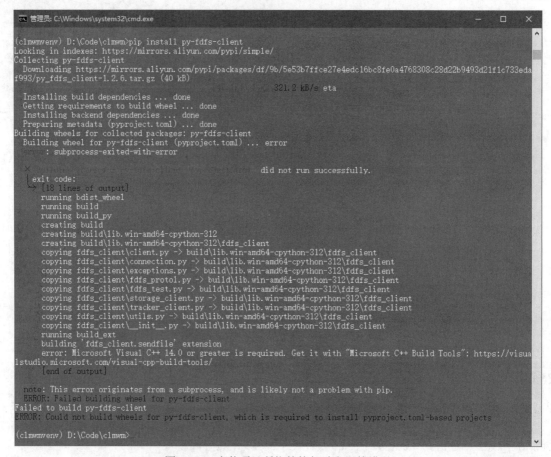

图 10.12　安装项目所依赖的包时出现的错误

（7）使用 migrate 创建数据表，命令如下：

| python manage.py makemigrations | # 创建迁移仓库，首次使用 |
| python manage.py migrate | # 创建迁移脚本 |

运行完成后，在 clmwm 数据库下会新增很多数据表，但是新增的数据表中数据为空，所以需要导入数据。

（8）使用 MySQL 命令行方式或 MySQL 可视化管理工具（如 Navicat）将 clmwm\clmwm.sql 文件导入数据库中。在命令提示符对话框中使用命令行方式时输入如下命令：

USE clmwm;
SOURCE D:\Code\clmwm\clmwm.sql;

在上面的命令中，D:\Code\clmwm\clmwm.sql 为数据库脚本文件的绝对路径，需要根据实际情况填写。

（9）搭建 FDFS 文件服务器，具体方法请参见资源包中的"FDFS 搭建"视频（视频位置：资源包\Code\附件\FDFS 搭建.mp4）。

（10）在 PyCharm 的菜单中选择 Run\Edit Configuration...菜单项，如图 10.13 所示。

（11）在打开的对话框中，单击左上角的+按钮，在弹出的快捷菜单中选择 Python，然后在右侧单击 Script path 右侧的文件夹图标，选择项目目录下的 manage.py，并且在 Parameters 文本框中输入 runserver，单击 Apply 按钮，如图 10.14 所示。

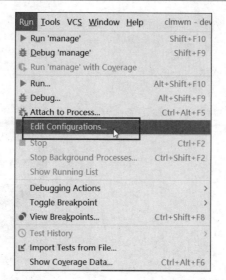

图 10.13　选择 Run\Edit Configuration…菜单项

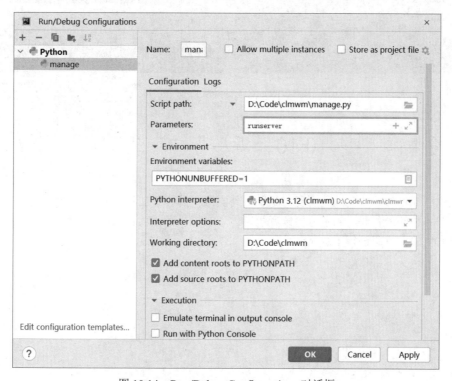

图 10.14　Run/Debug Configurations 对话框

（12）在浏览器中输入网址 http://127.0.0.1:8000，即可进入吃了么外卖网的首页，效果如图 10.15 所示。如果是买家，注册并输入买家身份的用户名和密码，登录网站后，可以选择店铺及商品并下单点外卖；如果是卖家，注册并输入卖家身份的用户名和密码，登录网站后，可以创建店铺并添加商品、跟踪订单、更改店铺状态等。

本章主要介绍如何使用 Django 框架实现"吃了么外卖网"项目。在本项目中，我们重点讲解了外卖平台买家选择店铺并下单和卖家店铺管理及订单跟踪功能的实现。通过本章内容的学习，读者可以了解外卖项目的开发流程，并掌握 Django 的 Web 开发技术，为今后项目开发积累经验。

图 10.15　吃了么外卖网前台首页

10.10　源码下载

本章虽然详细地讲解了如何编码实现"吃了么外卖网"的各个功能，但给出的代码都是代码片段，而非完整的源代码。为了方便读者学习，本书提供了该项目的完整源代码，读者可以通过扫描右侧的二维码进行下载。